Felix Publishing 2020
Email: info.felixpublishing@gmail.com
Print copies available from publisher.

Surviving Global Warming

Other non-fiction books by the author include:

Exploration Science (Field Geology and Mapping)
Changing the Surface (Erosion and Landscapes)
Rocks - Building the Earth
Fossils - Life in the Rocks
A Dangerous Planet (Earth Hazards)
Through Sea and Sky
Beyond Planet Earth (Astronomy)
Adventures in Earth Science (composite of the above books)
Adventures in Earth and Environmental Science Vol. I & II

2020 digital book release
ISBN: 978-1-925662-09-2
2020 Print Edition
ISBN: 978-1-925662-08-5

Author: Dr P.T. Scott
All illustrations, photographs and videos by the author unless otherwise stated.
Cover photo:

Registration:
Thorpe-Bowker +61 3 8517 8342
email: bowkerlink@thorpe.com.au

No part of this publication may be reproduced, stored in a retrieval system, or transmitted in any form or by any means, electronic, mechanical, photocopying, recording or otherwise, without the prior written permission of the publisher.

© All rights reserved Felix Publishing

SURVIVING GLOBAL WARMING:
A Guide for the Future

Dr. Peter T. Scott

First released 2020
© All rights reserved Felix Publishing

To my grandchildren and others who
Will have to live in a changing climate.

Foreword

This is a book written for the future but to be read now. I have attempted to put into it all of the knowledge and experience gained from over forty years of study and teaching. I have been lucky enough to have acquired a good and broad education in the Sciences, especially in the Earth Sciences in which I have done my own post-graduate research. I have also travelled to six continents to make my own observations and have seen the retreat of glaciers in New Zealand, Switzerland, Canada and Chile as well as some of the warming effects in Antarctica. I hope that the reader finds this book easy to read as it is not meant to be an academic thesis.

Most of the photographs and diagrams, unless stated, are my own work and I have included many links to my own videos and some useful websites. This should give more detailed information as well as to keeping this book down to an acceptable size. Many of these links are document files which should be downloaded as they may not stay on the web for too long. It is also meant to be a book encouraging reader action - peaceful rather than getting in everyone's' way in the streets.

There is a great need for action now concerning human-induced global warming as governments and some vested interests seem to be slow to realise the problems faced by our fragile world. I would also hope that some clever and innovative readers will take some of the ideas gained from this book and use their skills and initiative to add to the research and development already currently occurring behind the scenes that is not reported in the media.

Dr. Peter T. Scott
(March, 2020)

Table of Contents

Chapter 1: Truth and Reality	1
1.1 Personal Revelations	1
1.2 Who Should Read This Book	4
1.3 Credibility	6
1.4 Global Warming? - Says Who?	8
Chapter2: Climate Change	16
2.1 Introduction	16
2.2 The Ice Ages	17
2.3 Measuring Climate Change	24
2.4 Findings from the Research	29
2.5 Models & Natural Processes of Climate Change	38
2.6 Climate Change & Extinctions	51
2.7 The Gaia Hypothesis	54
Chapter 3: Humans and Climate Change	59
3.1 First Use of Energy & Resources	59
3.2: The Rise of Populations & Specialization	60
3.3 Renewable & Non-renewable Resources & Energy	63
3.4 The Greenhouse Effect	67
3.5 The Use of Fossil Fuels	72
3.6 The Effect of Land Clearing	74
Chapter 4: Fossil Fuels	79
4.1 The Need for Energy	79
4.2 Fossil Fuels - Coal	79
4.3 Fossil Fuels - Oil & Gas	88
4.4 Fossil Fuels - Coal-Seam Gas (or Coal-bed Methane)	93
4.5 Processing Fossil Fuels	95
4.6 Fossil Fuels - Advantages & Disadvantages	98
4.7 Fossil Fuels with Reduced or Zero Emissions	101
Chapter 5: Effects of Global Warming	113
5.1 Impact of Climate Change on the World's Systems	113
Chapter 6: Rain, Wind and Fire	**130**
6.1 Introduction	130
6.2 Extremes of Weather	130
6.3 Severe Thunderstorms & Tornadoes	135
6.4 Tropical Cyclones, Typhoons & Hurricanes	140
6.5 Floods	149
6.6 Wild Fires	159

Chapter 7: Monitoring the Environment — 175
7.1 Introduction — 175
7.2 Monitoring the Atmosphere — 177
7.3 Monitoring the Hydrosphere — 191
7.4 Monitoring the Geosphere — 202
7.5 Monitoring the Biosphere — 204

Chapter 8: The Nuclear Alternative — 213
8.1 Nuclear Energy — 213
8.2 Advantages & Disadvantages of Nuclear Power — 219
8.3 Towards Safer Nuclear Power — 224

Chapter 9: Alternative Energies — 231
9.1 Energy Production - Where to Now? — 231
9.2 Energy from Water — 233
9.3 Wind Power — 239
9.4 Energy from the Sun — 240
9.5 Storage of Renewable Energy — 245
9.6 Biofuels — 248
9.7 Geothermal Energy — 249
9.8 Advantages & Disadvantages in Using Renewable Energies — 251
9.9 Some Possible Home Applications — 255
9.10 Conclusion — 259

Chapter 10: Renewable Resources — 267
10.1 Introduction — 267
10.2 What are Renewable Resources? — 267
10.3 Ecosystems as Sources of Renewable Resources — 270
10.4 Replenishment of Renewable Resources — 276
10.5 Sustainability — 285
10.6 Sustainable Harvesting — 286
10.7 Human Impact on the Environment - Revisited — 290
10.8 Ecological Footprint — 294

Chapter 11: More About Water — 299
11.1 Introduction — 299
11.2 A Tale of Two Rivers — 302
11.3 Subsurface Water — 307
11.4 Desalination — 318
11.5 Water Recycling — 324
11.6 Hydroponics — 326
11.7 Future Consequences and Water — 330
11.8 Water Future in Australia - Bradfield Revised — 332
11.9 Diminishing Water in America — 335
11.10 Water Storage & Evaporation — 339

Chapter 12: The Way Ahead — 347
12.1 Introduction — 347

12.2 Back to the Future	347
12.3 The Urban Context	349
12.4 Living Space	350
12.5 Decentralization & Satellite Cities	354
12.6 Retrofitting Cities	360
12.7 Transportation – on Land	367
12.8 Transportation – at Sea	372
12.9 Transportation – in the Air	378
12.10 Construction & Materials	385
12.11 Permaculture	399
12.12 Waste and Recycling	403
12.13 Socio-political Consequences	413
Glossary & Index	**423**
About the Author	**438**
Other Books by the Author	**439**

Chapter 1: Truth and Reality

1.1 Personal Confessions

There have been a lot of words wasted on the topics of global warming and climate change. Many of the warnings about global warming and climate change are based on good research, current observations and wise knowledge. Unfortunately, some of the negative talk, especially that reported in the popular press, is based on fears, political and economic expediency and a lot of denial for comfort sake. The truth of the matter is that there are a lot of reputable scientists and scientific organisations which have been telling governments and the media for some time that our world is heating up due to the influence of greenhouse gas emissions from the uses of fossil fuels.

I would like to think that this book provides some of the scientific evidence for global warming and provides some suggestions on how its effects can be compensated for, or even reduced. Living with global warming, reducing its effects, or even reversing it can be assisted to some limited extent by private action, but any major reduction will have to come at the government level, especially on a global, cooperative level.

During my childhood, environmental issues tended to be concerned mainly with resources, such as forests and specific groups of animals which were then endangered. There were plenty of conservationists around, especially in places like Africa where the big game parks were being established, but a world-view of the environment was yet to hit the media. The media, like today, were more concerned with issues that sold newspapers or television time such as the progress of sporting teams, celebrity scandals and other such trivia.

Meanwhile, out in the suburbs, we still burnt our green waste in smoky backyard incinerators and ran sewerage into the sea. However, we did have milk, soda and most other liquids in re-useable glass bottles many of which could be returned to the retailer for cash. It was a good day when my friends and I could roam the local beach for a few bottles and then buy sweets with the money. When aluminium beer and soda cans came into use, they too could be traded in to be used again. Recycling is yet to be fully used in many of the big cities and waste dumping continues to be a major problem.

During my student days I became involved in limestone cave conservation and on one occasion, the locality in which we were mapping and studying the caves was being considered as a source of cement by a large, multi-national cement company. Rather than have a protest march through the city and chain ourselves to their doors, we took a quieter but more effective approach. A city protest would be useless for us as our caving society was small and the general public thought that caving was for 'weirdos'. Moreover, the media and the government were more interested in cement and jobs and there would be no competition against the large anti-Vietnam War protests then in vogue. Instead we looked at the legal aspects of mining in the region and the destruction that it would cause to the local farmers' land and water resources. Certainly, they would be compensated for the destruction of their land and perhaps gain lucrative royalties, but their land would become useless for farming due to the destruction of their waterways through the pollution from the fine quarry debris which came with limestone quarrying. Accordingly, a colleague and I visited all of the farmers concerned with a petition to turn their lands into a water reserve which they unanimously thought was a great idea. All of the official application forms were completed and we presented them on behalf of the farmers to the State Water Commission and received approval. Quarrying and mining leases were not permitted on water reserves so the caves and their rare bats still exist today.

Protesting and the freedom to complain has a place in our society provided that the people concerned remember to send the appropriate message to those who need to hear it. The use of disruptive and sometimes violent mass action targeted towards the general public often will have the reverse effect. Often, I feel that such protest movements underestimate the level of understanding of the majority of the populace. General informative protests to bring environmental issues to the public can be done in a legal and non-violent way which cause minimal disruption to the everyday activities of the general public - the very people who need to be able to agree with the message. Stunts such as forming human chains across public thoroughfares, disrupting public transport and hanging from bridges in weird costumes are a puerile self-indulgence, more likely to bring ridicule to the exhibitionists involved and public refusal to listen to the genuine concern that needs attention. The targets of mass or individual protest should be the appropriate private companies or government agencies and then they should be non-violent, legally arranged and legitimate.

I support genuine environmental group action against whole-scale exploitation of the environment, especially where there are endangered animal and plant species at risk, but I am also not completely against responsible resource mining and operations. Whilst it sounds like an oxymoron, a term which seems to contradict itself, responsible mining is carried out by many companies which extract the particular resource in a controlled manner with restoration of the environment during and after the mining operation. These companies do this under law and employ conservationists to study the region before mining and then maintain a program of restoration during and well after the operation has finished. Of course, this works well in theory and only in countries where there are strict environmental laws with honest and vigilant empowered people to monitor the process. Unfortunately, laws can be 'bent' or avoided by good corporate lawyers or ignored completely in 'developing countries' where governments or officials are also looking for revenue and the local people for jobs. Unfortunately, even in developed countries, these environmental laws often were not available at the time, or enforced when they were introduced, so there are still many old mine sites and waste lands yet to be rehabilitated.

We are going to need mining and the use of natural resources in the future; the difficulty will be in exploiting nature in a limited, controlled and restorative manner with minimal long-term effects. This will also involve the more efficient use and recycling of materials, especially in the use of waste materials as a resource rather than something to be dumped.

Figure 1.1: An old mine site near Queenstown, Tasmania which has not been rehabilitated.

My post-graduate field work also introduced me to the real world of investigative science rather than following someone's lectures or notes or simply being a follower of the popular press or social media. Along with my curiosity, I tried to get to the root cause of any scientific question, investigation or fact and my personal research taught me that Science as a methodology and philosophy has very strict balances and controls.

Provided that the original research and experimentation follow all of the rules required under this discipline, and that the data is given by reputable individuals or their organisations, then the data and the logical conclusions which comes from it can be believed with some certainty. Science, however is always a study involving change when new ideas or information comes to hand.

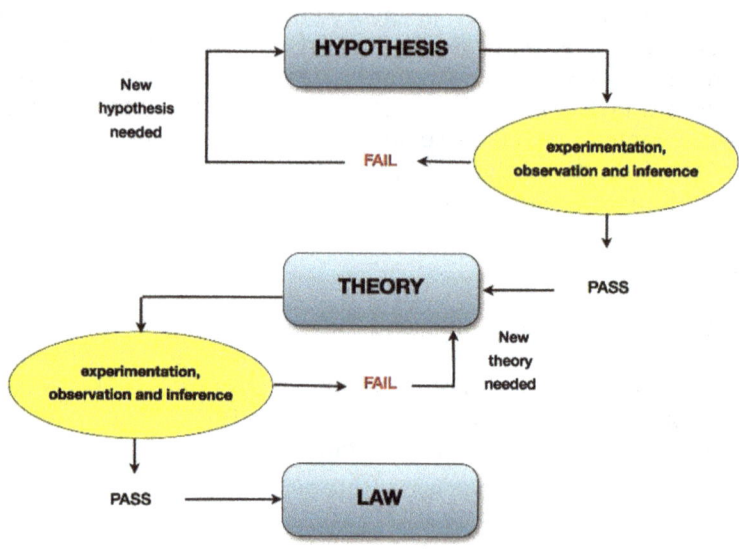

Figure 1.2: The Scientific Method

1.2 Who Should Read This Book

Everyone who is concerned about the future of this planet should read this book. This book is aimed at a general audience but the emphasis is on urban dwellers in developed and developing countries. Country folk are in a better position to do more at the personal level except where the effects of global warming is interfering extremely with their lifestyle. People in low-lying coastal regions of developing countries such as Bangladesh, in ocean nations such as Kiribati and in some inland arid regions are already experiencing problems caused by global warming and can do very little but appeal for assistance. For a large number of people who live in high-rise apartments, there will be little control over their environmental issues except at the governmental level. In cities where populations are growing, high-rise units will gradually replace low-rise within the urban expanse. Many of the 'applications' given at the end of each chapter in this book may be difficult for the apartment dweller to put into practice, but there may be other ways in contributing to personal and collective environmental conservation. For example, residents of an apartment complex may consider using the roof space and the surrounding grounds for beneficial and collective use such as: installing solar panels on the roof to supplement electrical supplies; developing rooftop and surrounds as collective gardens or recreational green space; using balcony space for personal hydroponic or other small gardens; and generally reducing waste and energy use within the complex. Certainly, the developers of such high-rise zones should consider the use of an appropriate amount of green space around such complexes.

Figure 1.3: One of the many high-rise suburbs of Xi'an, China now being built to replace low-rise housing.

I doubt that this book will directly influence national governments nor global organisations; it is aimed at the individual who will have to live with the effects of global warming and climate change currently and in the near future. Individuals can do a lot in their daily life to help themselves and their local communities to cope with climate change. This might mean a change in lifestyle such as: conserving electricity by reducing the number of appliances used; installing solar panels; walking or riding bicycles short distances instead of using a car; recycling wastes so that organic waste is used in the garden (another saving) and non-organic materials such as glass, aluminium and paper are recycled; and contributing time, goods or money aid to those who need it. Individuals can also vote for local, state and national politicians who show some willingness to combat global warming and not voting for those who don't, regardless of old voting habits. Individuals can also protest in a legal, non-violent and non-disruptive way to local government members and through reliable organisations such as those who volunteer to clean up local parks and protect wildlife and waterways.

The Applications section given at the end of each chapter may be useful on a personal basis or it may be useful in stimulating people who have influence in the local community or at the various government levels who can make a larger impact on the problems associated with global warming. Some of the applications come from personal experience and are workable and many of the photos showing home applications to global warming are of my own home. Others have solid basis in science and may be currently applied or are being developed. A few may be still at the idea stage, are possible and could be further developed by enterprising people with initiative. There is considerable amount of research and development going on around the world, especially in countries which are currently living with severe pollution, such as China and India, but the direct application on a large scale, depends on needs, motivation, government internal politics and certainly international cooperation. Governments seem reluctant to make the rapid changes suggested by scientists and creditable environmental groups and it may be up to private enterprise to lead the way by applying such new technology to replace the older technologies which have contributed to global warming and pollution. There will also have to be a massive shift in attitude and lifestyle in many of the urbanised and industrialised countries to cope with or reverse current destructive environmental trends.

1.3 Credibility

At the beginning of this chapter, it was stated that too much nonsense in the literal sense, has been written and spoken about global warming and climate change. With such a plethora of information and misinformation, one must carefully examine the original sources of the comments and the large amount of statistics given out by both doomsday environmentalists and climate change deniers. This, of course should also apply to this book.

There have been many short articles in magazines and newspapers and in books about the truth or not about global warming and climate change. How honest are these sources? What is the motivation of the people who write or speak about these and related topics? The range of people who make statements about global warming and climate change varies greatly from ill-informed individuals and organisations with little or no scientific understanding nor social consciousness to scientists and wise socially-minded people either individually or within creditable organisations who have done the research, have observed the consequences of the effects of humankind on the environment and can support their conclusions through results and data which have been verified by many other independent sources.

The media also varies in its approach. Printed articles in newspapers and magazines, television specials and even movies can provide a great deal of well-produced and dramatic influential material which is designed to change opinion and make a lot of money for the people who produced it. Some of it might even be true and give a good, balanced viewpoint of what really is happening but most of it often attempts to push a particular political, social or economic agenda and certainly to provide a good profit to the producers and sponsors.

Governments at all levels like to be seen to be doing the right thing for their constituents and other potential voters, but many politicians often have vested interests and certainly want re-election so adhere to their political party's interests. Some politicians may even be in politics for the good of their country and try to do the right thing regardless of their party's internal pressures. Unfortunately, representation of the people may not be their first priority and often the people are more concerned with local issues such as jobs. One should always analyse government-issued documents about the environment and look for the time frame of proposed actions, the funding which is to be available and the potential benefits to both the local community and to the nation. In some so-called Third World countries, governments may not see environmental issues as important compared to future development and prosperity of their people. Multinational companies, especially those involved in mining and agriculture may be prepared to overlook environmental concerns, especially if there is little or no environmental government legislation. On the other side, aggressive environmental activists may oppose development on vague environmental grounds without consideration for those companies and politicians who do have a responsible approach to the environment and the future. There will always be a certain element in society that will protest about something simply because they do not fit in or it is trendy and self-indulgent to do so. However, there is a place for genuine, concerned citizens to protest, personally or in groups, in a peaceful manner without defeating their own cause by outrageous puerile acts and seriously disrupting the rest of the community.

In this book, I have genuinely attempted to use my own experience, direct observations, and training in science, as well the information provided by individuals and organisations which have an international reputation for honesty and credibility in scientific exploration, data measurement and truthfulness of their findings. It must be pointed out that often scientists and large organisations can sometimes give misleading opinions because they too have agendas which may conflict with reality. This is because individuals can be pressured by their superiors or governments to provide conclusions which support the organisation's economic goals or political perspectives.

1.4 Global Warming? – Says Who?

Organisations which seem to have a genuinely scientifically-honest basis with excellent research facilities and staff giving sound results, data and with conclusions which have been rigorously examined by independent peers include:

- National Research Council (US);
- National Aeronautics and Space Administration (NASA);
- National Oceanographic and Atmospheric Administration (NOAA);
- United States Geological Survey (USGS);
- Geoscience Australia;
- Australian Academy of Science;
- Australian Bureau of Meteorology (BOM);
- Australian Institute of Physics;
- Australian Meteorological and Oceanographic Society;
- Academies, Associations or Royal Societies of many countries including those of: Belgium, Brazil, Cameroon, the Caribbean, China, France, Germany, Ghana, Indonesia, Ireland, Italy, India, Japan, Kenya, Madagascar, Malaysia, Mexico, Nigeria, New Zealand, Poland, Russia, Senegal, South Africa, Sudan, Sweden, Tanzania, Turkey, Uganda, the United Kingdom, the United States, Zambia and Zimbabwe;
- Commonwealth Scientific and Industrial Research Organisation (CSIRO);
- European Academy of Science and the Arts;
- Federation of Australian Scientific and Technological Societies; and the
- International Council of Academies of Engineering and Technological Sciences just to name a few.

The **Intergovernmental Panel on Climate Change (IPCC)** is an internal organization of the United Nations (UN) set up in 1988 by the **World Meteorological Organization (WMO)** and **United Nations Environment Program (UNEP)** to collate and provide scientific information and give advice through its various reports. In 1992, it produced the **United Nations**

Framework Convention on Climate Change (UNFCCC) as an initial step in looking at the climate change problem and today there are 197 countries that have ratified the convention. Soon after, there was an international effort to strengthen the global response to climate change.

In 1997 the member countries adopted the **Kyoto Protocol**. This agreement was an attempt to legally bind developed countries who are signatories, to agree on targets to reduce greenhouse gas emissions of carbon dioxide (CO_2), methane (CH_4), nitrous oxide (N_2O), hydrofluorocarbons (HFCs), perfluorocarbons (PFCs,) and sulphur hexafluoride (SF_6). The protocol's first commitment period started in 2008 and ended in 2012. In 2009 unfortunately, the hacking of emails from the Climate Research Group (CRG) at the University of East Anglia and the subsequent litigation was leaked to the press and taken up as support for the inaccuracies of the scientific evidence of climate change. This Climategate affair and some real inaccuracies in data analysis damaged the credibility of the IPCC and was seen as a smear campaign before the 2009 **Copenhagen Summit** on climate change. Later, in defence of the scientific findings which were criticised, several scientific bodies including the American Association for the Advancement of Science (AAAS), the American Meteorological Society (AMS) and the Union of Concerned Scientists (UCS) came out in support of the scientific consensus that the Earth's mean surface temperature had been rising for decades and that this was most likely caused by human activity. Further investigation later found no evidence of fraud or scientific misconduct by the IPCC.

After the Kyoto Protocol, the second commitment, known as the **Doha Amendment**, began on 1 January 2013 and is set to end in 2020. Under this amendment, 37 countries agreed to have binding targets. These countries include: Australia; the European Union and its 28-member states; Belarus, Iceland, Kazakhstan, Liechtenstein, Norway, Switzerland and Ukraine. However, since then Belarus, Kazakhstan and Ukraine have stated that they may withdraw from the Kyoto Protocol or not put into legal force the amendment with second round targets whilst Japan, New Zealand and Russia, which had participated in the first-round of the agreement, have not taken on new targets in the second commitment period. Other developed countries which have not committed themselves to second-round targets are Canada and the United States which withdrew from the Kyoto Protocol in 2012.

In 2015, the parties to the Framework Covenant (UNFCC) met in Paris for the 21st Conference of the Parties. Here they agreed to combat climate change and to accelerate all actions and financial assistance needed to develop a sustainable low carbon future. The **Paris Agreement** attempted to bring all nations to the common cause in the undertaking of ambitious efforts to combat climate change and to support to assist developing countries also to do so. To do this, the members of the Paris conference developed the central aim to keep the current global temperature rise well below 2 degrees Celsius above pre-industrial levels. Furthermore, it hopes to limit temperature increase down to 1.5 degrees Celsius in the future. On 22nd April 2016, 175 world leaders signed the Paris Agreement at the United Nations Headquarters in New York and this number has grown to 184 countries. In December 2018, the 24th Conference of the Parties to the **United Nations Framework Convention on Climate Change (or COP-24)** met in Katowice, Poland with 197 countries signing off on a set of guidelines needed to implement the agreement aiming to limit global warming to well under 2°C this century. This will send a signal that governments are serious about addressing climate change.

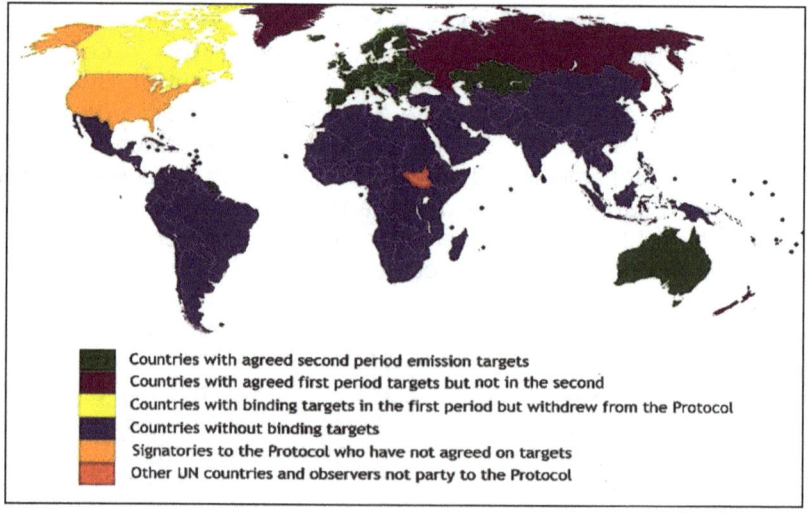

Figure 1.4: Map showing various UN countries and their response to the Kyoto Protocol (Photo: Public Domain)

In September 2019, further efforts were made by the United Nations in combating global warming at the Climate Action Summit in New York. The current Secretary-General António Guterres, urged support by both government and the private sector in combating climate change. There was also a passionate plea from Greta Thunberg, the sixteen-year-old climate activist who urged governments to listen to the pleas of the youth of the world who would inherit the problems of a future world. The final press release stated that 65 countries and major economies such as the

State of California were committed to cut greenhouse gas emissions to zero by 2050. In addition, 70 countries announced they would boost their national action plans by 2020 or commence new action plans to reduce gas emissions. Over one hundred business leaders delivered concrete actions to align with the Paris Agreement targets, and accelerate their organisation's transition towards a green economy. Many smaller countries, including small island developing states and least developed countries, were among those who made the biggest pledges, despite the fact that they have contributed the least to the problem. UN Secretary-General António Guterres, in closing the Summit, said "You have delivered a boost in momentum, cooperation and ambition. But we have a long way to go." This meeting was designed to focus further attention on climate change problems for the next UN climate conference in 2020 and there was general optimism that reduction in greenhouse gas emissions will occur in the future and meet previous set goals despite recent findings that the rate of increase of global warming has been faster than previously anticipated (UK Meteorological Office report in Scientific American.

See also:

https://blogs.scientificamerican.com/observations/scientists-have-been-underestimating-the-pace-of-climate-change/).

The Australian Government claims to have outperformed its first target under the Kyoto Protocol and is on track to meet the second amendment's agreement commitment to reduce emissions by 13% below 2005 levels by 2020 with further reductions to 26-28% by 2030. This target represents a 50-52% reduction in emissions for each person in the country and a 64-65% in emissions generally on the 2005 levels by 2030. Currently there is still some debate as to the ability of this government to do so.

See also:

https://www.aph.gov.au/About_Parliament/Parliamentary_Departments/Parliamentary_Library/Browse_by_Topic/ClimateChangeold/governance/international/theKyoto
(Australian Government summary of the Kyoto Protocol)

http://www.environment.gov.au/system/files/resources/c42c11a8-4df7-4d4f-bf92-4f14735c9baa/files/factsheet-australias-2030-climate-change-target.pdf
(Australia's 2030 Kyoto Protocol target)

http://classic.austlii.edu.au/au/journals/UNSWLawJl/2001/42.html
(Another opinion about Australia's approach to Kyoto)
http://climatecollege.unimelb.edu.au/files/site1/docs/9834/Paris%20Agreement%20-%20Fair%20share%20for%20G20%20countries.pdf
(The Paris agreement and Australia)

https://cop24.gov.pl/news/news-details/news/success-of-cop24-in-katowice-we-have-a-global-climate-agreement/
(Details of COP-24, December, 2018)

However, one must be wary of some international bodies which may have members with a political or environmental bias. This occurs when powerful or influential persons or countries on these bodies will push an agenda against the truth for their own personal aggrandizement or political and economic motives. In most cases, such agendas are publicly seen as incorrect and the person, report or organisation is then exposed. One must be able to report the reality of the environment openly, honestly and with good scientific support. There are many individuals and organisations who seek personal attention through climate change activities, press releases and other media outlets, including the media itself. Moreover, many individuals, politicians and governments find it easier to deny or delay action against global warming because it is easier to maintain the *status quo* than it is to take positive action. Politicians at all levels of government and across many countries and political systems seem to prefer actions which keep them in office and they find it difficult to make courageous decisions about climate change. Individuals and progressive organisations seem to be making the most headway into combating or living with global warming and climate change and there is often a ray of hope when an environmental organisation or individual is shown to be making a positive contribution.

In this book I have briefly outlined some data and the consensus from many of the most influential scientific bodies and have taken the assumption that global warming has occurred for some time due to both natural and human processes and that there is a need for change in the way energy and processed materials are produced. I have seen at first hand the recent effects of melting of glaciers in the European Alps, the Andes, the Rockies, the New Zealand Alps and Antarctica as well as experiencing extreme heat wave, drought and wildfires in Australia and believe that change is needed now.

Final Remarks

1. Global warming is occurring and has been increasing for a long time. We are now starting to bear the effects of global warming;

2. Many reputable scientific bodies doing research in this area suggest that global warming is caused by humankind's use of fossil fuels since the start of the Industrial era;

3. Global warming has become a major source of international debate and media comment;

4. There has been a considerable number of mistruths and general nonsense given to the media by both deniers of global warming and doomsday environmental prophets;

5. There has also been considerable scientific research done about climate change in the past and this is actively ongoing by many reputable scientific bodies and individuals with development of new technologies;

6. When statements are made about global warming and climate change, the reliability of the source and their motives behind their statements should be closely examined.

More Applications

Each Chapter will have some applications which are available, being researched or still being planned. Some are just my ideas in the hope that they will be developed. More Applications are some smaller ideas for the individual or wider application.

1. Research the local environmental groups in your local area and ask for their environmental policies. Examine the validity (truth) of their statements and find out what positive steps they are taking against climate change.

2. Use the list of scientific organisations from section 1.3. to look at their websites and make a list of their conclusions about climate change and what plans they have for the future.

3. Also compare these statements with those of section 1 above.

4. Find the names and addresses of your local, state and national government representatives and ask them what they (not their political party) are doing for climate change. There may be some specific question such as how the local waste dump is used, how the local water supply is managed or steps being taken to prevent coastal damage etc,

5. Do a preliminary energy/resources audit of (a) your living space and (b) your work environment to see what energy and waste can be reduced/recycled.

Further References

Australian Government, 2016: Climate Change in Australia.
https://www.climatechangeinaustralia.gov.au/en/

Australian Government Bureau of Meteorology, 2019: Climate Change - Trends and Extremes.
http://www.bom.gov.au/climate/change/#tabs=Tracker&tracker=timeseries

Climate Council, 2019: What is Climate Change and What Can We Do About it?
https://www.climatecouncil.org.au/resources/what-is-climate-change-what-can-we-do/

Hulme, Mike, 2009: Why We Disagree About Climate Change - Understanding Controversy, Inaction and Opportunity. Cambridge University Press; ISBN 978-0-521-72732-7.

National Aeronautics and Space Administration (NASA) 2019: Global Climate Change.
https://climate.nasa.gov/

Scott, P.T., 2016. Exploration Science. Brisbane, Felix Publishing. ISBN: 978-0-9946432-8-5

Wikipedia, 2019: Scientific Consensus on Climate Change:
https://en.wikipedia.org/wiki/Scientific_consensus_on_climate_change
(This has an extensive list of organisations and their scientific opinions)

Chapter 2: Climate Change

2.1 Introduction

The world's climate changes throughout time by natural processes and such changes are studied by the science of **climatology**. Of late, there has been considerable debate about climate change, so it is important to understand that climates have been changing since the day that the Earth was formed. The current debate is about **anthropogenic** climate change, that rapid and extreme set of processes fueled by the activities of humankind which are changing aspects of the climate, especially due to global warming, to threatening levels.

In the past, say about sixty to seventy years ago, climate change was considered by many people to be only a natural process; there was plenty of air and the oceans were wide. Moreover, despite local industrial pollution, many of the abuses of the 19th century had been reduced through more efficient industrial processes and more attention to health and social wellbeing. Waste was still being burnt or levelled into landfills or dumped into the sea as was nuclear waste from the new atomic power stations.

Climate change discussion traditionally centred around discoveries about past ice ages and it was considered that the current long, warm part of climate history was only the latest **interglacial period** or time between ice ages. It was just another part of the natural cycle and that soon the Earth would again gradually start to become colder. It did not happen and the idea soon developed that perhaps this was not going to happen when it became apparent that the Earth was still getting much warmer and too quickly. It was also thought that this warming may have to do with human activity. However, the Earth has experienced around 20 ice ages with several cycles of advance and retreat of large continental ice sheets, so exactly how Human activity has influenced global climates and how this activity has affected the large-scale natural cycles, remains to be seen.

Climatologists are concerned that changes in the climate may reach a **tipping point** i.e. a threshold of change which is abrupt, often unnoticed but an irreversible change after which a return to previous conditions would be uncertain. The belief is that with global temperature rise, there may be a crossing of multiple tipping points which would permanently damage the world's environments. For example, the continued melting of the Greenland ice sheet may get to the tipping point where the melt of the entire ice sheet will become inevitable even though it may take thousands of years to complete. Some climatologists, such as James Hansen (American: 1941 -) hold the view that the world has already passed the tipping point for atmospheric carbon dioxide levels in going from 350 parts per million (ppm) to now over 400 ppm. Whether or not this value can be reversed remains to be seen.

2.2 The Ice Ages

As an undergraduate studying glaciation in the seventies, it was thought that our long current warm period would soon be replaced by a new ice age. Previous studies of Earth's global environment centred on the various ice ages and the periods of warmth between them and there have been a number of defined ice ages in the geological and climatological history of Planet Earth.

An **ice age** is an extended period of time when the Earth's poles are covered with continental ice sheets and such events are called **glacial periods**. These occur when global cooling causes periods of extreme low temperatures over the entire Earth's surface. This produces an expansion of the continental ice sheets at both poles as well as alpine glaciers in both the Northern and Southern Hemispheres.

Between these glacial periods, the temperatures of the Earth's surface warmed and are known as **interglacial periods** or simply interglacials. Currently, the Earth is in the latest interglacial period which has been called the Holocene Interglacial. This began about 11,500 years ago and only the Greenland, Arctic, and Antarctic ice sheets remain.

See also:
https://www.atmos.washington.edu/1998Q4/211/project2/moana.htm
(A good summary of most recent glacials and interglacials)

Interest in the past climates of the world began in the Swiss Alps where it was noted that there were many large rocks sitting in the middle of glacial valleys. In 1815, the position of these boulder **erratics** in the middle of the valley were explained by a local hunter Jean-Pierre Perraudin (Swiss: 1767-1858) as being due to glaciers which had carried them down the valley and then left them behind as the glacier retreated. Such explanations could also be found in other parts of the world and when the naturalist Ernst von Bibra (German: 1806-1878) visited the Chilean Andes in 1849, he also was told by the local people that these boulders had come from the glaciers.

In 1824, the geologist Jens Esmark (Danish-Norwegian: 1762-1839) proposed that changes in climate were the cause of these ancient glaciations which had come and gone and had left the **erratics** in the valleys. He suggested that the changes in climate which caused these glaciations originated from changes in

Figure 2.1: A glacial erratic in the Andean valley of the Morado Glacier, Chile.

the Earth's orbit. Considerable debate on glaciation continued throughout the early 19th century, as the established view was of an Earth which was gradually cooling down from its original molten state. The concept that there had been a sequence of cooling in which glaciers and ice sheets extended followed by a warmer period in which they retreated was at first unpopular. However, further observations and proposals from more open-minded scientists such as geologists Jean de Charpentier (Swiss-German: 1786-1855) and Jean Agassiz (Swiss-American: 1807 - 1873) meant that by 1875, most of the scientific community had accepted that the Earth had gone through several ice ages. There have been at least five major ice ages in the Earth's history:

- **Huronian Ice Age**, also called the Makganyene glaciation, is named from the glacial rock formations of the Huron Supergroup deposited between 2.5 and 2.2 billion years ago, near Lake Huron in North America. This glacial period followed the **Great Oxygenation Event (GOE)**, when there was increased atmospheric oxygen with a decrease in methane which was oxidised into carbon dioxide and water. It is the oldest and longest of the ice ages and occurred at a time when only simple, unicellular life existed on Earth. During this event, glaciers covered most of the continents, extending from the poles to low latitudes and reaching down to sea level. This was followed by a short interglacial period of global warming.

- **Cryogenian Ice Age** which began about 850 million years ago, although some researchers think that it started later at about 720 million years ago. It followed the breakup of the ancient supercontinent **Rodinia,** which was in the Southern Hemisphere, which contained much of the landmass at that time. Rifting of the seafloor and the rise in temperature of the crust at that time lifted landmasses upward and also produced more evaporation of the oceans which in turn produced a considerable amount of rain which then cooled down the planet. Ice caps and glaciers extended all the way from the poles to the middle of equatorial areas, covering all or most of the planet and creating a phenomenon known by some as 'Snowball Earth'. It consisted of two short glacial periods, the Sturtian and Marinoan glaciations, separated by a very short interglacial period. This most extreme ice age lasted until about 630 million years ago and was followed by a longer interglacial period.

- **Andean-Saharan Ice Age** took place between 450 and 420 million years ago and its name is derived from the change in pattern of the glaciation which seemed to shift as temperatures dropped from what is now present-day Sahara Desert across Morocco, West Africa, and Saudi Arabia to present-day South America and into the Andes Mountains. The majority of the ice and glaciers were concentrated over Africa and the eastern region of present-day Brazil. Carbon dioxide levels in the **atmosphere** were abnormally high about this time so the cause of this ice age is probably not climatic but may have been due to geographic or solar radiation changes. Certainly, at this time, the supercontinent of Gondwanaland was moving southwards towards the pole and there was considerable mountain building in what is now the Appalachian region of North America. The onset of the Andean-Saharan glaciation set about a chain of events that brought about the Ordovician-Silurian mass extinction of living things. Because of its short duration, many geologists consider that the Andean-Saharan Ice Age may have been only a minor period of glaciation.

- **Karoo Ice Age,** also called the Late Paleozoic Ice Age, occurred from 360-260 million years ago. It is named after the glacial tillite found in the Karoo Basin of South Africa. It is thought that major mountain building or orogeny, associated with the formation of the supercontinent of Pangea disrupted warm currents in the ancient oceans causing cooling as the two separate regions of glaciation, one in Africa, the other in South America, joined together. There was also probably a later episode of glaciation in both India and Australia which also combined. This glacial period is also thought to have assisted in the rise of atmospheric oxygen and the evolution of larger land animals and fire-resistance in many plant species.

- **Quaternary Ice Ages** started about 2.58 million years ago at the beginning of the Pleistocene Epoch of the Quaternary Period. This led to the spread of ice sheets from the Artic into Europe as far as France and Germany and into North America past the Great Lakes. The corresponding drop in sea level also saw the migration of many animals, including early Humans, across land-bridges, especially between Asia and America and Asia and Australia.

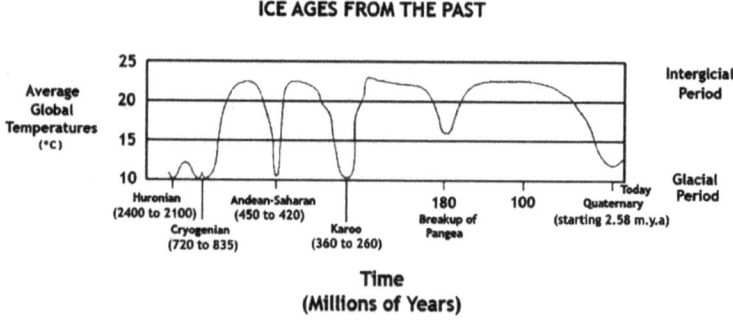

Figure 2.2: Graph showing the major Ice Ages.

Since that time there have been many cycles of glacial and interglacial periods with ice sheets advancing and retreating. The Earth is currently in an interglacial period, with the last glacial period ending at the start of the Holocene Epoch, the second epoch of the Quaternary Period, about 11,700 years ago. The ice continental sheets of Greenland and Antarctica are all that remains of the last glacial period. Because of the accessibility of more recent data from such diverse sources as ice cores and tree rings, this period of glaciation has been studied in great detail. Several minor cooling – warming episodes have been identified throughout its length so far.

When ice ages are mentioned, it usually is those of the Quaternary Period, especially those of the Pleistocene, the first epoch of this period. The Pleistocene Epoch, the first of the two epochs of the Quaternary Period, had a series of glacials interrupted by short warming periods with at least 20 cycles of such advances and retreats of the ice masses. During this time, global average annual temperature was over four Celsius degrees colder than today but temperatures become colder further into the Pleistocene.

During the glacial periods, a large part of the world's water was locked up in gigantic ice sheets. The cycle time between cold periods and warm periods was about 41,000 years and the temperature differences between cold and warm periods were also about 4 degrees or less. However, about 800,000 years ago the climate became colder, the glacial cycles lengthened to about 100,000 years, and the temperature difference between the glacials and interglacials increased to about 9 degrees. Thick glaciers moved down across Europe and North America to about 40 degrees north latitude and much of North America, South Scandinavia and Northern Germany were covered by ice. The Northern European ice sheet, called the **Fenno-Scandinavian Ice Shield,** extended down and covered much of Great Britain. The North American ice sheet, called the **Laurentide Ice Shield,** covered most of Canada and extended into the United States along the valleys of the Mississippi and Ohio Rivers. In the Southern Hemisphere glaciers existed in Tasmania and in the mountains of southern Australia. In Argentina, New Zealand and Antarctica where glaciers still exist, glaciers were much more extensive and deeper.

In the latter part of the Pleistocene, several major glacial cycles have been identified. In studies in Germany and the European Alps these were named after German rivers such as the **Günz, Mindel, Riss** and **Würm** glaciations. Today however, **palaeoclimatologists** prefer to simply use the numerical dates for periods of glaciation because these often varied from country to country making it difficult for comparison between the names used.

Near the end of the Pleistocene, roughly between 24,000-18,000 years ago, the last of most recent glacial events occurred. This has become known as the **Last Glacial Maximum (LGM).** It was a time when ice sheets and glaciers were at their thickest and the sea levels at their lowest. During this time, ice sheets covered much of northern Europe and North America with the permanent Summer ice covering about 8% of Earth's surface and sea levels were about 120 and 135 metres lower than they are today. Currently about 3.1% of Earth's surface is covered in year-round ice. Evidence for this glaciation can be found in sediments deposited by sea level changes all over the world and from glacial scouring across the North American plains. During at least part of the great ice ages, the continents of Asia and North America were joined in a land called **Beringia**, named after the current Bering Sea by the Swedish botanist Eric Hultén (1894-1981) in 1937.

The following web site shows an animated gif of the Asia-North America region from about 21,000 years Before Present to modern times:

https://en.wikipedia.org/wiki/File:Beringia_land_bridge-noaagov.gif

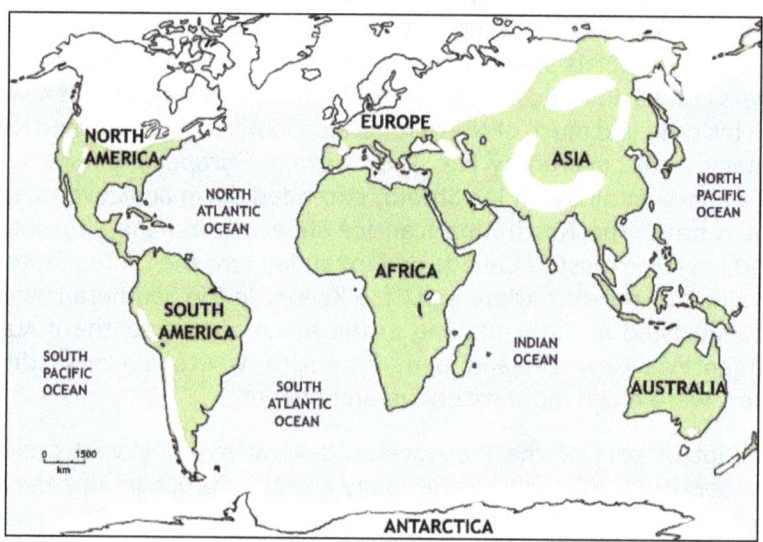

Figure 2.3: A simplified map of the extent of the Pleistocene ice.

It is thought that the land bridge between modern-day Alaska and Asia allowed for the interchange of some animal species between Asian and the America, especially after the joining of the two American continents and after the closure of the Isthmus of Panama which closed about 2.8 million years ago. This closure also stopped the flow of water between the Pacific and Atlantic Oceans and so altered the climate in the North Atlantic as warmer seawater from the Caribbean moved north. It is also thought that humans migrated across this land bridge or came by sea across more narrow waterways.

In the Southern Hemisphere, the Australian mainland, New Guinea, Tasmania and many smaller islands were joined together as a single land mass sometimes called **Sahul**. To the north-west of this continent were an archipelago of islands called **Wallacea** and beyond that the peninsula of **Sunda**. Between what are now the islands of Bali and Lombok in Indonesia is the **Wallace Line**. This was a boundary line drawn 1859 by the naturalist Alfred Wallace (Welsh: 1823 - 1913) and named by biologist Thomas Huxley (English: 1825 -1895), which separates the animal ecologies of Asia and Australia, with the islands of what was Wallacea as a transition zone. Plant species do not follow the Wallace Line as clearly as animal species except for *Eucalyptus* species which are confined to the Australian mainland. Evidence of human habitation in Australia suggests that the first indigenous Australians probably crossed the relatively short sea distance between Sunda and Sahul about 50,000 to 65,000 years ago. To the south-east of Sahul, the two main islands of New Zealand, along with associated smaller islands, were joined as one landmass and all of the Southern Alps of New Zealand were under permanent ice.

Figure 2.4: An estimation of the land bridge of Beringia during the Wisconsin Ice Age somewhere between 75,000 to 11,000 years ago. (After data from the USGS)

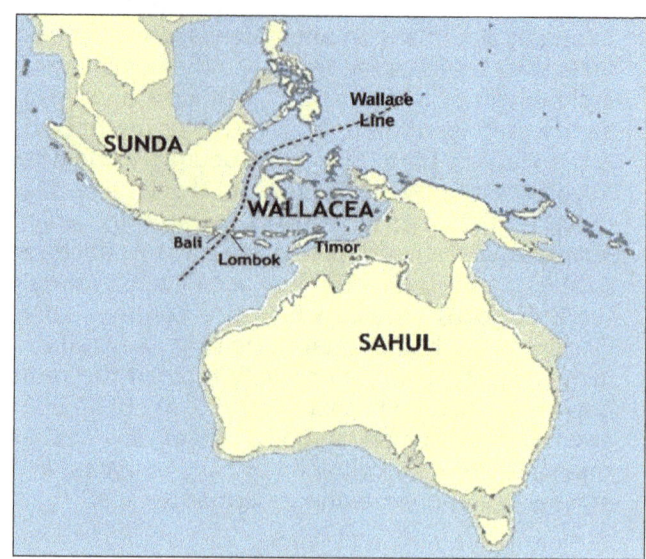

Figure 2.5: A map of the Australasian region during the lowered sea level of the Pleistocene ice.

See also:

https://pubs.usgs.gov/gip/ice_age/ice_age.pdf
(Ice age summary from the USGS)

http://www.dandebat.dk/eng-klima2.htm
(Good summary of some of the older ice ages)

http://gogreencanada.ucoz.com/index/ice_age_timelines/0-52
(Another good summary of older ice ages)

2.3 Measuring Climate Change

There is considerable evidence about how climates have changed over time and palaeoclimatologists have used a variety of methods, scientific disciplines and areas of study to obtain data about how climates have changed through time. A few of the methods used by researchers include:

- **Fossils,** which are the remains of living things which existed a long time ago and have been preserved as impressions in rocks, as actual remains in fossilized tree sap or amber, in tar pits and in the permafrost soils of northern latitudes. **Palaeontology**, the study of past life has been a well-tried science for centuries. Coupled with more recent discoveries in radioactive **isotopes** and radiometric dating for a variety of isotopes, palaeontologist now can put fairly accurate and absolute dates for the age of many rocks, minerals and fossils.

Figure 2.6: Fossil plant: *Dicroidium odontopterpoides* collected by the author showing that the climate in parts of Australia was once colder and wetter supporting extensive fern forests.

- **Pollen grains** as microfossils in some surface rocks, drill cores and ice cores sometimes give a very good indication of the type of vegetation which existed when the rock or ice was formed. **Palynology** (from the Greek: *palunō* meaning 'to sprinkle') is a precise science which studies small particles such as dust and pollen. Certain pollen is indicative of specific plants which have distinct conditions of temperature and moisture. Ice cores in particular give a great deal of detail from the pollen spores they contain as many of them are the same as or related to modern species whose ecological parameters are well known.

- **Foraminifera** are tiny marine, single-celled animals which secrete calcium carbonate shells or tests. These live in the ocean today and are very sensitive to temperature change i.e. they live within a very specific range of temperature in the sea. Foraminifera are also found as microfossils in marine sedimentary rocks because as free ocean plankton, they fall to the bottom of the ocean when they die and mix with the sea-floor muds. Thus, a collection of foraminifera fossils can give a very precise indication of the temperature of past times.

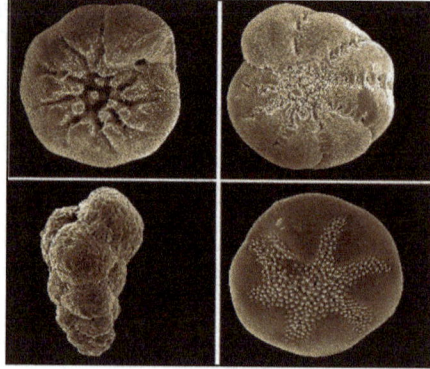

Figure 2.7: Photomicrographs of foraminifera (Photo: USGS)

- **Rock types** also give indications of past lithification or rock-forming environments. Igneous rocks now found in very stable parts of the Earth, such as Australia, show that in the past, the areas were volcanically active. The last volcanic event in Australia occurred at the Blue Lake at Mount Gambier in South Australia no less than 6000 year ago. Sedimentary rocks show their environment of deposition by wind, water and ice and these events can be easily dated both in relative time by the principles of stratigraphy such as the Law of Superposition and by radiometric dating of carbon fragments and ash layers within them.

Figure 2.8: A jointed marine sandstone rock platform typical of coastal regions now several thousand metres above sea level in the Andes of Chile.

The shapes and orientation of the grains in sedimentary rocks and any other sedimentary structure within it can also indicate the type of environment in which the sediments were deposited. Glacial conglomerates in hot, dry parts of Australia indicate a much colder climate. Limestones with fossil corals can be found well inland in parts of Australia and other countries. This indicates that once these areas were below warm, tropical seas. These observations assume that the **Principle of Uniformitarianism** is correct i.e. that the processes observed today in the formation of rocks and their structures were the same that caused them in the past. For example, corals are now found in warm tropical seas, so coralline limestone well inland suggests that these areas were once under tropical seas. Moreover, chemical signatures such as the magnesium to calcium ion ratios in the mineral calcite, both in limestone rocks and spelaeothems, such as stalactites, can be used to determine past temperatures. Other indicators of past environments come from:

- **Ice cores** which are pulled out of drill sticks or hollow cylinders when ice is drilled. As with ice being formed today, the ice taken from deep drill cores in places such as Antarctica and Greenland contain bubbles of air. Within the drill core ice, these bubbles contain air which may have existed thousands or even millions of years ago. By thorough laboratory analysis of this gas, the composition of the air in ancient times can be found. In the atmosphere, there are several isotopes, or types of the one element within its gases. Oxygen gas (O_2) can have atoms of its most common isotope, Oxygen-16 or it can have another of its isotopes Oxygen-18 present in its molecule. The ratio of these oxygen isotopes in the samples of the air in the ice bubbles changes with the air temperature at the surface of the ocean where the water freezes. Water molecules containing the heavier Oxygen-18 evaporate at a higher temperature than water molecules containing the normal Oxygen-16 isotope and so the ratio of Oxygen-18 to Oxygen-16 will be higher as temperature increases. There are other factors which can be taken into account, but these oxygen isotope ratios in air bubbles in ancient ice can give a very accurate idea of the air surface temperature in the past. The layers in the ice core can also be used to determine the relative age of the core samples and their thickness can be used to determine changes in precipitation and temperature when the ice was formed.

Figure 2.9: Ice core drilling in Greenland (Photo: NASA)

- **Relic landscapes** also show evidence of past climates. U-shaped glacial valleys in Tasmania, for example, show that once there as a much colder climate on the island but now all permanent ice has disappeared and warmer climates have prevailed. Sandstone rocks with internal lines of **current bedding** can indicate whether the past climate in which they were formed was a hot desert, a mountain stream or an ocean beach. Moreover, the angles of these structures can give an indication of the speed and direction of the wind or water currents which formed them.

Figure 2.10: Cradle Mountain and Dove Lake in central Tasmania. Australia. Both are relics of past glaciation.

- **Dendroclimatology** or the study of tree growth rings, can be useful, as trees respond to climatic changes with different amounts of growth each year by having different amounts of thickness for that growth ring.

Using older wood that is further within the trunk can give a more accurate record with some tree-ring records being able to be dated back a few thousand years. Carbon-14 **radiometric dating** can also give an accurate absolute age. A tree-ring record can be used to produce information regarding precipitation, temperature, soil moisture as well as some environmental information such as old wildfire events and even insect types found within sap.

Figure 2.11: Annual growth rings of an old tree (Photo: Public Domain)

- **More recent indications of climatic change** over the last two hundred years can also be obtained from younger ice cores, tidal records and weather station records of daily temperatures, pressures, wind speeds and direction, as well as atmospheric composition such as oxygen and carbon dioxide content. In addition, satellite technology has allowed scientists to get a complete global view of climate changes. Precise measurements can be made of ice sheet and glacial changes, natural and man-made changes to vegetation, sea level heights and fluctuations in weather patterns. There is also a great deal of information from historical records as well as current observations from astronomy. These include very detailed observations about the nature of Earth's orbit and of its position in the Solar System and Milky Way Galaxy as well as the activity of the Sun and the amount of solar radiation reaching the Earth's surface.

2.4 Findings from the Research

Collected data from a wide range of disciplines, scientific organisations and individual research have all shown that over the last few hundred years the global climate is changing. Moreover, these changes have become more pronounced in modern times and many changes impact on the global environment in such a way that they then precipitate other changes e.g. an increase in ocean temperature will melt sea ice and raise sea level.

The most dramatic change since the advent of the industrial age has been the increase in levels of carbon dioxide and other **greenhouse gases** in the atmosphere. This increase has been monitored extensively around the world both in urban, industrialised areas which would have an expected higher concentration due to Human activity, and also in areas well away from such obvious sources to gauge carbon dioxide increase in the global atmosphere. The **National Oceanic and Atmospheric Administration (NOAA)** regularly monitors carbon dioxide levels from atop the snow-covered extinct volcano Mauna Loa on the island of Hawai'i. Their readings are perhaps some of the most scientifically accurate of the world's monitoring stations.

This data, and that obtained from ice core air bubbles, carbonate mineral data and other sources allow other organisations such as the **National Aeronautics and Space Administration (NASA)** to extrapolate and represent data of pre-history levels. Data from carbon dioxide levels have shown an alarming increase in this gas in the last sixty years and it is thought that this increase is anthropogenic i.e. caused by human activity. There are additional consequences of this increase; carbon dioxide (CO_2) molecules along with water vapour (H_2O) and methane gas (CH_4), the other natural greenhouse gases, absorb infra-red radiation coming from the Sun as sunlight. These molecules have at least three loosely-bound atoms in their molecules and these loosely bound structures are efficient absorbers of the long-wave infra-red radiation which reflect off the Earth's surface. This is a simplified view of how the greenhouse effect causes global warming. Carbon dioxide gas is also very soluble in water and has been absorbed by the oceans since their formation, but recent increases in carbon dioxide gas also makes the oceans more acidic which has both good and bad effects on marine life. It is the increase in the temperature of the atmosphere and oceans which is the most dramatic result of increased carbon dioxide levels. This in turn has other follow-on effects such as the rise in sea level due to melting ice sheets in Greenland, the Arctic and to a lesser degree, the Antarctic. There has also been a corresponding decrease in the amount of glacial ice which impacts wildlife in alpine environments and of all of the communities which depend upon glacial meltwater for their freshwater supplies. Global snow cover also has also decreased, especially in the Northern Hemisphere where the vast areas of snow help to regulate the Earth's temperatures by its **albedo** or its reflectivity of solar radiation.

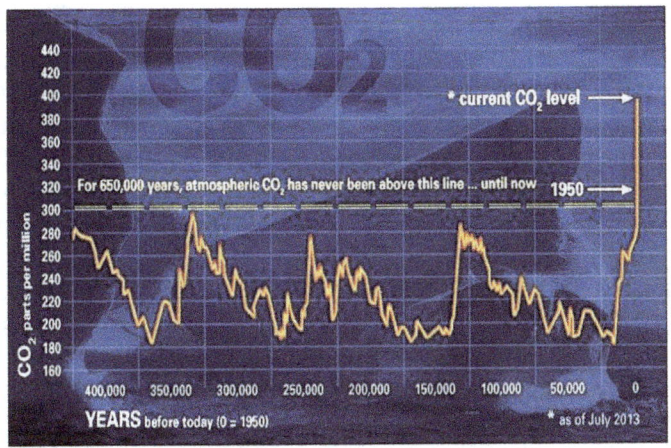

Figure 2.12: Graph showing increased carbon dioxide levels in the atmosphere over the last 400,000 years (Photo: NASA)

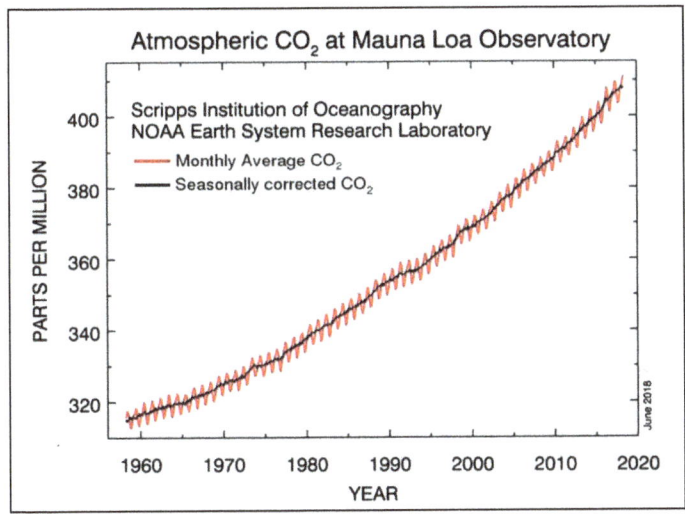

Figure 2.13: Graph showing increased carbon dioxide levels in the atmosphere in more recent times (Photo: NOAA)

Some of the effects of climate change due to global warming are shown in the following graphs:

- **Atmospheric temperatures** have increased. Temperature recordings of the weather over the last 150 years have provided extensive data and the last decade has been the warmest since these records have begun. These temperatures are those of global averages with considerable variation at different localities. Because of local factors such as El Niño and La Niña weather events, volcanic ash eruptions, the influence of aerosols, reduced solar activity and changes in local ocean currents, short-term fluctuations with a reduction in local temperatures can temporarily mask the overall effect of the increase in global average temperatures. These occurred in the 1970's when media speculation about the effects of aerosols predicted global cooling during an episode from 1998 to 2012, which was called the **Global Warming Hiatus**, when there seemed to be a pause in the increase in global air temperatures. However, ocean heat storage throughout this period continued to steadily increase, and since that time surface temperatures have again risen.

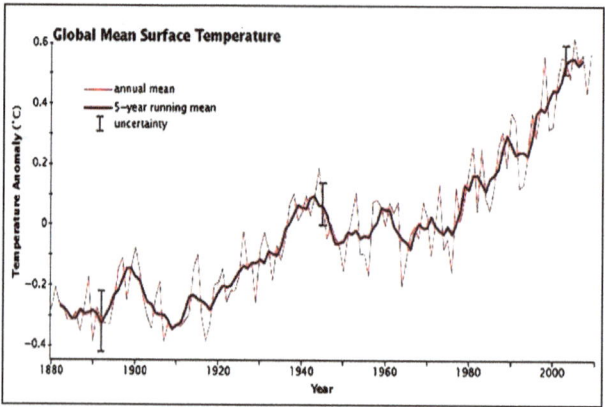

Figure 2.14: Graph showing increasing global heat content of the Earth's surface (Photo: NOAA)

- **Ocean heat content**, for which records go back over half a century. More than 90% of the extra heat content from global warming is going into the oceans, contributing to a rise in sea level through sea-ice melting and water expansion. Much of the measured increase in global temperatures is noticed at lower altitudes over land, as oceans and ice have great capacities to absorb heat producing increased ocean temperatures and melting of ice, especially in the Northern Hemisphere where there is more exposed land mass and because the Arctic ice cap is mostly sea ice which melts faster.

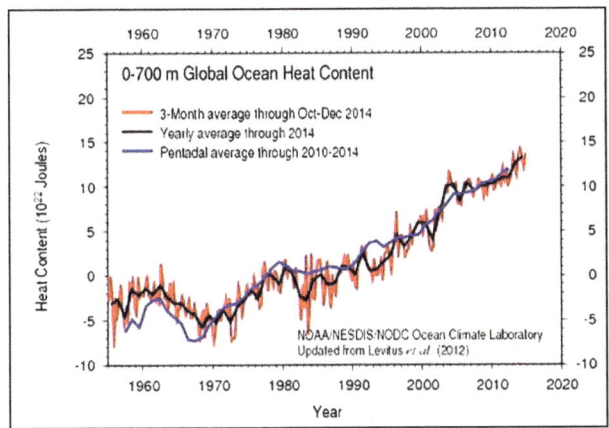

Figure 2.15: Graph showing increasing global heat content of the oceans (Photo: NOAA)

- **Measured sea level changes** from tidal gauge records going back to 1870, showing that sea level has risen at an accelerating rate due to general temperature rises in the atmosphere and oceans. Prior to the mid-19th century, global sea level changes were small, at only about a few centimetres per century. However, once the industrial age rapidly developed and more greenhouse gases were emitted, the rate of rise has increased dramatically such that in the past 20 years, satellite and coastal sea level data have indicated that the rate of rise has increased to about 3 centimetres per decade.

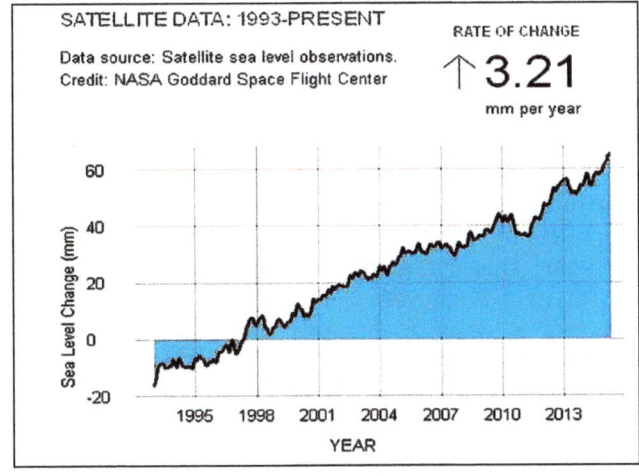

Figure 2.16: Graph showing recent sea level changes (Photo: NASA)

- **Glacial retreat** for many of the world's glaciers with an overall net loss of ice from glaciers worldwide is another consequence of atmospheric temperature rise. Glacial retreat over the last 150 years has been rapid and well-documented in many countries. Freshwater from the Summer melting of glaciers is often the only supply for some mountain communities such as those in the Andes of South America, the Himalayas of Asia, the Rocky Mountains of North America and in the alpine regions of Europe and many other countries. The overall retreat of mountain glaciers, provides evidence for the rise in global temperatures as glaciers often do not recover sufficiently from one season to the next. Glaciers advance when there is good snowfall at their head and less **ablation** or melting at their snout.

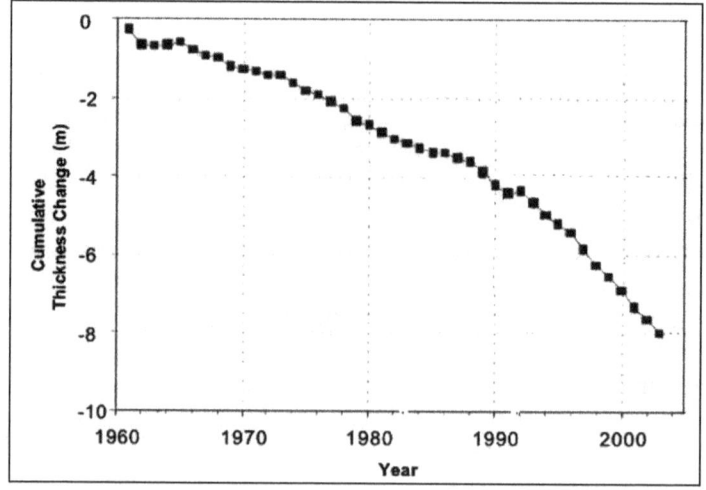

Figure 2.17: Graph showing global cumulative mass change (Modified from that of Mauri Pelto with data from the World Glacier Monitoring Service)

The retreat of glaciers in warmer months is a natural consequence of the seasonal increase in temperature but if temperatures become warmer over the entire year, this cycle is altered and the glaciers generally suffer additional retreats each year. In New Zealand, several of their short, coastal glaciers actually advanced more than they retreated due to an abnormally high rate of snowfall which the South Island often receives. However, since 2011 all of the glaciers have suffered greater retreating phases of their natural cycle.

Figure 2.18: The snout or front of the Franz Josef glacier in South Island, New Zealand. Apart from a short growth between 1983 and 2008, glaciers in New Zealand are rapidly retreating like many in the world.

- **Northern Hemisphere snow cover** has also decreased in recent decades with Spring snows disappearing earlier in the year with the total amount of area of snow generally decreasing. Snow cover has a significant role in the global temperature of the planet. Because it has a high albedo i.e. it is highly reflective, a considerable amount of heat in sunlight is reflected from snow surfaces back into space. Without snow cover, the ground absorbs about four to six times more of the Sun's heat energy. The presence or absence of snow controls patterns of heating and cooling over Earth's land surface more than any other single land surface feature. Satellite data has confirmed that average snow cover has decreased, especially in the Spring and Summer of northern Europe, Russia and Canada which accounts for much of the Earth's snow cover.

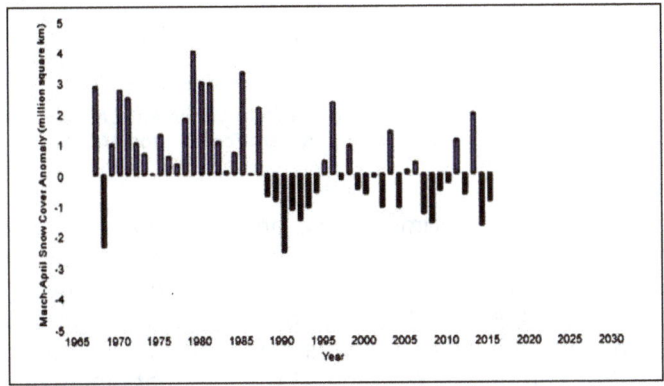

Figure 2.19: Graph showing average area of Northern Hemisphere Spring snow cover (Photo: NOAA)

- **Arctic and Antarctic ice cap** reduction is the most dramatic change of all. Satellite measurements from 1979 and shipping records from 1953 show a dramatic loss of sea ice from the high latitude oceans and ice from Greenland and Antarctica. Satellite observation has shown that the extent and thickness of the sea ice in the Arctic area has been in decline for several decades and it is speculated that it may cease to exist sometime later in this century. In 1988, older ice which was a permanent feature of the Arctic Ocean that was at least 4 years old accounted for about 26% of the Arctic's total sea ice. By 2013, this type of permanent ice had shrunk to only about 7%. In addition, oceanographers in the region have noticed that the open Arctic Ocean has started to show an increase in more violent wave action, something new to this part of the world. This is significant because violent wave action also breaks up the ice and allows for faster melting. In January 2016, satellite-based data showed the lowest overall extent of Arctic sea ice of any Winter since records begun.

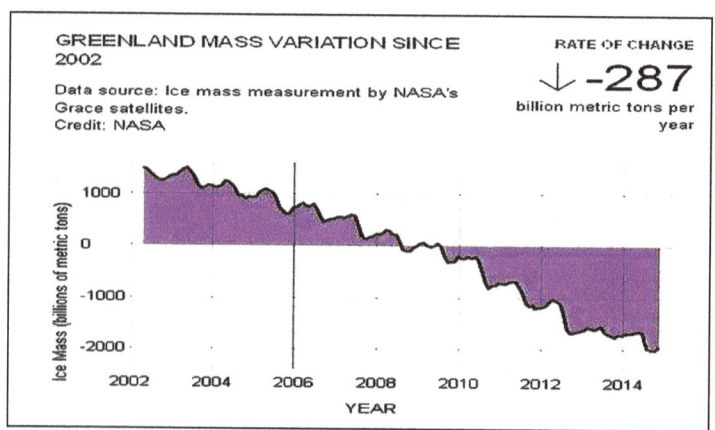

Figure 2.20: Graph showing ice loss from Greenland. A similar but slightly lower rate is occurring in Antarctica (Photo: NASA).

At times Antarctica sometimes shows an overall gain in the extent of its ice cover. It has different processes to that of the Arctic which is more confined and consists of sea ice which is more subject to melting. Antarctica has a greater thickness of several kilometres of ice above mostly land, although it has a surrounding region of sea ice which extends out into the ocean during Winter and then retreats in Summer. It is also surrounded by water which can provide more precipitation as snow when conditions are right. East Antarctica had a slight growth in its ice over the period 1992-2011, but overall, the Antarctic ice sheet in this region has lost on average approximately 70 billion tonnes of ice per year and West Antarctica and the Antarctic Peninsula have lost slightly less. In addition, the transient floating ice shelves have also become progressively thinner so from 1992 to 2015, the ice loss of Antarctic ice has contributed approximately 5 mm on average to the global sea level. However, a recent study of Antarctica suggests that when new snow accumulation and ice loss is taken as a part of the whole global ice sheet extent, the amount of ice being lost altogether, especially from the Greenland ice sheet and the Arctic region has become significant.

- **Changes to the patterns of storms** would be the expected result of increased ocean surface temperature because more water would evaporate as humidity in low-pressure areas from the rising warm air mass. Furthermore, with global oceanic warming, the regions of potentially severe low-pressure storms, including tropical cyclones, also called hurricanes or typhoons, would also probably extend beyond the current tropical regions. A warmer ocean would mean a higher amount of moisture in the air, or its humidity, due to greater evaporation.

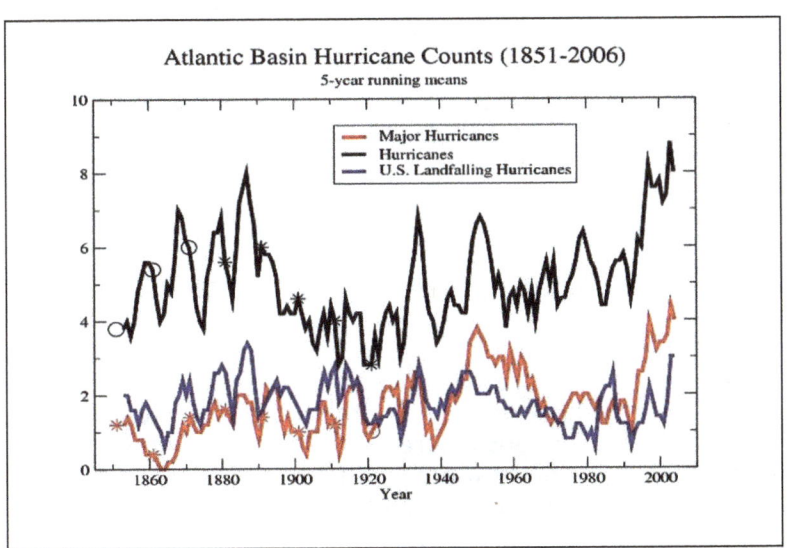

Figure 2.21: Graph showing frequency of hurricanes in the Atlantic Basin (Photo: NOAA).

Data analysis and computer modelling by the National Oceanographic and Atmospheric Administration (NOAA) suggests that in the future, tropical cyclones may not increase in number but may increase in intensity and with more rain. With a global rise in sea level this will also mean that storm surges, or water blown onto land, may also be more intense. Their modeling suggests that an increase of about 10-15% for rainfall rates for a 2°C global warming could occur and that tropical cyclone intensities globally will likely increase from 1 to 10%.

See also:

https://www.climatecouncil.org.au/uploads/3ca765b1c65cb52aa74eec2ce3161618.pdf
(Good summary about future storm development in the world with reference to Australia)

https://www.climatecouncil.org.au/uploads/1b331044fb03fd0997c4a4946705606b.pdf
(Extreme weather events in Australia)

https://www.climatechangeinaustralia.gov.au/media/ccia/2.1.6/cms_page_media/176/AUSTRALIAS_CHANGING_CLIMATE_1.pdf
(Australia's changing climate. Good summary with data)

https://www.zurich.com.au/content/dam/australia/general_insurance/risk_engineering/australian-storms-and-floods-white%20paper.pdf
(Excellent summary of natural disasters and climate change with reference to Australia)

2.5 Models and Natural Processes of Climate Change

Climate change generally refers to variations in average weather conditions over time taken in the broad context of changes in the atmosphere and the **hydrosphere**. It also involves the two-way interaction with the **biosphere** and **geosphere** as well longer-term average climatic conditions. Scientists often use computer-generated models which take into account the current trends and possible variations so that predictions can be made about the future. In the last two hundred years and especially in the last fifty years, there have been many hypotheses, theories and models to explain some or all of the changes which have occurred to the world's climate. Influences which can cause climate change are called **climate forcings** and they can be both natural and man-made.

More recently, **climate-change modelling** has been used to examine and predict what will happen to the climate in the future. These models are used because there is huge amount of data coming from many complex and interactive processes involved in all of the Earth's natural spheres. Modelling uses advanced computer **algorithms** to analyse this data and give trends or describe confidence levels – values which indicate the probability if some event will occur given certain input data. Algorithms include the basic laws of physics and chemistry as well as known relationships between the Sun's radiation and its interaction with the Earth. These modelling methods are usually complex and often are used by linking data from several different data sources such as sea ice and energy transfer or radiant heat and land use. Models are usually tested or validated by using known data from the past- a method known as **hindcasting**. Using well-established data from past events and then applying it to the specific computer model, the output predictions of the model can then be matched to climatic events which did occur and have been observed or measured directly in the field. Moreover, such testing models can be modified so that changes to input parameters can be made to see whether or not the model will agree with current trends. For example, within the computer program, the input values for carbon dioxide levels can be increased and then the output predictions can be matched to current trends in such observations as atmospheric temperature increase, sea level rise, glacial melting and the like.

In addition, current interactions within the Earth systems can be tested to see what variations in natural processes can affect climate change. There are some natural events or situations which can alter factors which cause climate change and these can be tested by computer models then compared to observed outcomes over a short time as a sort of cause-and-effect process. Such natural processes which can increase or decrease climate change and give predictable results are known as **feedbacks**. Usually such feedbacks simply modify direct warming and cooling effects without changing the planet's temperature directly. This is much like a brief covering of the Sun by a dark cloud will change how the Sun's heat is felt on the skin without actually changing the daytime temperature. Feedbacks that increase climate changes are called positive feedbacks and feedbacks which reduce climate changes are called negative feedbacks. Such feedbacks include changes in surface reflectivity, clouds type and cover, water vapour, and changes in the Carbon Cycle. Water vapour causes the most important positive feedback because as the surface warms, the rate of water evaporation and humidity both rise, which leads to an increase in global warming as water, as a greenhouse gas, traps more heat. Melting of sea ice is another example of positive feedback. For example, as the Arctic sea ice melts, it reduces the albedo effect of the planet, so less heat is returned to space through reflection and the exposed water then can absorb more heat giving more evaporation. Melting of permafrost in northern latitudes due to global warming and the release of its stored methane is another example of positive feedback. Negative feedback can come from the formation of some types of clouds formed by sea-warming evaporation which can increase the amount or reflectivity of sunlight back into space, cooling the surface of the planet.

See also:

https://www.climate.gov/maps-data/primer/climate-models
(Some of the basic principles of climate modelling from NOAA)

https://www.climatechangeinaustralia.gov.au/en/climate-campus/modelling-and-projections/
(Good summary of climate change and its modelling in Australia)

Some of the other suggested models or theories of the causes of climate change by natural processes include:

- **Changes in the Earth's orbit** which occur as a natural process in several long-term cycles initially suggested by Jens Esmark (Danish: 1763-1839) in 1824, but confirmed in more detail mathematically in 1920 by Milutin Milankovitch (Serbian: 1879-1958) and now known as the **Milankovitch Cycle.** Milankovitch was able to show that the three variations in the way the Earth travels around the Sun were very similar to the cycles measured and plotted earlier for the most recent sequence of glacial and interglacial periods in the Northern Hemisphere. These orbital variations are the: Earth's orbital eccentricity or changes to the elliptical shape of the Earth's orbit; its obliquity or the tilt of the Earth's axis, which is currently decreasing from 23.44 degrees; and its precession, or the way that the Earth's axis wobbles around its axis through its poles. This correlation suggested strongly that Earth orbital variations were a major contributing factor in causing global temperature variations which led to this cycle of glacial and interglacial periods.

In Milankovitch's model, the position and orientation of the Earth in relation to the Sun will naturally change the amount radiation falling upon the Earth's surface, the angle at which it strikes and the length of time which a location will receive it. The **eccentricity** of the Earth's orbit would mean that when the Earth was at its **perihelion** or position nearest to the Sun, the Earth would receive more solar radiance than when it was further away at its **aphelion** position. However, this is not a constant state. These values for the perihelion and aphelion change over time because the Earth's orbit is always changing its shape slightly in regard to how much it varies from a circular orbit. This is its eccentricity and it seems to have a major cycle every 413,000 years with some internal variation every 95,000 and 125,000 years. This eccentricity is probably due to the gravitational pull of Jupiter and Saturn and it effects the semi-minor axis of the orbit, that is the shortest distance or the vertical distance between the Sun and its orbit. This changes the amount of solar irradiation by up to 26%.

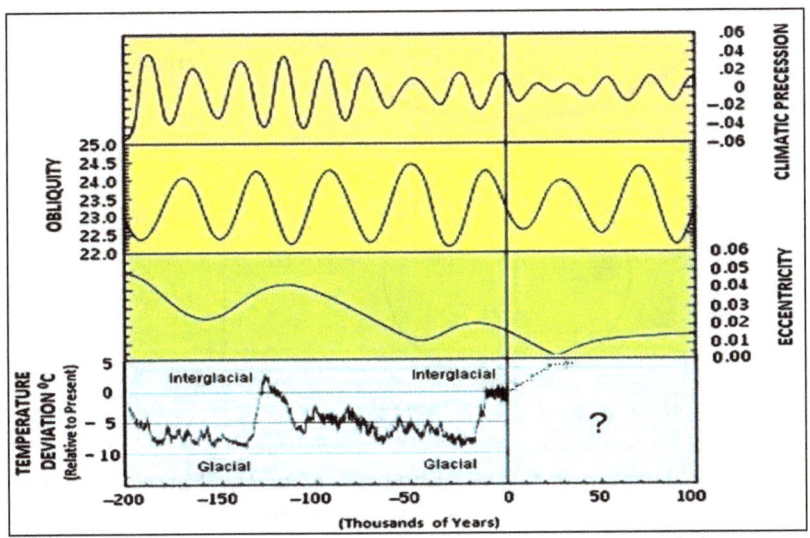

Figure 2.22: Graph showing Milankovitch Cycles for the Earth's orbit and Ice Age estimations. (Modified after Berger & Loutre, 1991, Petit et al., 1999 and others)

Obliquity is the angle at which the Earth's axial tilts in relationship to its orbital plane or the flat surface prescribed by its orbit. This angle varies from 22.1° to 24.5° over a 41,000-year cycle. The last time that the tilt was at its maximum was 10,700 year ago and it is now tilted at 23.4° and is decreasing to its minimum in another 9,800 years. An increase in tilt increases the angle at which the Sun's rays strike the Earth's surface during each season with more solar radiation in Summer and less in Winter. It also increases the total annual solar radiation at higher latitudes, and decreases the total closer to the equator. A decrease in tilt may promote colder temperatures in Summers and in the higher latitudes, increasing the chances of more snow and ice in these latitudes.

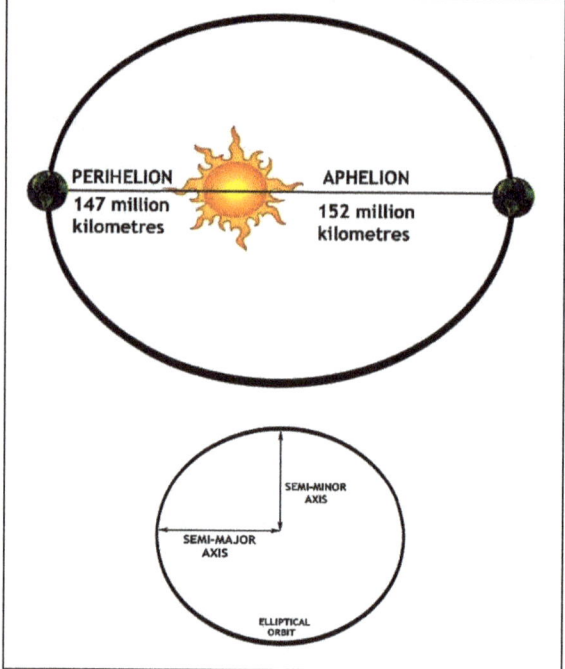

Figure 2.23: The elliptical or eccentric orbit of the Earth around the Sun

Axial precession is the wobbling effect of the Earth's axis and has a period of about 25,700 years. This also changes the angle of the Earth's axis and so adds to or subtracts from the effects of the obliquity of the planet. It is likely that orbital variations influence the Earth's climate and there have been several correlations between geological data and these orbital forcings, but there have been some proven variations which suggest that other factors may also be involved. For example, climatic changes observed on the 100,000-year cycle are not large enough, by themselves, to explain the observed cooling and warming based only on the amount of solar radiation reaching the Earth.

- **Solar output,** i.e. the radiation given off by the Sun, changes naturally would affect the amount of solar radiation reaching the Earth however the effects of the Sun's irradiance on global climate change is not fully understood and seems to be insignificant when compared to other forcings of climate change such as the greenhouse effect. Attempts to correlate the 11-year sunspot activity with climate change also seems to be inconclusive. References in the literature are often made to the **Maunder Minimum**, a period between 1645 and 1715 during which sunspot activity was generally exceedingly rare, as was then noted by many of the active astronomers of those times. The term was named in honour of the solar astronomers Annie Maunder (Irish: 1868-1947) and her husband, Edward Maunder (English: 1851-1928), who studied how sunspot latitudes changed with time. During this time period and for some time thereafter, Europe and North America experienced exceptionally cold Winters, a period incorrectly referred to as the **Little Ice Age (LIA)**. However, following more recent analysis of temperature data for that period, the correlation between low sunspot activity and cold Winters was most likely a local seasonal effect and not a global one.

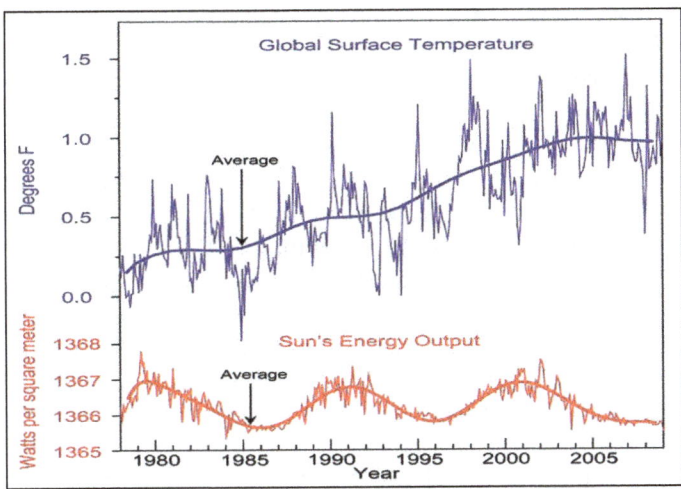

Figure 2.24: Graph comparing solar output and global temperature (Photo: NOAA)

More recent satellite data since 1978 onwards suggests that the variation in solar irradiation is minor at about 0.07% of the total yet Earth's temperature has risen much more during that time.

 **Online Video 2.1:** Solar variation and global temperature rise. Go to:
https://www.climatecommunication.org/climate/natural-factors-video/

See also:

https://www.imperial.ac.uk/media/imperial-college/grantham-institute/public/publications/briefing-papers/Solar-Influences-on-Climate---Grantham-BP-5.pdf
(Good summary of the effects of solar irradiation and climate change)

https://www.climate.gov/news-features/understanding-climate/climate-change-incoming-sunlight
(Sunlight and climate change)

- **plate tectonics** or the theory that the Earth's surface is covered by many large, moving plates, is also thought to have an influence on global temperature changes from the time of the Earth's formation to the present day. The great movement events in plate tectonics such as the breakup of supercontinents such as Laurasia in the Northern Hemisphere and Pangea in the Southern Hemisphere and the opening which formed the ancient oceans would also have climatic effects. Collisions of continental plates produced mountain chains through the processes of orogeny, the result of which would be the uplifting of large amounts of land surface, cooling these regions and the formation of snow fields with greater albedo adding to the cooling effect. The winds and their moisture contents would also be changed affecting the amount of rainfall with increased rain on the windward side of the mountains and less rain on the opposite side due to the rain shadow effect of the mountains. The opening of new landmasses would also provide additional carbon dioxide to warm the atmosphere.

One of the most rapid periods of warming in Earth's history, the **Palaeocene-Eocene Thermal Maximum (PETM)**, occurred as Greenland pulled away from Europe and the associated volcanic events put a considerable amount of carbon dioxide into the atmosphere. A similar, albeit smaller process can be seen today along the East African Rift. This is a region near the horn of East Africa where the eastern part of Africa is pulling apart from the rest of the continent. In this volcanic and earthquake prone region stretching down the eastern interior of Africa, substantial amounts of carbon dioxide gas are seeping from the soils due to the rifting. When new oceans are formed by the breakup of old supercontinents and when new oceans are made by sea-floor spreading at divergent boundaries, sea levels usually rise, and the climate generally becomes warmer.

One problem that must be considered when looking at localized climate changes and their relationship to plate tectonics, is that more recent movements of the continents may have changed the nature of the local climate since the time that the evidence was created. For example, plant fossils of the fern of genus *Dicroidium* are known from the Late Triassic sedimentary rocks of Australia, Antarctica, South America, South Africa and India. Citing the existence of these plants as evidence for a climate change from say, the original cool, wet forests of the Late Triassic to the cold wastes of Antarctica or the arid plains of India without understanding that these continents have changed position as well as climates by plate movements is poor reasoning.

See also:

http://www.indiana.edu/~g105lab/1425chap13.htm
(Revision of plate tectonics)

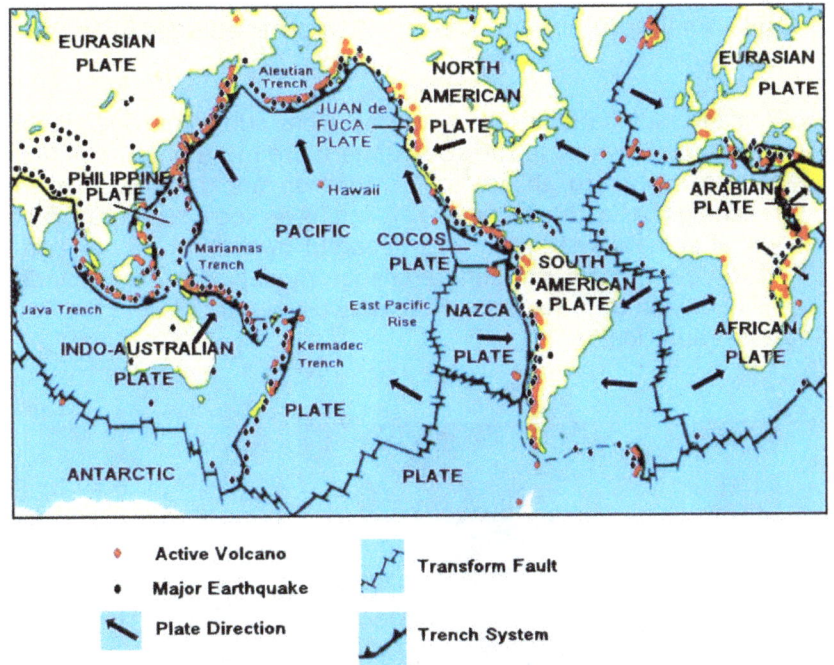

Figure 2.25: The Earth's major tectonic plates.

- **Volcanic eruptions** are another result of plate tectonics, but unlike slow-changing events such as the breakup of supercontinents, large eruptions of volcanoes have been suggested as having an immediate cooling effect on global climates. Eruptions on a massive scale can release large amounts of ash, aerosols, carbon dioxide, sulfur dioxide and other gases into the atmosphere. Ash can be carried long distances around the Earth within high altitude currents obscuring sunlight, lowering the temperature of the Earth to the point where some plants and animal species cannot survive – an event often erroneously called a **nuclear winter,** assuming that such events would be like the theoretical model of a cloud-covered world following a nuclear war.

In the long term, the volcanic gases ejected into the atmosphere, especially carbon dioxide and water vapour can add slightly to the total amount of greenhouse gases and contribute to further global warming but their amount is small compared to other sources such as those from human activities according to the **United States Geological Survey (USGS).** Water vapour is usually the most common volcanic gas but this often falls as rain and has little effect. Carbon dioxide is also emitted but usually this is small compared to other sources of the gas in the atmosphere.

Volcanic sulphate aerosols formed from sulfur dioxide dissolving in atmospheric water in the stratosphere are more of a problem because they are able to absorb solar radiation and reduce solar radiation reaching the Earth's surface. This can lower local temperatures and the global climate system for a few years. However, the long-term effects of large eruptions on climate-change models is not fully understood. Volcanic ash from such eruptions may temporarily shield the Earth from solar radiation.

Figure 2.26: A huge eruption cloud from Sarychev Volcano, (Kuril Islands) as seen from the International Space Station. Note the pyroclastic flow at the bottom right. (Photo: NASA)

Volcanic ash containing high concentrations of iron minerals may also modify climate by dumping these iron compounds into the ocean and stimulating phytoplankton growth. Phytoplankton are a major absorber of oceanic carbon dioxide which they use for photosynthesis and this could also reduce the global temperature. It has been estimated that during the past century major eruptions such as Mount Pinatubo in the Philippines on June 15, 1991, caused a decline in the average temperature at the Earth's surface of less than two degrees Celsius for periods of one to three years. It caused the largest aerosol disturbance of the stratosphere in that century, although it was probably smaller than those from eruptions of Krakatau, also known as Krakatoa, in 1883 and Mount Tambora in 1815. Volcanic eruptions however have other associated effects on the world's climate through such disruptions as the destruction of plant life and pollution of waterways.

See also:

http://climate.envsci.rutgers.edu/pdf/ROG2000.pdf
(Scientific paper on the effects of volcanoes and climate change)

http://www.pastglobalchanges.org/download/docs/magazine/2016-1/PAGESmagazine_2016%281%29_29_Toohey.pdf
(Good summary including some of the major volcanic events of the past)

- **Meteorite or asteroid impact** with Earth, such as that suggested by the Chicxulub crater of the Yucatán Peninsula in Mexico would also cause immediate global climate change. Large impacts such as this cause massive destruction on land and huge tsunamis would have swept across the sea. In addition, a huge cloud of dust and aerosols would have quickly spread across the Earth's surface and high into the atmosphere. Similar to a large volcanic eruption but on a much larger scale, the coverage of such a large dust cloud would prevent sunlight from getting through and cause global temperatures to fall. This model has been suggested as a possible cause of the extinction of the dinosaurs at the end of the Mesozoic era 66 million years ago. The global cooling below this cloud cover could last for several years but it would eventually dissipate as the dust fell back to Earth. After the dust had fallen back to Earth, the greenhouse gases such as carbon dioxide, water vapour and methane and many of the aerosols caused by the impact with the surface would remain in the atmosphere. This would then cause global temperatures to increase for probably decades.

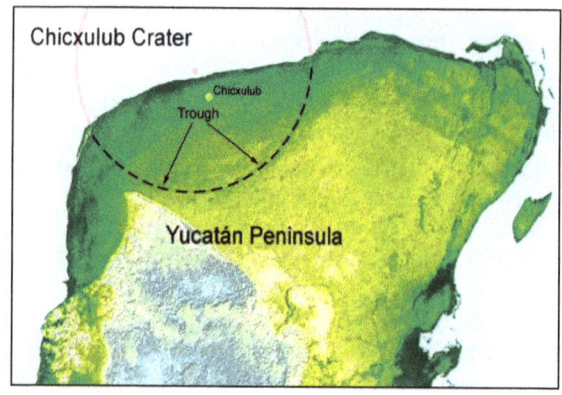

Figure 2.27: Map showing the location of the Chicxulub crater in the Yucatán Peninsula of Mexico (Photo: Modified after NASA)

- **Ocean currents** play a major role in the world's climates. Today warm currents such as the Gulf Stream give warm climates to countries such as Ireland and Great Britain which would normally have much colder temperatures. Countries with about the same latitude which do not receive this current such as eastern Canada have a much colder climate. Similarly, cold currents such as the Peru or Humboldt Current which runs parallel to that country's coastline do not bring much rain to the coast. Ocean currents also have a three-dimensional nature with warm, surface currents sinking down when they reach colder latitudes where they become colder and denser and so sink to form sub-surface deep currents which move across the sea floor and flow back into the ocean basins. This great conveyor belt of ocean currents is called the **Thermohaline Circulation (THC)**. The warm **Gulf Stream**, for example flows northwards from the warm waters of the Caribbean up into the North Atlantic, bringing warmer air above it to the coasts of Western Europe. Continuing further north it then cools when it reaches the Arctic Ocean where it sinks and flows at depth back towards the south as the **North Atlantic Deep Water (NADW)** were it warms, becomes less dense and rises again off the tropical coast of Africa to reform again as a warm, surface current.

Ocean currents distribute the heat of the oceans around the world. Disruption of these currents and their sub-surface counterparts could bring localized climate changes in various parts of the world and no doubt this also could have occurred in ancient oceans, especially with the movement of the world's tectonic plates. In future times, for example, if the Gulf Stream were to be interrupted, say by mid-Atlantic uplift at the mid-ocean ridge, then temperatures in Europe would drop.

Some data suggests that during the last glacial period, the sea level has dropped as much as thirty metres as seawater became ice, mostly Arctic ice sheets. With this reduction in sea level, current flow through the shallow Bering Strait between Siberia and Alaska was reduced. Consequently, over on the other side of the Arctic, the North Atlantic from the North Atlantic into the Arctic Ocean. This melting of the Arctic ice released new water which then raised the sea levels again, restoring the balance of water flow in the Bering Strait. In 1958, oceanographer Maurice Ewing (American: 1906 - 1974) and meteorologist William Donn (American: 1918-1987) published a theory using a similar mechanism to account for the various ice ages which had occurred in the late thermohaline circulation was altered and warmer water was able to flow Pleistocene. They proposed that ice ages were caused by such interruptions to the North Atlantic thermohaline current allowing warm water to intrude into the Arctic Ocean which would then melt the ice sheet. With the ocean now free of ice, more water would evaporate, rise and then freeze again in the higher and colder air producing snow which would then fall over a wide area. If it fell on the Arctic Ocean, its heat content would melt the snow, but if it fell on the land of what is now northern Europe, Russia and Canada, then this snow would accumulate over time forming ice. With successive Winters and further accumulation of snow on the land, ice sheets would spread out further south into Europe and North America and also across the Arctic Ocean.

This cycle could be repeated with new interruptions of the North Atlantic currents. It sounded like a good theory at the time, but Ewing and Donn could then not find a mechanism which would interrupt the ocean currents in the North Atlantic allowing warm water to flow into the Arctic Sea. Proposals of fault movements on the sea-floor across the North Atlantic did not fit the model and so it was generally discarded. At that time, the Theory of Plate Tectonics was in its infancy. Today, it is interesting that general global warming has once more caused a great reduction in the Arctic Ocean and Greenland ice sheets. If this trend continues then perhaps some of the ideas of Ewing and Donn might once more suggest a new ice age scenario. Periods of excessive snowfalls over several seasons in Russia and Canada would be an indication that such an event could be happening.

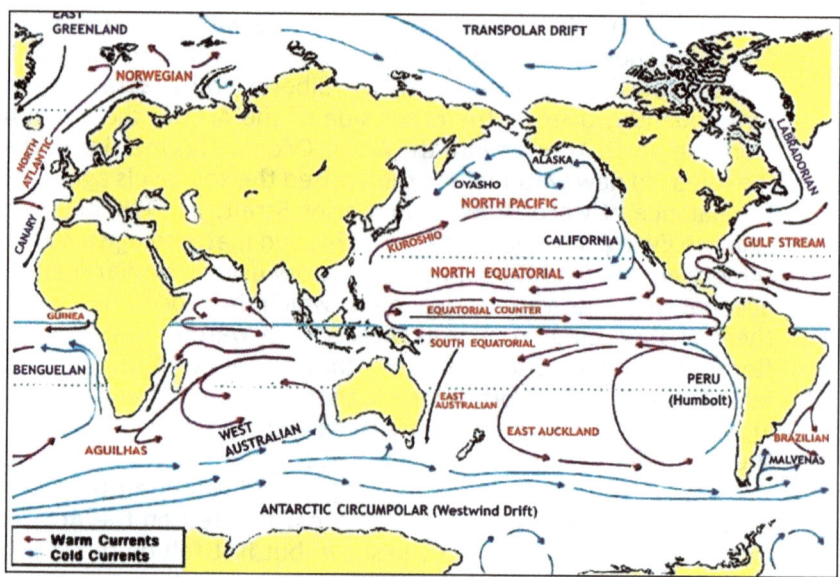

Figure 2.28: A map showing some of the world's major surface currents.

- **Surface temperature changes** are also caused by ocean current change. Such changes are involved in the **El Niño - La Niña** events across the equatorial region of the Pacific Ocean, the Indian Ocean Dipole in the Indian Ocean and the complex Tropical Atlantic Ocean Dipole. More on these weather effects later.

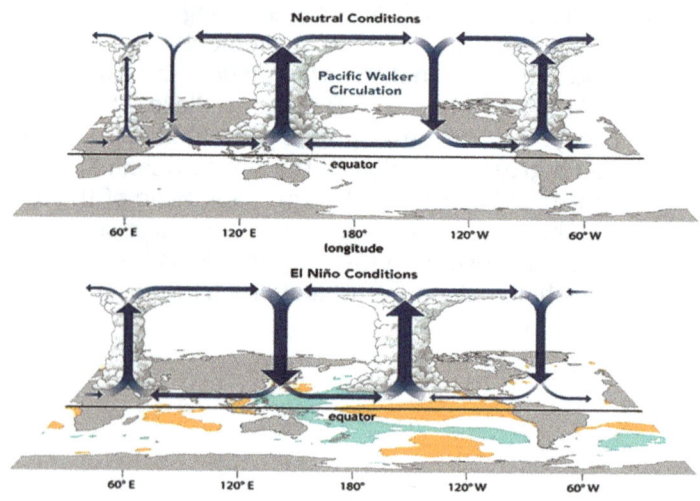

Figure 2.29: A map showing some of the world's major surface currents.

See also:

http://www.pik-potsdam.de/~stefan/Lectures/ocean_currents.html
(Good summary of ocean currents and climate change)

https://www.climate.gov/enso
(NOAA's El Niño page with many links)

http://www.bom.gov.au/climate/enso/index.shtml#tabs=Overview
(ENSO alert page from the Australian Government with many links)

Both in the past, when computer modelling did not exist and today when computing power is excellent for handling complex interrelationships, researchers have made use of physical models to explain the processes which caused observed climate change trends. One of these models is the greenhouse effect which is thought to be enhanced by man-made or anthropogenic activities, especially the increased production of gases such as carbon dioxide, methane and oxides of nitrogen and other gases. This is the dominant model for the cause of global warming and all of the other associated changes which come from it. In summary, most of the world's scientific organisations and individual scientists support this concept that emissions of greenhouse gases since the start of the industrial age has led to a global warming of both the atmosphere and the oceans.

2.6 Climate Change and Extinctions

Species of living things and evolved, some becoming **extinct** and some remaining as **extant** organisms still surviving today. In examining the fossil record, palaeontologists have noted that at certain times, some species have suddenly disappeared. This could be due to lack of sedimentary processes which form the remains of once-living things as fossils, but these sudden disappearances have been more extensive and widespread than any localized effect could have had on simply the lack of fossilization. Most scientists now agree that these sudden disappearances have been most likely due to catastrophic events which were sudden, and involved large parts of the entire Earth. They are referred to as **mass extinctions**.

Scientists have identified five time periods which may have been mass extinctions:

- **Ordovician-Silurian** extinction event which occurred at the end of the Ordovician Period. It may have consisted of two separate events which destroyed a large number of marine species, especially trilobites, graptolites and brachiopods. During this time there was a sudden drop in sea level as glaciers formed, followed by a time when there was a sudden increase in sea- level as glaciers melted.

- **Devonian-Carboniferous** extinction event which occurred at the later part of the Devonian Period, when a series of extinctions eliminated a significant number of species, especially those of shallow seas such as many of the coral species. It is possible that global cooling and some sea level reduction may have triggered these extinctions.

- **Permian-Triassic** extinction event which occurred at the end of the Permian Period was possibly the Earth's largest extinction, killing a majority of all species. This included both terrestrial and marine species, including many insect species and the mammal-like reptiles. It did, however, allow for the rise of the archosaurs – the ancestors of the dinosaurs, birds and crocodilians in the Triassic Period.

- **Triassic-Jurassic** extinction event which occurred at the end of the Triassic Period and eliminated most of the archosaurs, therapsids, and large amphibians. This enabled the dinosaurs to become the dominant animal group on land. About this time there was a great production of flood basalts – eruptions of great volume with little ash which spread over large areas and contributed to an increase in global temperature. It is possible that this may have triggered this mass extinction.

- **Cretaceous-Lower Tertiary** extinction event at the end of the Cretaceous Period. This is the well-known K-T Extinction, from the German, *Kreide*, meaning chalk, and *Tertiary*. It is the extinction which removed all of the non-avian dinosaurs and assisted in the rise of the mammals and birds as the dominant vertebrate animals.

There have been several theories put forward for the cause of this event; most notably that of an asteroid collision with supporting evidence coming from the amount of the chemical element **iridium**. This is a natural chemical element which is found only in the Earth's mantle and extra-terrestrial objects such as meteors and asteroids. It is also found world-wide as a fine layer within rocks of the time of this extinction. Moreover, several large impact craters have been found, one 180 km in diameter dating to the Late Cretaceous buried beneath sediments of the Yucatán Peninsula near Chicxulub, Mexico and a second, smaller and slightly younger crater, at Boltysh in Ukraine.

Tektites, small broken fragments typical of such impacts, have also been found in deposits associated with the time of this extinction. As well as the indications of impact, including those caused by the huge fireball and tsunami which would have occurred at that time, such a collision would have also caused a major reduction in the amount of sunlight needed for plants and therefore the rest of the food-chain.

Whilst this impact theory gives a good explanation of why such extinction could occur, some palaeontologists have doubts that the time of impact at these sites actually coincides with the time of the extinction event. Another theory suggests that large-scale eruptions of basalt in the Deccan of India, would have poured considerable amounts of carbon dioxide into the atmosphere and produced a greenhouse effect, which would have warmed the entire planet.

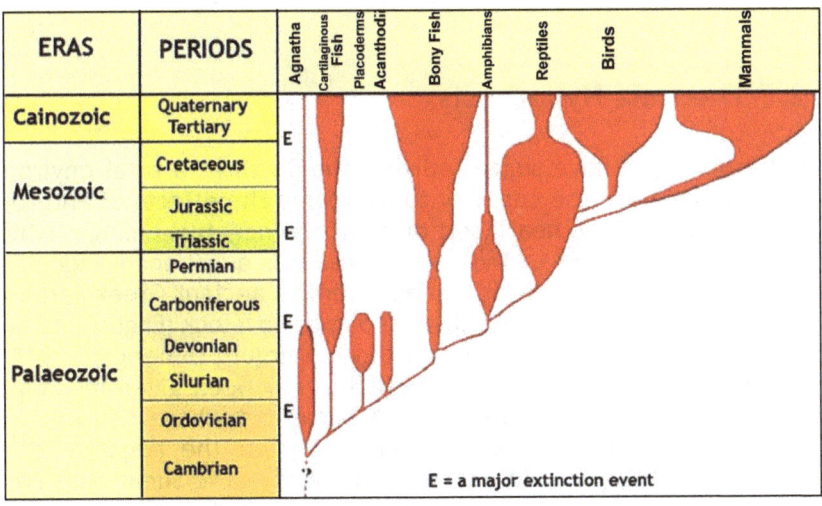

Table 2.1: A simplified chart for some major animal groups showing their development and some extinctions. Red areas are an approximation of the size of each major group.

A third theory suggests that sudden movement of the Earth's crustal plates could have cause major disruption to the habitats of the dinosaurs as well as a major climate change in some localities. It is conceivable that there may have been several factors at this time with the asteroid impact possibly being the tipping point which caused the Earth's climate to change to such an extreme that mass extinctions occurred.

In addition, some scientists have referred to a new, sixth extinction period called the:

- **Holocene extinction** for the current modern era since about 10,000 BC, mainly due to human activity. It has also been suggested that the latter part of our current epoch be renamed the **Anthropocene Epoch,** because of the destructive influence of humankind. The large number of extinctions spans numerous animal and plant species with over 800 extinctions being documented by the International Union for Conservation of Nature and Natural Resources between 1500 and 2009. This extinction possibly includes the disappearance of large mammals starting at the end of the last Ice Age. Such disappearances might be the result of the rise of modern humans and continues into the 21st century and the consequences are yet to be fully understood.

2.7 The Gaia Hypothesis

Some optimists have suggested that the Earth as a total environmental system may have the capacity to overcome the effects of anthropogenic climate change and heal itself of all of the negative changes which have occurred due to humankind's activities. This questionable idea has been called the **Gaia hypothesis** named after the ancient Greek Earth goddess and was formulated by the chemist James Lovelock (English: 1919 -) and co-developed by the microbiologist Lynn Margulis (American: 1938 - 2011) in the 1970s. Whilst this hypothesis has not been given much scientific credibility, there have been some occasions where natural processes seemed to have had some minor effect on the negative effects of humankind. For example, it has been noted in the Summer of 2012, that some icebergs in Antarctica, which had melted due to increased local temperatures have deposited iron oxides which they contained causing large blooms of algae on the ocean's surface. The iron oxides have come from wind-blown dust from distant land masses and dropped into the snow which eventually formed the glaciers which spawned the icebergs. The iron oxide then has acted as a fertilizer to increase the population of marine diatoms which remove carbon dioxide from the sea and air. There also has been some limited success in artificially seeding the ocean using iron sulfate fertilizer. When this is done, there is a sudden increase in the local surface algae population which then quickly dies and sinks to the floor of the ocean. Whilst this has been promoted as a method of sequesting, or burying, atmospheric carbon dioxide, there are still some doubts as to the long-term effects of large amounts of dead ocean-floor algae which is able to be returned to the surface after a few hundred years by upwelling ocean-floor currents.

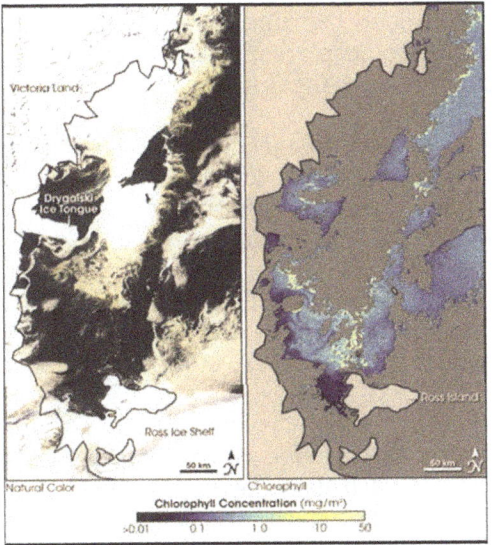

Figure 2.30 Algal blooms observed in Antarctica from NASA's Aqua satellite (Photo: NASA)

Final Remarks

1. The Earth's climate has always been in a state of natural change due to changes in: its orbit around the Sun; energy output of the Sun; movement of its tectonic plates; volcanic activity; number of meteorite impacts; and changes in ocean currents;

2. Previous studies of ancient climates (palaeoclimatology) were more concerned about the several great Ice Ages which the Earth has experienced which suggested that modern times represented an extraordinary long interglacial period;

3. More recent studies have suggested that the Earth's atmosphere and oceans have been warming due to the activities of humankind since the start of the Industrial Revolution over 200 years ago;

4. This anthropogenic warming is largely due to the Greenhouse Effect cause by the release of such gases as carbon dioxide, methane and others into the atmosphere largely due to the burning of fossil fuels, coal and oil products;

5. Evidence for the changes in climate and global warming due to these gases has come from a variety of thorough research with consistent and reliable data.

More Applications

1. Use local weather data to find out how climate has changed in the local region over (say) 100 years;

2. Look at the various evidence for natural climate change and analyse their limitations as scientific indicators;

3. Take a sceptical look at what is being said in the local media about climate change. Can you detect any bias?

4. Research the Internet about how super volcanoes can reduce global warming and find out where such volcanoes are located;

5. The old theory of Ewing and Donn predicted a new Ice Age from a clearing of the Arctic Ice. Such an event would follow massive successive snow falls in countries bordering the Arctic Ocean. Find out how much of the Arctic Ocean is free of ice and research snowfall data from Canada and northern Europe.

6. Look at the various graphs used as evidence for climatic changes and list the trends and percentage changes from each.

Further References

National Academy of Science (US) and The Royal Society (UK). Climate Change: Evidence & Causes.
http://dels.nas.edu/resources/static-assets/exec-office-other/climate-change-full.pdf
(A good summary of climate change)

National Aeronautics and Space Administration (NASA), 2019: Climate Change. How Do We Know?
https://climate.nasa.gov/evidence/

National Geographic, 2019. Is Global Warming Real?
https://www.nationalgeographic.com/environment/global-warming/global-warming-real/

National Oceanographic and Atmospheric Administration (NOAA), 2019. Ten Signs of a Warming World.
https://cpo.noaa.gov/warmingworld/air_temperature_over_ocean.html

National Research Council. 2012. Climate Change: Evidence, Impacts, and Choices: PDF Booklet. Washington, DC: The National Academies Press.
https://doi.org/10.17226/14673.

Scott, P.T., 2017. Fossils - Life in the Rocks. Brisbane, Felix Publishing. ISBN: 978-09946432-4-7

The Royal Society (UK), 2019. The Basics of Climate Change.
https://royalsociety.org/topicspolicy/projects/climate-change-evidence-causes/basics-of-climate-change/

Chapter 3: Humans and Climate Change

3.1 First Use of Energy and Resources

When trying to understand global warming one usually looks at the changes which have been happening since the start of the Industrial revolution in the 18th century. To put that in true perspective, however one must go well back to an earlier time when humankind first began to use energy and resources on a collective scale. The need for energy and resources has always been a prime concern for humankind and it will continue to be so in the future.

Natural resources have been used by humankind from the dawn of time. The first resources used were of stone for tools such as axes, animals for hides and food, timber for fire and native grains for food. The first of our ancestors were hunters and gatherers who roam specific regions following animals and seasons of plant growth. They hunted specific animals for their meat, hides for clothing and shelter, sinews for stitching and bone a variety of uses. Stones, especially hard igneous rocks were used as axe heads, grinding implements and obsidian was chipped with other rocks or teeth to make knives and spear tips. These resources came from the local environment and the energy to obtain or produce useful articles from the resources came from the use of fire and human muscle.

Figure 3.1: Flint arrow heads from early North America (Photo: US NPS)

Figure 3.2: A stone axe of metamorphic greenstone made recently by a hill tribe of Tana Island, Vanuatu.

With time, family groups amalgamated into tribes and settled in areas which provided regular supplies of natural resources without the need to wander vast territories. Animals became domesticated and wild grains were cultivated as crops. Other natural resources such as timber, stone and water were in plentiful supply if climatic seasons were good. Conservation of resources were practiced by both the hunter - gatherers and the first settlers. Nomadic tribes of the past and present had their tribal lands on which to hunt and specific animals and plants to use. Others which were in short supply or used by other local tribes were considered taboo and not taken. Some indigenous peoples such as those in Australia often carried out practices to further enhance the natural environment such as regular burning of the forest to encourage further growth of plant species which could only germinate after fire.

3.2: The Rise of Populations and Specialization

With the rise of the great civilizations in Mesopotamia (now Iraq), China, Greece, Rome, Africa and the Americas, larger numbers of peoples gathered together and built cities for mutual protection. They used the resources at hand such as timber, other plants for medicine and for weaving cloth and rope, water and wind for powering granaries to turn their grain into flour and eventually metals such as copper and tin for bronze and eventually iron. Eventually these settlements became specialized in how they used their local resources, the raw materials they processed and the products they made from them. These were traded with other settlements and in return other materials not available to the city were imported. Some important cities developed because of their specialization e.g. Salzburg (salt town) in Austria became a prosperous city from the salt mines began by local Celtic tribes over 3000 years ago. Salt was an important commodity used in preserving food.

Online Video 3.1: Travel underground into the ancient Celtic salt mine at Hallein, near Salzburg, Austria.
Go to https://www.youtube.com/watch?v=QCj7CjsYqHw

Metals such as copper, gold, silver, lead, mercury and iron have been used since ancient times with tin and zinc being alloyed with copper to form bronze and brass respectively as these are harder than copper. Gold, silver and copper, often found in the native state, have been prized since the **Neolithic Age** as metals of trade and adornment. Later they were used to great effect as crafted ornaments, utensils and coins. Lead was smelted in ancient Egypt. It was later used by the Greeks and Romans as a coating in their copper pots and Romans used lead pipes as plumbing to carry water into their homes. The old Latin name for lead is *Plumbum* and the Greeks and Romans also suffered from the toxicity of the metal as lead poisoning. Other important metals such as nickel, manganese, molybdenum, tungsten, titanium and chromium were not isolated as separate metals until the 18th Century.

Figure 3.3: The Pyramids at Giza, Egypt are made of huge sandstone blocks which were covered with a highly polished layer of white limestone, some of which still remains at the top of the pyramid at right

A simple development of humankind is shown in the following table:

AGE	TIME	EVENT	ENERGY
Stone Age (Neolithic Age)	Prehistoric times lasting approximately 3.4 million years and ending between 8700 Before Common Era and 2000 BCE in most continents - later in some remote areas.	Stone was widely used to make implements with an edge, a point, or a grinding surface.	Mostly human and the invention of fire.
Bronze Age	From about 3300 BOE to 2100 BCE following a brief time when some civilisations were able to smelt copper. The harder bronze was smelted using copper and alloying it with tin, arsenic, or other metals and was widely traded.	Rise of many of the great civilizations in Egypt, Asia, Mesopotamia, Europe and South America. Many forms of written script also occurred at this time.	Human, beasts of burden (horses, camels, llamas, donkeys, oxen). Animals to turn grinding wheels. Early development of water and wind power. Wheels (except in the Americas)

Iron Age	Varies from place to place but usually from about 1200BCE to about 800CE in Europe. Iron continued to be made as the main metal for tools and weapons through the Medieval times and into modern history.	Smelting of iron ca steel containing about 2% carbon using iron ore, charcoal and forced air using a bellows.	Humans and beasts of burden used to power mills and hoists. Windmills and water wheels.
Industrial Revolution	This was the dynamic transition to the new manufacturing processes in Europe from about 1760 which continued into the 19th century and then on to the modern era of manufacturing, mining and industry.	Transition from hand-operated machinery in home-based industries to large factories and mines using large-scale machinery and production methods	Development of steam power in the 18th century and later in the 19th century of electrical power.

Table 3.1: A brief outline of the development of resources and energy

Aluminium, perhaps one of the most common elements on Earth was not isolated as a metal until 1827 and uranium was not prepared as a metal until 1841 although it had been known as the oxide as a yellow pigment since the later 18th Century.

Figure 3.4: A Persian gold Daric coin from about 420 BCE (Photo: Wiki Commons)

Energy use became more efficient as humankind developed machines. Initially, the only source of energy was the strength of groups of workers but later the power of the wind and of water were put to use to grind grain into flower, lift heavy weights in construction and mining as well as to lift water for irrigation. The use of water wheels to raise more water for inland irrigation was invented by the ancient Egyptians whilst the Greeks and later the Romans became masters of using water wheels for domestic and construction purposes. The use of water and wind as a source of cheap power continued through the Middle Ages to modern times with today's hydroelectrical systems and wind turbines.

Figures 3.5 and 3.6: A horizontal water wheel and its grinding stone on the floor above at the Molino de Sabandia, near Arequipa, Peru, constructed in 1785.

3.3 Renewable and Non-renewable Resources and Energy

In the Neolithic era, humankind in most of Europe were hunter gatherers who hunted within well-defined areas of forests and grasslands and relied on the natural resources of stone and timber as well as native animals and plants. Energy needs were simple and usually made use of fallen timber and human power for simple construction and transportation. Resources and energy were both renewable and sustainable for the relatively small population.

With the development of larger villages and cities and a rapid increase in the world's population, the need to mine and smelt metals, to quarry resources such as stone, gravel and clay and to cut timber became more of a necessity. By about the 16th century, Europeans had begun to clear so much forest for their construction, energy and agricultural needs that they had difficulties in acquiring adequate food and energy supplies. With the depletion of wild animals following deforestation and inadequate numbers of domesticated livestock and inefficient agriculture, they were faced with a nutritional decline. They were only saved by new improvements in agriculture following the demise of the feudal system and the importation of new foods such as the potato and maize from the New World.

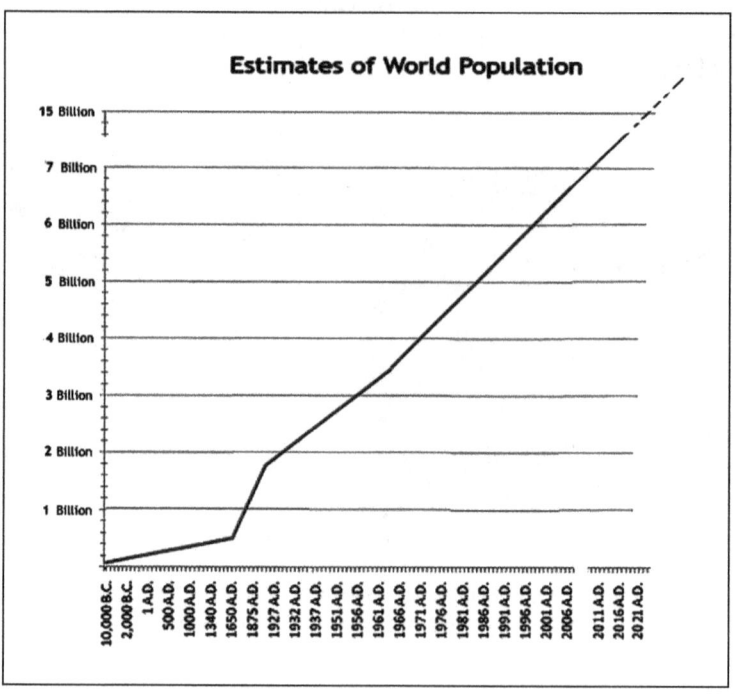

Figure 3.7: Graph showing an estimate of world population increase (Data: UN and US Census)

Massive deforestation in Europe and especially Britain was also assisted by the use of **charcoal** on an industrial scale for smelting and shaping of metals and metal products. Charcoal is produced by the slow heating of wood in the absence of oxygen, a process called **pyrolysis**, by firing the wood and then covering it or burning the timber in special kilns. Charcoal consists largely of carbon and is a condensed fuel which can be easily transported, stored and used in a variety of functions such as cooking, smelting metals and re-working metals in a blacksmith's forge. The advantage of using charcoal instead burning wood is that in charcoal, a considerable amount of water, ash and other components in timber have been removed. This allows charcoal to burn at a higher temperature with less smoke and ash remains. Charcoal and other natural renewable resources are still used today, especially at the rural village level in many countries.

Figure 3.8: Old charcoal kilns in Idaho, United States (Photo: US NPS)

In India, the Middle East and Africa and other places where timber is difficult to find, animal dung from cows and camels is often heaped in piles to dry. In India, some of these small piles are covered with plastic sheeting and a small tube is used to carry methane gas from the pile's decomposition into the home as a source of fire for cooking.

The development of the Industrial Revolution required a larger supply of raw material in the furnaces of the new steam engine boilers. This led to the large-scale use of coal, as the steam engine took over from the water wheel. Whilst coal had been known for centuries and had been used as a fuel in China, it was not until the beginning of the 18th century that large-scaled mining of coal for industrial use began to make use of this resource. Coal has also been used for the production of steam to run turbines to produce electricity since the 1880's and hydroelectricity began to be used from about the same time.

Oil from below the ground had been used for centuries, but it was only obtained from natural soaks and tar pits where the oil had come to the surface. Later in the mid-19th century, a more efficient way of pumping oil from beneath the surface led to its use as a fuel. Natural oil obtained from plants and animals, especially from the whale, had been used for a great number of uses such as lubrication, heating and light but now there was a more convenient source.

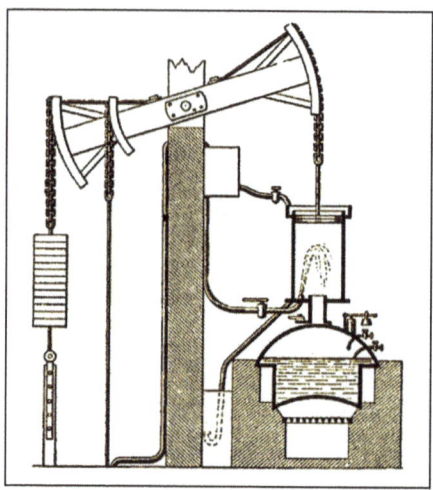

Figure 3.9: A plan of Thomas Newcomen's 1712 Beam Engine powered by steam to pump water from mines.

Before and during the Industrial Revolution, the conservation of natural resources and the concept of renewable and non-renewable resources and energy were little understood. It did not fit into the mode of thinking that these natural resources were to be used regardless of the consequences and that their depletion and the waste products they produced would one day be a major world-wide problem. To some extent this attitude still persists amongst some of the developers in industrialised nations. The need for resources and cheap power has become a major priority and the consequences of their depletion and of the increase in waste and pollution from their use is still ignored by many because this is the way things have always been. It is only in modern times that there has been an awareness that most of the world's resources and energy come from non-renewable sources and that sooner or later they will have to be replaced by more sustainable renewable sources. The difficulty here is that with an ever-increasing demand for resources and energy, renewable sources cannot meet these demands without appropriate changes in attitude to their use, their costs as well as to the removal of waste and pollution.

In summary, resources and energy have and continue to be obtained mainly from non-renewable sources. These resources include metallic and non-metallic material resources, water and energy resources all of which have a limited life-time at their current use. Renewable resources such as crops, vegetables and domestic animals have been used in the past, sometimes excessively and some renewable energy sources have been exploited in those countries having available hydroelectricity and geothermal power. With current, traditional resources in material resources and energy resources becoming depleted, more attention is needed in establishing permanently sustainable substitutes. In addition to the development of these new resources through scientific and industrial innovation, there will also need to be many social changes in how the world's population uses and manages its resources and limits the amount of waste and pollution caused by resource use and production. The chart below lists some of the major material and energy resources:

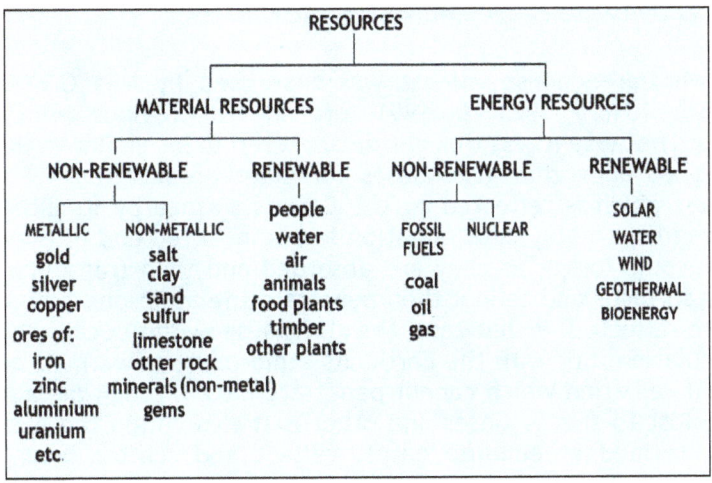

Figure 3.10: A chart showing the main types of renewable and non-renewable

3.4 The Greenhouse Effect

Since the start of the Industrial Revolution, when the burning of coal for steam became the dominant source of energy, the levels of carbon dioxide gas in the atmosphere has increased. Whilst early attempts in environmental control were centred on cleaning up the fumes and smoke of the cities and of the pollution in the waterways, little attention was given to the increase atmospheric warming. Throughout the 19th century, scientists such as Joseph Fourier (French: 1768-1830), Eunice Newton Foote (American: 1819-1888) and Svante Arrhenius (Swedish: 1859 -1927) argued that the increase in carbon dioxide would heat up the atmosphere by a similar process as observed in green houses used in plant propagation.

This **greenhouse effect** is another natural process in climate change and has been around since the development of an atmosphere containing carbon dioxide, methane water vapour and to a lesser extent oxide of nitrogen. The problem since the advent of the Industrial Age is that the increase of the emissions of these gases have greatly increased the retention of heat within the atmosphere causing an alarming increase in global temperatures. It should be remembered that due to the heat capacity of the ocean particularly, but also the gases in the atmosphere and the Earth's surface generally, a rise in only a few Celsius degrees represents a huge amount of retained heat which can have catastrophic consequences for the natural processes of the weather and seas and in turn for life on the planet. Such changes to the environment caused by human activity are called **anthropogenic** changes.

The term 'greenhouse effect' was first used by Nils Gustaf Ekholm (Swedish: 1848 – 1923) in 1901 referring to previous work of other scientists because it was thought (incorrectly) to be similar to the process which goes on within glasshouses for plant propagation. Of the solar radiation which is reflected by the Earth's surface by its albedo, some wavelengths, notably heat radiation in the infra-red end of the spectrum have become longer as they are absorbed and then transmitted by the Earth's surfaces and cannot then penetrate the atmosphere above where they are reflected. It that case, the glasshouse warms because of reduced air circulation, but with the Earth, its atmosphere is warmed because of the heat radiation which cannot penetrate back through the atmosphere and be lost to space. Gases and other particles within the atmosphere, loosely termed greenhouse gases, reflect and scatter heat radiation coming from the surface and so add to its overall temperature. The primary natural greenhouse gases in Earth's atmosphere are water vapour (H_2O), carbon dioxide (CO_2), methane (CH_4), nitrous oxide (N_2O) and ozone (O_3). It is thought that without these greenhouse gases, the average temperature of the Earth's surface would be much lower, perhaps as low as about –18 °C rather than the present average of 15 °C. The high surface temperature and dense atmosphere of Venus, which contains mainly carbon dioxide as well as other greenhouse gases, may have come about because of a runaway greenhouse effect.

Since the Industrial Revolution from about the late 18th century, particles and gases from industrial activities and subsequent pollution has added an enormous amount of these gases to the air as well as others such as sulfur hexafluoride (SF_6) and carbon monoxide (CO). Carbon monoxide has only a minor effect by itself but it reacts with hydroxyl (OH^-) radicals in the atmosphere, reducing their abundance. Hydroxyl radicals reduce the amount and therefore the lifetime of strong greenhouse gases like methane and so carbon monoxide indirectly increases global warming.

Also see a good animation at:

https://climate.nasa.gov/causes/

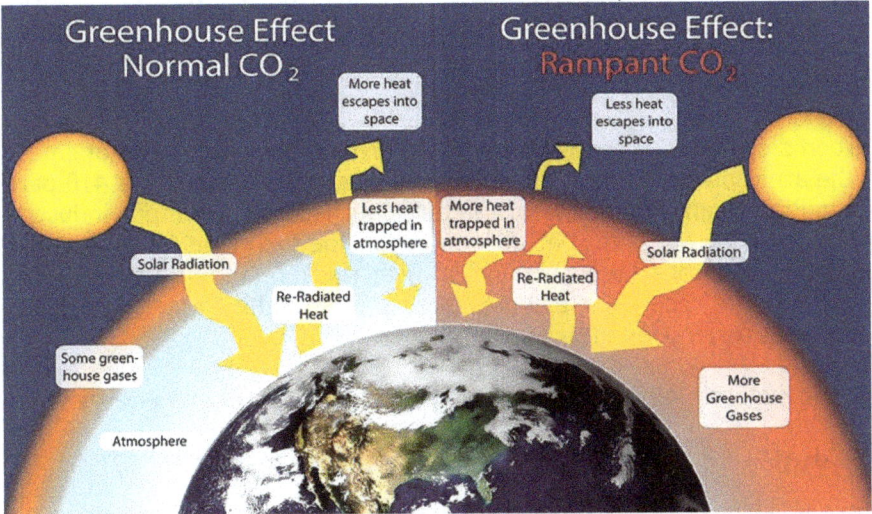

Figure 3.11: Diagram showing a simplified view of the Greenhouse Effect (Will Elder US NPS)

For a very good set of interactive charts with data from separate countries see:

https://ourworldindata.org/co2-and-other-greenhouse-gas-emissions

Other gases and pollutants which have been produced as a result of human activity and which also add to an increased greenhouse effect will be outlined later. Of the natural greenhouse gases, their contribution to overall warming is:

- water vapour, 36-70%
- carbon dioxide, 9-26%
- methane, 4-9%
- ozone, 3-7%

Water vapour enters the atmosphere through the processes of the Water Cycle, including evaporation, ejection of volcanic gases, water loss from the leaves of plants by **transpiration**, and from respiration of all organisms. Carbon dioxide is also a product of decay, volcanic eruption and respiration as well as from the various mineral reactions. There is a two-way equilibrium between this gas dissolving in the sea and rivers and it being released to the atmosphere.

Carbon dioxide (CO_2) and methane (CH_4) are the two main gases of carbon in the atmosphere. Carbon dioxide gas makes up about 0.04% or 410 parts per million (ppm) by volume of the air. This has risen from pre-industrial levels of about 280 ppm. Natural sources of the gas come from volcanoes and associated thermal events and from chemical reactions between acids and carbonate rocks such as limestone, marble and dolomite. It is also a product of respiration in living things:

simple sugars	**oxygen**		**carbon dioxide**		**water**		**energy**
$C_6H_{12}O_6$	+ $6O_2$	→	$6CO_2$	+	$6H_2O$	+	ATP

(ATP is the biochemical adenosine triphosphate which is stored in living cells as a source of energy)

Carbon dioxide is also very soluble in water (1.45 g/L at 25 °C, 100 kPa) and so occurs naturally in all water and ice environments. Data for CO_2 levels in climate change studies often come from air bubbles in drill core samples of polar ice caps and glaciers. It is also present in deposits of petroleum and natural gas.

Naturally - occurring methane is also produced by the breakdown of organic matter by microorganisms by the process of **methanogenesis** used by these microorganisms as an energy source. The net reaction is:

carbon dioxide	hydrogen ions	electrons		methane		water
CO_2	+ $8 H^+$	+ $8 e^-$	→	CH_4	+	$2H_2O$

Methane gas is also found both below ground and under the sea floor within sedimentary rocks where it was formed by the decomposition of organisms and as a gas of excretion in many organisms. It is a major gas found in coal, petroleum deposits and trapped in frozen soil in the Arctic tundra of Canada, Alaska and Russia. With the warming of these regions, additional methane gas would be released to the atmosphere. In the atmosphere, its concentration has increased by about 150% since 1750, and whilst it accounts for only 20% of greenhouse gases it has about 28 times more heat absorbing power of heat than carbon dioxide. However, the methane molecule (CH_4) in the atmosphere and has only a short lifespan and is removed by its reaction with the hydroxyl radical (OH^-). Methane gas levels are normally held in check by the hydroxyl radical (OH^-), which removes most of it out of the atmosphere within a decade.

Unstable hydroxyl group		methane		water		unstable methyl group
OH^-	+	CH_4	→	H_2O	+	CH_3^-

The methyl (CH_3^-) radical then reacts rapidly with oxygen to form the methylperoxy radical (CH_3O_2) and by further reaction forms the more stable formaldehyde molecule (CH_2O), which has a lifetime of about 5-8 hours in sunlight.

The hydroxyl radical is formed in the presence of sunlight by water vapour and pollutants like ozone and nitrogen oxides but it persists for just a short time in the air before it reacts with other agents. The increase in methane levels in the atmosphere however, could be due more to increased emissions from industry, agriculture and previously trapped sources such as the permafrost. Studies from NASA suggest that the reaction between methane and the hydroxyl radical appears to have remained constant.

See also:

https://Earthobservatory.nasa.gov/images/144358/detergent-like-molecule-recycles-itself-in-atmosphere

Carbon dioxide is removed from the atmosphere mainly by photosynthesis in green plants on land and in the sea. It also is removed from the air by dissolving in the waters of the hydrosphere. Dissolved in water, carbon dioxide forms weak carbonic acid (H_2CO_3), which contributes to ocean acidity and dissolves out carbonate rocks.

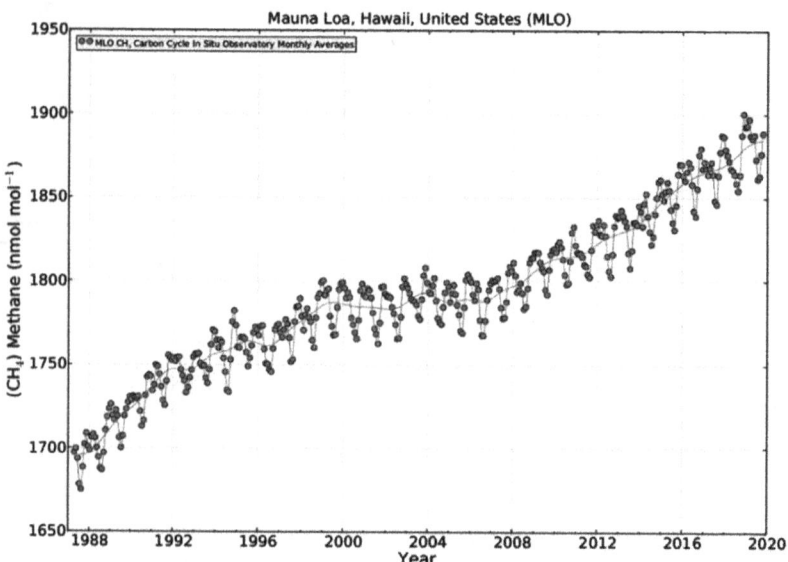

Figure 3.12: Atmospheric methane levels (Data: NOAA)

The ocean is considered as a sink or storage trap of carbon dioxide gas. The absorption of carbon dioxide in the ocean is increased by its ability to simply dissolve the gas by the marine carbonate buffer system, which transforms CO_2 into carbonate ions (CO_3^{2-}). However, this buffering capacity of the ocean becomes saturated as the CO_2 concentration in the surface ocean increases, and the ability of the ocean to act as a CO_2 sink will therefore decrease in the future. This will cause an increasingly large fraction of CO_2 emissions from human activity (anthropogenic sources) to stay in the atmosphere. Also important is the amount of photosynthesis by phytoplankton taking place in the upper waters of the sea and if this rate of marine photosynthesis increases, there may be some decrease in CO_2 absorption by the oceans.

3.5 The Use of Fossil Fuels

By far the dominant cause of the increase in carbon dioxide gas in the atmosphere is by the use of fossil fuels such as coal from near the end of the 18th century and later petroleum from the mid-19th century. At the beginning of the Industrial Era from the late 18th century, coal was seen as the ultimate source of energy to power transport, operate factories, heat buildings and later generate electricity in coal-fired power stations.

With the discovery of large reserves of oil from the mid-18th century, it became the dominant fuel source in transportation. Until relatively recent times, there was little control over the emissions from coal-burning factories and power stations and industrial cities soon became centres of a sooty, smog-ridden atmosphere. Even in the 1950's, the author recalls layers of black soot settling on the washing hanging outside.

In simple terms, much of the carbon dioxide and other pollutants come from combustion from fossil fuels:

- **Combustion of carbon in coal:**

$$\text{carbon} \quad \text{oxygen} \quad \text{carbon dioxide}$$
$$C + O_2 \rightarrow CO_2$$

(Burning of coal also gives a large amount of soot and flue gas which contains mostly carbon dioxide but also carbon monoxide, water vapour and oxides of sulfur and nitrogen which are also harmful).

- **Combustion in internal combustion engines** e.g. octane in gasoline (a mixture of C-H-O compounds of low carbon value):

$$\text{octane} \quad \text{oxygen} \quad \text{carbon dioxide} \quad \text{water}$$
$$2C_8H_{18} + 25O_2 \rightarrow 16CO_2 + 18H_2O$$

(This is a simplified equation as there is usually incomplete combustion with carbon monoxide being produced as well as oxides of nitrogen, unburnt hydrocarbons and other products. Much of these gases are reduced using catalytic converters on car exhausts. Diesel engines burn oil with a larger carbon number and give similar results).

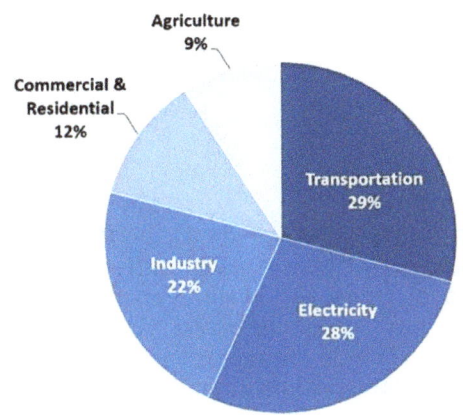

Figure 3.13: Greenhouse gas emission sources in the United States in 2017 (Source: US EPA)

Coal and oil have many other uses apart from producing energy by combustion. Much of the synthetic material created over the last two hundred years have some origin in these fossil fuels. It is most probable that in time, the use of fossil fuels as an energy source will be limited to those fitted for zero-emissions or discarded in favour of renewable sources. However, unless renewable alternatives can be found, many of the current synthetic resources may be continued to be manufactured, hopefully with zero or very limited waste. A few of the useful products from coal and oil are given in the following table:

COAL AND ITS WASTE ASH	OIL PRODUCTS OTHER THAN LUBRICANTS AND FUELS
coke for iron & steel	solvents
dyes	plastics
plastics	artificial rubber
graphite	inks
abrasives	synthetics (threads such as nylon, rayon etc.)
baking powder	
batteries	pharmaceuticals
chalk	dyes
concrete (small pieces)	tar and bitumen
fertilizer	anaesthetics
insulation	waxes and candles
linoleum (some)	ammonia
mothballs	refrigerants
paint	detergents & soaps
perfumes	glycerine
synthetic rubber	antifreeze liquids
chemicals e.g. sugar substitute	paints

Table 3.2: Uses of coal and oil other than for fuel

3.6 The Effect of Land Clearing

Another anthropogenic problem is the effect of land-clearing. Over the last two hundred years many countries have suffered massive deforestation as land is cleared of trees and major vegetation for the purposes of mining, agriculture, grazing and urbanisation. This greatly reduces the amount of carbon dioxide removed from the air by photosynthesis, the process by which green plants make sugars for energy:

$$6CO_2 + 12H_2O \xrightarrow[\text{Chlorophyll}]{\text{light}} C_6H_{12}O_6 + 6O_2 + 6H_2O$$

(Carbon dioxide + water → glucose + oxygen + water)

(This is also a simplified combination of several reactions within the green chloroplast bodies in plant cells)

Vegetation also assists in retaining topsoil and soil moisture by its root systems and evaporation and transpiration (**evapotranspiration**) from plant leaves. The latter process sees water being drawn up through the soil, into the roots of plants and then exiting through the stomata in the leaves. This also increases the humidity of the air above and ensures a cooler, wetter climate. Removal of vegetation limits these processes and also increases the amount of bare land which will then absorb solar radiation and reemit this as heat which reduces the biodiversity of the area. As a consequence of land clearing, the air above becomes dryer, the soil moisture is lost, topsoil can then be eroded, local streams are reduced by evaporation and clogging with debris and the local climate becomes dryer. CSIRO modelling has suggested that there is a strong correlation between climate change and loss of vegetation since European settlement, with an average Summer temperature increase in eastern Australia of from 0.4°C to 2.0°C, and a 4% to 12% decrease in Summer rainfall. The modelling also showed that there has been an increase in the intensity and duration of droughts in south-eastern Australia as a result of large-scale clearance of native vegetation. In addition, the running of sheep, cattle and other hooved animals on cleared land in countries such as Australia where there were no hooved animals before European settlement means that the soil is broken up more than without grazing. Also, the soils become more enriched with nitrogen-based products from grazing which break down and add to the methane gas produced by the animals themselves.

See also:
https://www.climatecouncil.org.au/uploads/c1e786d5d0fe4c4bc1b91fc200cbaec8.pdf
(Land clearance in Queensland)

https://environment.des.qld.gov.au/wildlife/threatened-species/documents/land-clearing-impacts-threatened-species.pdf
(Impact of land clearing)

Final Remarks

1. With increased gases such as carbon dioxide and methane in the atmosphere, more heat radiation from the Earth's surface fails to be reflected back into space and so is trapped and heats up the lower atmosphere. This is known as the Greenhouse Effect;

2. The main source of the additional greenhouse gas emissions in the atmosphere in modern times is the burning of fossil fuels such as coal and petroleum products;

3. Any combustion of carbon-based compounds such as coal or hydrocarbons will produce large amounts of carbon dioxide as well as smaller amounts of carbon monoxide (unstable) and the oxides of sulfur and nitrogen from impurities as well as other compounds and particulate matter;

4. Land-clearing for agriculture or grazing greatly reduces the amount of carbon dioxide which can be removed from the atmosphere by plants. Moreover, the open land now radiates more heat back to the atmosphere and much less water vapour.

More Applications

1. Reduce the amount of combustion of fossil fuels by supporting renewable energy campaigns, shopping around for home energy deals which use renewable energies, and being more conservative in the use of energy in the home, office and in industry;

2. Reduce the amount of unnecessary use of automobiles by walking/riding bicycles short distances, taking public transport; and car-pooling for work;

3. Plant more trees and shrubs in the home area. Apartment dwellers can have window/balcony gardens or combine with fellow tenants to have roof gardens or garden/vegetable plots around the apartment buildings;

4. Advocate the reduction in land clearance. Those on the land can plant more fast-growing trees (e.g. eucalypts) along fence lines and roads or as specific animal shelter clumps;

5. Support the removal of native forests by forestry unless there is selective logging with extensive re-planting. Consider soil characteristics when planting fast-growing plantation species.

6. Try to find alternatives for or reduce the amount of plastics and synthetics in the home.

Further References

Andrews, Timothy; Betts, Betts; Booth, Ben; Jones, Chris; et al. (2016). Effective radiative forcing from historical land use change. Climate Dynamics. **48** (11-12): 3489-3505. doi:10.1007/s00382-016-3280-7

Fiske, S.J., Crate, S.A., Crumley, C.L., Galvin, K., Lazrus, H., Lucero, L. Oliver-Smith, A., Orlove, B., Strauss, S., Wilk, R. 2014. Changing the Atmosphere. Anthropology and Climate Change. Final report of the AAA Global Climate Change Task Force. 137 pp. December 2014. Arlington, VA: American Anthropological Association.
https://s3.amazonaws.com/rdcms-aaa/files/production/public/FileDownloads/pdfs/cmtes/commissions/upload/GCCTF-Changing-the-Atmosphere.pdf

Powell, James Lawrence (2017-05-24). The Consensus on Anthropogenic Global Warming Matters. Bulletin of Science, Technology & Society. **36** (3): 157-163. doi:10.1177/0270467617707079.

Skeptical Science, 2019. Empirical evidence that Humans are causing global warming.
https://skepticalscience.com/empirical-evidence-for-global-warming.htm

Chapter 4: Fossil Fuels

4.1 The Need for Energy

The development of humankind from simple hunter-gathers to urbanized farmers also saw the development of technology in building, agriculture, transportation and then industrialization. Prior to the Industrial Era or the **Industrial Revolution**, humankind relied upon renewable sources of materials and energy such as wood from forests, animals, water and wind power. In rapidly developing countries such as Britain, wood as a fuel source soon meant that many of the great forests there had been depleted. The discovery that coal could be used as a convenient fuel source soon led to this new revolution.

The Industrial Revolution, was the rapid development of new manufacturing processes in Europe and the United States from about 1760. This transition included going from hand production methods to machines in factory systems using steam-power generated by burning coal. Later, many of the coal-fired machines were replaced by those burning petroleum products including natural and coal-seam gas. The Industrial Revolution also led to an unprecedented rise in the rate of population growth with the problems of over-crowding and waste production in the new industrialised cities.

4.2 Fossil Fuels – Coal

Coal is formed from the breakdown of large volumes of plant matter underwater without air, that is under **anaerobic** conditions. This usually occurs when plant material such as the ferns, palms and other simple plants of past times have fallen into the quiet waters of freshwater swamps or **paludal** environments and lakes or **lacustrine** environments.

Coalification is the process which then turns compressed vegetation into coals with the loss of water (H_2O), carbon dioxide (CO_2) and methane (CH_4). The extent and rate of this process depends upon:

- type of original vegetation

- depths of burial

- temperature and pressure at these depths

- length of time

With time and the temperature and pressure associated with burial, compressed vegetation begins to transform into different **ranks** of coal with varying amounts of components such as:

- fixed carbon which is the useable organic fuel component as carbon;

- ash which is the silica content which will be left over when the coal is burnt and which also contains some toxic materials such as the salts of arsenic, lead, selenium, chromium and mercury;

- moisture which is the remains of the water expelled by the coalification process which hinders burning but is liberated as steam during burning; and

- volatiles, except for moisture, are usually a mixture of short and long chain hydrocarbons, aromatic hydrocarbons, nitrogen compounds and sulfur.

A generalised list of the chemical components of bituminous coal would be:
>carbon (C), 75-90%;
>hydrogen (H), 4.5-5.5%;
>nitrogen (N), 1-1.5%;
>sulfur (S), 1-2%;
>oxygen(O), 5-20%;
>ash (mainly silica SiO_2), 2-10%; and
>moisture (H_2O), 1-10%.

Generally, the rank of coal increases with depth if the thermal gradient or increase in temperature with depth, is entirely vertical. This relationship is called **Hilt's Law**. However, there may be some changes in rank due to other external changes, such as the heating effects of local igneous intrusion or pressure by large scale rock folding. The main ranks of coal are:

- **Peat** is often considered only a precursor to coal as it consists of considerable vegetable matter with little coalification, high percentage of water, silica and clay (as soil). It is found extensively in parts of northern Europe in large areas of marsh and bog. It has, however, been a source of poor fuel for centuries and is still dug for local consumption in Ireland and Finland.

- **Lignite** (or brown coal) is the lowest true rank of coal, having about 25-35% carbon but still high in moisture content and amount of silica (or ash). It is useful for generating heat in power stations but is considered environmentally dirty.

- **Sub-bituminous coal** has 35-45% carbon content, about 15-30% moisture and considerable ash content. It is not suitable for making coke, the solid fuel made from distilling coal, but it is useful as a fuel in steam boilers. Together with lignite, this rank makes up the largest reserve of the world's coal.

- **Bituminous coal** is the typical hard, black coal used in making coke for industry and most high-grade steam-generation boilers. It contains bright and dull bands and contains 45-86% carbon with some moisture and ash.

- **Anthracite** is a very hard, often glossy in lustre, and is very brittle. It has 86-98% carbon and is considered to be a transition towards pure carbon as graphite which forms after further changes by heat and pressure. It is used mainly in metallurgy.

Figures 4.1 & 4.2: Lignite or brown coal showing a woody texture (left) and black coal with bright bands of vitrain.

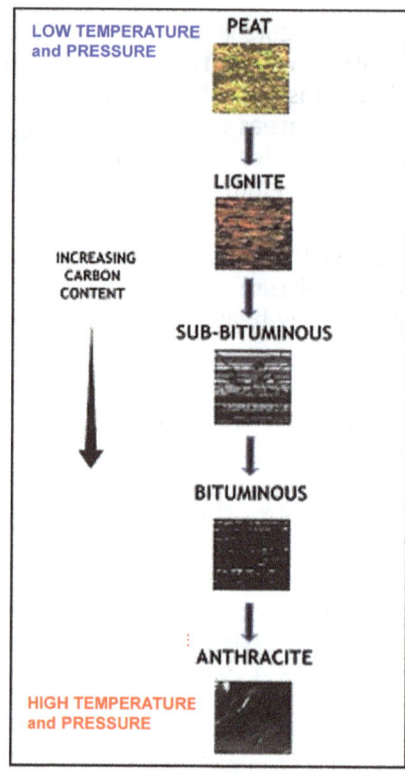

Figure 4.3: The ranks of coal formed by increasing temperature and pressure with depth.

About three metres of compressed vegetation will eventually form one metre of black coal, and because of the nature of the deposition of the vegetation into freshwater lakes and swamps, the coal is formed in layers called **seams**. These are found within the originally horizontal sequence of sedimentary rocks, usually with shales with sandstones and conglomerates as coal measures. Seams can be many metres thick and the area of the seam may be several hundred square kilometres. Coal measures may contain several seams, representing different periods of deposition. Much of the world's coal was formed in the Carboniferous Period (359.2 to 299 million years ago), although considerable coal formation occurred in more recent times in the Permian Period (299 to 251 million years ago) and the Triassic Periods (251 million and 199 million years ago), especially in the Southern Hemisphere.

Coal can be mined by several methods, and coal mining has been a major industry since the 18th century in Europe and America. The main techniques of coal mining are:

- **Strip mining** by mining on the surface as huge open-cut strips if the coal seams are close to the surface, are mainly horizontal, and cover a sizeable area. In responsible mining, extensive environmental work is undertaken before the mining commences. Native vegetation is sampled and seedlings are grown in greenhouses or generally stored and corridors for the movement of the local fauna are retained. Only then does true mining take place with the removal and storage of valuable topsoil and then the overburden from the mine itself. Water quality in the natural water table as well as in and around the mine is usually monitored constantly by the professional environmentalists employed by the mine. As mining continues across the lease, the unwanted debris and topsoil are replaced into the exhausted pit by back-filling. The surface re-sculptured and re-planted with native vegetation and fauna are encouraged to return to the area.

Figure 4.4: Dragline (right) at an open pit coal mine, Queensland, Australia

See also:

https://www.im4dc.org/wp-content/uploads/2014/01/Mine-rehabilitation.pdf
(General outline of mine rehabilitation)

https://www.industry.gov.au/sites/default/files/2019-04/lpsdp-mine-rehabilitation-handbook-english.pdf
(A good outline of mine rehabilitation from the Australian Government)

Unfortunately, good practice in some mining operations, especially open pit coal and metallic resources, has not taken place in the past or in some areas at present. Rehabilitation and environmental legislation either were not in place or enforced. This is especially evident in some Third World countries where environmental issues were not major concerns of the government despite opposition from local people. In countries where mining has been on a large scale, there are many areas which have not been rehabilitated in the past. Apart from the destruction to the surface environment, these abandoned areas also contain the toxic waste products of the mining operation and of the processing of the raw material extracted.

See also:

http://siteresources.worldbank.org/INTOGMC/Resources/notoverwhenover.pdf
(A good general overview of mining rehabilitation problems from the World Bank)

https://s3.treasury.qld.gov.au/files/better-mine-rehabilitation-in-qld-discussion-paper.pdf
(a very general paper about mine rehabilitation in Queensland, Australia with some ideas as to how the Public can make comment)

https://www.epa.gov/superfund/abandoned-mine-lands
(Good site by the Environment Protection Agency of the United States with many general and specific links).

https://www.abandonedmines.gov/
(Another good site from the US about abandoned mines with good links and data).

https://www.undp.org/content/dam/undp/library/Sustainable%20Development/Extractives/UNDP-MMFSD-LowResolution.pdf
(Mining rehabilitation in developing countries from the United Nations)

- **Sub-surface mining** such as the bord-and-pillar method which is used when the seams are deep, close to centres of habitation or greatly tilted and faulted. Tunnels or rooms (bords) are cut leaving pillars to support the roof. These are also then cut allowing the roof to cave in from the back of the mine to the entrance. Another sub-surface technique is the **Longwall technique** in which a special tracked runway is placed parallel to the face of the seam and a large toothed excavator runs along the track excavating the coal which is removed by a conveyor belt. The track is then moved closer to the seam, being protected by hydraulically-operated roof supports.

Figure 4.5: Longwall excavator – notice the hydraulic ram supports at top right (Photo: Peabody Australia)

 Online Video 4.1: Visit an open-pit coal mine in operation in Queensland, Australia. Go to https://youtu.be/9Z9a02oGfBQ

The main coal producing countries, the United States, Russia, China, Australia and India still regard coal as a major fuel and source of other manufactured materials. However, there is a global attitude that coal as a fuel source should be dramatically cut back due to the global warming effects of the massive amount of carbon dioxide, water vapour and heat emitted by coal-fired power stations.

Country	Anthracite & Bituminous	Sub-Bituminous	Lignite	Total	Percentage of World Total
United	108,501	98,618	30,176	237,295	22.6
Russia	49,088	97,472	10,450	157,010	14.4
China	62,200	33,700	18,600	114,500	12.6
Australia	37,100	2,100	37,200	76,400	8.9

Table 4.1: Proved recoverable coal reserves at the end of 2011 (in millions of tonnes) (Data from World Energy Council survey, 2013)

Whilst the environmental impact of subsurface mines is considerably less than open pit mines, there still are major concerns about mine rehabilitation. Often, the spoil or unwanted component of the mine is dumped in close proximity to the surface exits. This has produced many large dumps of solids and mine water which can sometimes be poorly contained. The issue of unstable tailing dams, especially has been a major problem in some countries where their failure has led to environmental and urban damage and sometimes deaths. An extensive list of such failures can be found at:

https://www.wise-uranium.org/mdaf.html

Moreover, many of the extensive workings of subsurface mines, especially coal mines have been built over by towns and cities which came from the mine development or were there before the mining operations started. The collapse of tunnels, the danger of old shafts at surface and the risk of toxic processing materials now buried under habitable areas are just a few of the environmental problems associated with irresponsible subsurface mining.

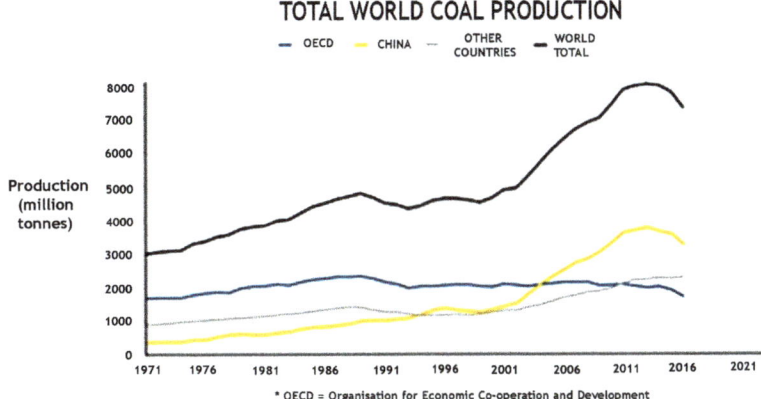

Figure 4.6: Total World Coal production (in millions of tonnes) (Data the International Energy Agency, 2017)

See also:

https://web.mit.edu/12.000/www/m2016/finalwebsite/problems/mining.html
(General environmental problems of mining)

https://www.americangeosciences.org/critical-issues/faq/can-we-mitigate-environmental-impacts-mining
(General notes about mining and the environment from the American Geosciences Institute)

Some useful sites about coal and its future can be found at:

https://www.eia.gov/beta/international/data/browser/#/?pa=00000000000000000
00000000000000000000000000000000000001&c=ruvvvvvfvtvnvv1vrvvvvfvvvvvvfvvvo u20evvvvvvvvvvvvuvo&ct=0&tl_id=1-A&vs=INTL.1-6-AFG-MST.A&vo=0&v=H&start=2013&end=2015
(Interactive map showing the future coal reserves of most countries)

https://www.iea.org/coal2017/
(International Energy Agency report - gives the current demand for coal)

https://www.worldcoal.org/file_validate.php?file=uses_of_coal(01_06_2009).pdf (How coal is used around the world)

https://coaltransitions.files.wordpress.com/2018/09/coal_australia_final.pdf (Transition of coal in Australia)

https://web.mit.edu/coal/The_Future_of_Coal_Summary_Report.pdf (Good summary of the future of coal)

4.3 Fossil Fuels – Oil and Gas

Crude oil and the natural gas which often comes with it are formed from the vast quantities of microscopic dead marine organisms which fall into and accumulate in the muds or ooze, of the deep oceans. In time, this is covered by more marine sediments, mainly muds and sand, and the organic debris goes through complex changes under anaerobic conditions to become oil.

The geological environment also plays an important part in the organic development of oil and gas. If the organic matter contained in the marine muds has been buried to only a relatively shallow depth, then the temperature and pressure conditions will favour the development of oil. If the burial has been deeper with greater temperature and pressure, then these conditions will tend to vaporise any oil into gas. For example, in Australia, oil wells in the Bass Strait between the mainland and Tasmania produce oil because this strait has a sea floor of shallow marine sediments. However, far to the north west between Australia and Timor, the sea and the sediments are much deeper so marine organic material buried here in the North West Shelf are subject to higher temperatures and pressures and only produce natural gas.

If the ocean floor is uplifted to become land or shallow sea basins, the oil, along with gas and trapped water will migrate upwards from its source rock until it reaches the surface, where it might form tar pits, or become trapped by an impermeable cap rock in a variety of geological structures collectively known as **oil traps**. A common form of oil trap is the **anticlinal** or dome trap in which the fluids are trapped within an uplifted section of sedimentary strata and other traps include those associated with faults, changes in the rock layers and intrusion of salt domes.

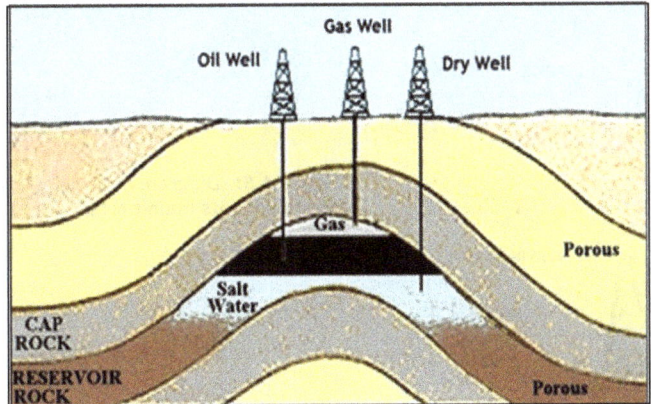

Figure 4.7: An anticlinal oil trap - in reality it is a 3-dimensional dome

Oil rigs either on land, or on drilling platforms at sea, are essentially the same - a tall derrick up which the sections of pipes, called **sticks**, are hauled so that they can be attached to the drill bit at its base and then passed through a hexagonal or square pipe called a **kelly**, which is attached to a rotating table driven by a motor. Drilling mud is also pumped down the drill pipe casing to act as a lubricant and coolant. As the bit drills down into the rock, more sticks are added by screwing them into the thread at the top of the previous pipe. When oil or gas is reached, it is usually under pressure so the drill pipe is then attached to a series of pipes and taps called a **Christmas tree**. As the pressure reduces, the oil or gas, stops flowing under its own pressure and must then be pumped out by injecting water down into the well.

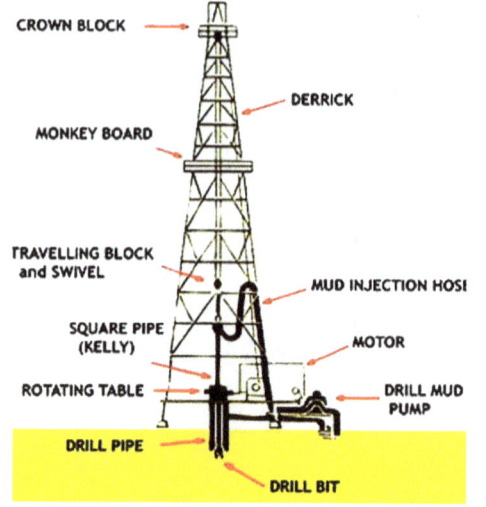

Figure 4.8: Diagram of an oil drilling rig showing its main parts

Australia has about 0.3 per cent of the world oil reserves. Most of Australia's known remaining oil resources are condensate and liquefied petroleum gas (LPG) held in giant offshore gas fields in the Browse, Carnarvon and Bonaparte basins. In addition, oil resources are also found in the Perth, Canning, Amadeus, Cooper/Eromanga, Bowen/Surat, Otway, Bass and Gippsland basins. Whilst producing oil and gas, Australia still imports most of its petroleum products.

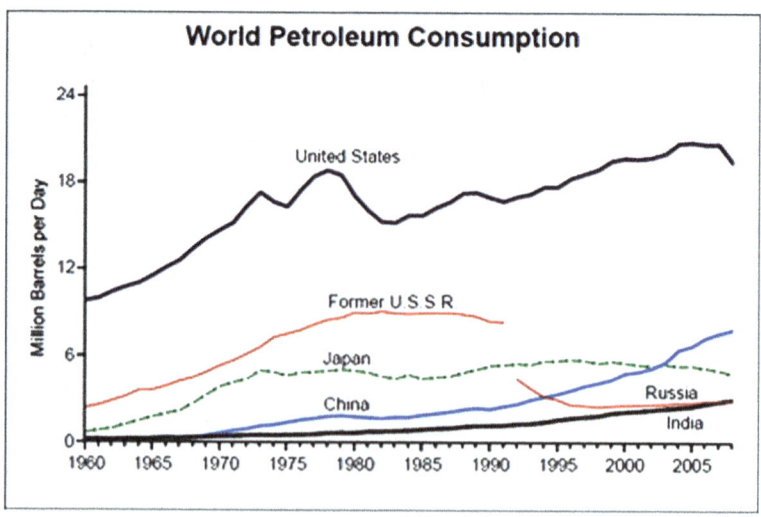

Figure 4.9: Graph showing World Petroleum Consumption to 2018 (Data from the International Energy Agency)

The main oil and gas producing nations are:

Rank	Country	Proved Reserves (billion barrels*)
1	Saudi Arabia	259.9
2	Canada	175.2
3	Iran	137.6
4	Iraq	115.0
5	Kuwait	101.5
6	Venezuela	99.4
7	United Arab Emirates	97.8
8	Russia	60.0
9	Libya	44.3
10	Nigeria	37.2
11	Kazakhstan	30.0
12	Qatar	25.4
13	China	20.4
14	United States	19.2
15	Brazil	12.8
16	Algeria	12.2
17	Mexico	10.4
18	Angola	9.5
19	Azerbaijan	7.0
20	Norway	6.7
	Top 20 Countries	1,281.5
	Rest of the World	72.2
	World Total	1,353.7

*One barrel is approximately 159 litres or 42 gallons. The energy contained in one barrel of oil is equivalent to 1700 kWh.
Data from U.S. Energy Information Administration (EIA).

Table 4.2: Major oil-producing nations in billions of barrels (Data from the International Energy Agency)

See also:

http://energyatlas.iea.org/#!/profile/WORLD/AUS

and

http://energyatlas.iea.org/#!/tellmap/-1920537974
(An excellent set of statistics for all countries from the IEA)

Oil sands are loose sands or broken sandstones containing a mixture of sand, clay, and water, saturated with a viscous form of petroleum (bitumen). It is also commonly and incorrectly called tar, and large quantities are found in many countries including Russia, the USA and Venezuela, and especially the Athabasca oil sands located in north-eastern Alberta, Canada. Canada has the largest industry of extracting and refining the oil sands on an economical basis. It is thought that this petroleum was formed in much the same way as in conventional oil deposits but here the product of the marine environment moved to another location and was trapped in coarse river sands. The lighter petroleum products evaporated away and the sand-clay-bitumen deposit was covered in new sediment.

See also:

https://www.api.org/~/media/Files/Oil-and-Natural-Gas/Oil_Sands/OIL_SANDS_PRIMER_MAY_2011.pdf
(Canadian oil sands summary)

For general information on oil and gas, see also:

https://www2.deloitte.com/content/dam/Deloitte/ru/Documents/energy-resources/ru_er_vision2040_eng.pdf
(Future trends for oil and gas)

http://www.kgslibrary.com/cms/Oil%20in%20the%2021st%20Century%20and%20Fut ure%20Challenges%20April%2015%202018.pdf
(Oil in the 21st Century)

https://www.eia.gov/outlooks/aeo/pdf/0383(2017).pdf
(Oil use in the future)

https://www.bhp.com/-/media/bhp/documents/investors/reports/2003/australiafutureoilandgas.pdf
(Australian oil and gas)

4.4 Fossil Fuels - Coal-Seam Gas (or Coal-bed Methane)

Coal-seam Gas (CSG) is found within the cracks and cavities of sub-surface coal seams. With the future threat of depletion of oil, this has become a new source of fuel which can be extracted from the surface without extensive mining. Coal seam gas contains about 95-97% methane gas (CH_4) with a little carbon dioxide, ethane and nitrogen. As a comparison, natural gas from oil wells also contains methane along with heavier hydrocarbons such as ethane C_2H_6, propane C_3H_8, butane C_4H_{10}, pentane C_5H_{12} and longer carbon molecules. It may also contain some hydrogen sulfide (H_2S) and carbon dioxide which must be expensively separated or removed. Methane gas has always been an explosion hazard in underground coal mines where it has been called coal mine methane CMC or firedamp, from the German: *dampf* for vapour. It is released during the mining process and often extracted by large ventilation systems to limit the hazard of mine explosions. Methane gas is a product of the breakdown of the organic matter within coal and is stored within the coal matrix as a liquid adhering to the walls of the pore spaces, a process called adsorption. The open fractures in the coal, called **cleats**, can also contain free gas or can be saturated with water.

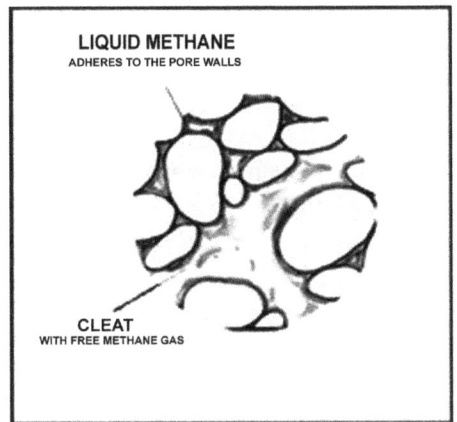

Figure 4.10: Diagram showing methane within the pore spaces of rock

Potentially commercial coal seam gas, containing mostly methane gas, is usually found in seams saturated with water. The underground water pressure also holds the methane to the pore walls. The CSG is extracted by means of wells which are drilled down into, and along the coal seams of depths of up to 1000 metres. When water is pumped out of the coal seams the confining pressure is reduced, leading to the evaporation of the gas so that it can be collected. Gas compressors are then used to pump released gas to a central processing facility where it is dried and transported along a high-pressure pipeline to shipping points.

Where the gas is tightly held within the coal seam, hydraulic fracturing (**fracking**) is used to crack up the coal seam and surrounding rock to release the gas. This process involves the high-pressure injection of sand-water slurry and sometimes a range of chemicals, into the coal seam. This not only fractures the rock but also holds the fractures open, thus releasing the gas. Some of the chemicals used in the extraction are potentially toxic, such as **BTEX** which is a mixture of benzene, toluene, ethylbenzene and xylene with sodium hypochlorite, hydrochloric acid, cellulose, acetic acid and disinfectants. The water extracted from these wells includes large amounts of natural sub-surface water and some of the water and waste chemicals which were injected down the well. The water which is extracted is pumped into surface ponds for treatment. The CSG extracted is then dried and pumped by long pipelines to storage facilities at market points.

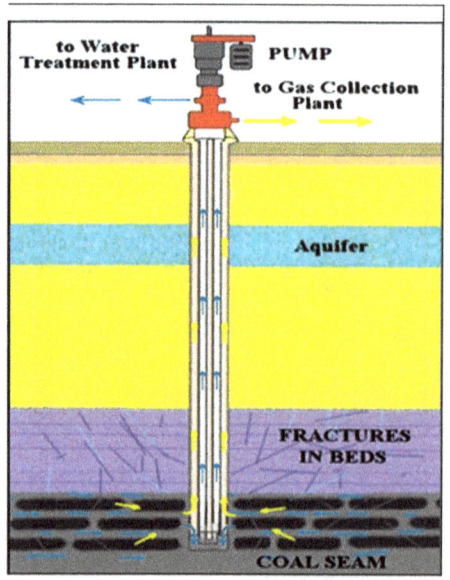

Figure 4.11: Coal seam gas well showing extraction and fracturing of beds due to fracking.

Environmentalists and farmers are concerned with the drilling of CSG in agricultural areas because of the potential damage to the environment and disruption to agriculture. They are concerned about:

- depletion and contamination of underground **aquifers** and surface water;
- methods of disposal of the large quantities of extracted polluted water brought to the surface;
- leakage of inflammable methane from wells, pipelines and storage facilities;
- harm to humans and animals from air, water and soil pollution;
- loss of agricultural land and native vegetation from the large surface area required for CSG operations; and
- risk of localised seismic activity from fracking and aquifer re-injection.

Each well and field has its individual features and properties, and mining companies and governments claim that the engineering and scientific controls through effective monitoring would prevent these problems. There still is considerable debate about the pros and cons of CSG, but governments faced with depletion of energy reserves and the attraction of CSG exports have welcomed CSG exploitation. Further development of environmentally-friendly energy sources, such as solar-electric power would be desirable not only to reduce the environmental impact but because CSG is also non-renewable.

4.5 Processing Fossil Fuels

In general, coal and oil are unsuitable for use when they are extracted from below the surface. Coal requires a simple process of washing before it is graded by size and type before it is used locally or exported. At the end of the supply chain, it can then be used as a fuel to generate power or used to manufacture steel, coal gas and a vast range of synthetic materials including fibres, medicinal drugs, plastics, useful chemicals, coke and tar. In many countries, the distillation of coal is still used to manufacture coal gas and sold to consumers as **town gas** although it has largely been replaced by **liquid petroleum gas (LPG)** and coal seam gas (CSG). Synthetic gas or **syngas**, a mixture of carbon monoxide and hydrogen is made by combining powdered coal with oxygen and steam while being heated and pressurized. During the reaction, oxygen and water molecules oxidize the coal into carbon monoxide (CO), while also releasing inflammable hydrogen gas (H2). This process has also been used in the production of town gas which was piped to customers to burn for illumination, heating, and cooking:

coal	oxygen gas	water	hydrogen gas	carbon monoxide gas
3C +	O_2 +	H_2O →	H_2 +	3CO

This can be used as a fuel or further converted into lubricating oils by the **Fischer-Tropsch process**. This is a series of chemical reactions which converts the mixture of carbon monoxide and hydrogen into liquid hydrocarbons as synthetic lubricating oil or petrol. These reactions require the use of metal **catalysts**, typically at temperatures of 150-300 °C and pressures of one to several tens of atmospheres.

Petroleum, when extracted as **crude oil** from oil wells, is an impure mixture of many useful products which must be separated out by **fractional distillation**. This is a process whereby the mixture is heated at the base of tall fractionating towers at oil refineries and the vapours rise up the tower. At different heights up the tower each petroleum product or fraction boils off at its specific boiling point, condenses and is removed from the tower. In the tower, some larger molecules are also broken down into more useful, smaller molecular fractions in a process called **cracking**. At the top of the tower, the smaller, lighter molecules are removed as gases and then liquefied as Liquefied Petroleum Gas.

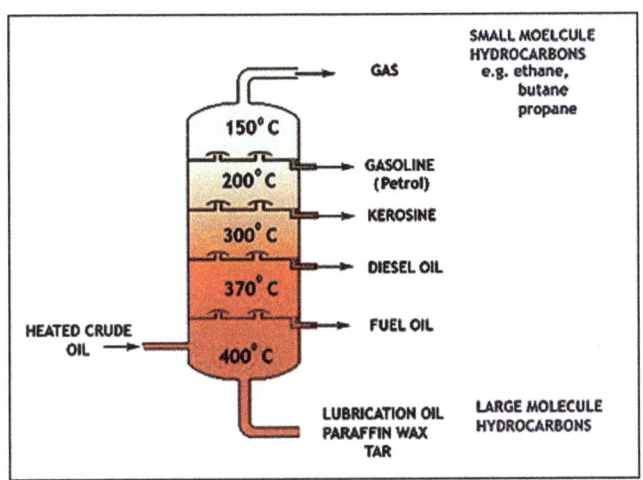

Figure 4.12: Diagram showing the parts of a fractionating tower

Most oil refineries, oil and gas wells and landfills will often have an associated gas flare combustion system to release unwanted gas pressure from time to time. These are seen as a tall stack with a large flame burning at top. Large amounts of carbon dioxide are formed from such combustion, contributing about 1.2% of the annual global emissions of this gas. When not operated efficiently, flares may emit methane, a gas with more greenhouse warming potential, as well as other volatile organic compounds, sulfur dioxide and other sulfur compounds, which can cause asthma and other respiratory problems. Other emissions from improperly operated flares may include, aromatic hydrocarbons such as benzene, toluene, xylenes, and benzo pyrene, which are known to be carcinogenic. However, much of the waste methane gas produced from an efficient flare system will convert most of the methane to carbon dioxide.

Figure 4.13: Fractionating columns in a refinery (Photo: USGS)

 Online Video 4.2: Animation showing how a fractionating column works. Go to https://www.youtube.com/watch?v=wq-bU08QlO4

See also:

http://www.petroleum.co.uk/hydrocarbons
(A good summary of types of hydrocarbons and their uses)

4.6 Fossil Fuels - Advantages and Disadvantages

Humankind has long used fossil fuels as a source of energy. From the pre-industrial revolution to today, nations have used timber, coal, oil and gas to provide energy for heating, processing of raw materials and production of manufactured materials including synthetic materials directly produced from oil and coal, and fuels and oils for transportation. At a time when many are looking for alternatives, one must consider the advantages and disadvantages of these fuels.

The **advantages** of fossil fuels have always been:

- **readily available** - humankind has relied on fossil fuels since the dawn of time, and there has been a very long and extensive development in gathering and using these resources. Moreover, the industrial revolution evolved to the present-day using machinery and technology for coal and oil. Countries which are rich in fossil fuels have large infrastructures designed and made for the extraction of fossil fuels and their societies have also developed assuming the continued use of these fuels.

- **easier to find and to extract** - fossil fuels are relatively very easy to find using simple geological exploration. They occur all over the world, usually in very rich seams or oil traps and that once committed to getting coal or gas out of the ground or out from under the sea, nations have access to a considerable amount of energy resources.

- **extremely effective** - fossil fuels are extremely effective in producing energy with a known and generally reliable technology. This means that they can generate huge amounts of energy which cannot yet be matched by any renewable energy.

- **easier to transport** – fossil fuels such as coal and oil are relatively easy to transport in bulk by rail, road and by sea with minimum risk compared to the transportation of nuclear fuel. Renewable fuels cannot be transported although their energy can in the form of electricity. This can be a problem when considering long distance or from country to country, whereas fossil fuels can be transported to local power stations.

- reliable - unlike renewable energy sources which have a dependency on sun, wind and water, fossil fuels are readily available. A fossil fuel plants can be set up anywhere in the world using existing technology and human resources.

- **cheaper** than most other sources – fossil fuels have a long-established technology and supply and are cheaper than most other energy sources. However, there are on-going costs of fuel supply and use, whereas hydroelectricity is more economic once the power plant has been constructed. This is also true of many of the other renewable sources but one has to consider the total amount of energy produced on a cost/unit basis. The problem with the argument that fossil fuels are cheaper fails when one has to consider the cost to the environment and what these fuels when the price of carbon-capture from them is added to production costs. When this is done, alternative fuels become more competitive.

- **a major source of employment** - fossil fuels generate hundreds of thousands of jobs every year and many localities and economies rely directly or indirectly on the production of these fuels or their processing, transportation and use.

- **convenient raw material** for a wide range of synthetic materials, drugs and industrial products.

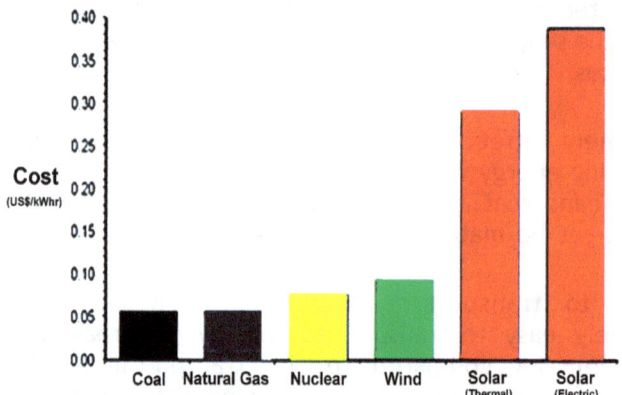

Figure 4.14: Chart showing some of the basic production energy costs of various sources without carbon-capture (Data: IEA/NEA)

The **disadvantages** of fossil fuels are:

- **environmental pollution** - this is the main disadvantage of using fossil fuels. The problems that are caused as a direct result of burning fossil fuels are well documented by scientists throughout the world. Carbon dioxide that is released into the air has been directly linked to global warming and its subsequent problems of sea level rise, species extinction and social upheaval.

- **extensive land clearance** for open cut mining with the disruption of natural fauna and flora, especially large-scale vegetation removal. There also may be long-term destruction of valuable farming and grazing land as well as problems with infrastructure such as roads and water resources.

- **extended problems** well away from mining or drilling sites with the need for additional rail lines, stock piles and processing plants and refineries and the need for port facilities with the added problem of ocean pollution.

- **continued resources are needed** - power stations require large amounts of coal and the transport and other industries require continuous supplies of oil and gas. These must be mined or extracted and so require continual exploration by geologists and large amounts of land use in their mining, processing and stockpiling. Land use for food supply as well as the sensitive issue of land rights by traditional and other land-owners is constant source of conflict.

- **safety-** this has always been an important factor in the extraction of fossil fuels, especially in coal mining. This has been a dangerous occupation and more personnel are killed directly in coal mining and over time with lung diseases than in the nuclear industry.

4.7 Fossil Fuels with Reduced or Zero Emissions

There is considerable on-going research in the United States, China, India and Australia on various systems to overcome the problems associated with the use of fossil fuels. The burning of petroleum products in vehicles and that of coal in power stations are the major sources of greenhouse gases. Burning fossil fuels primarily produces carbon dioxide (CO_2) and water vapour (H_2O). Other major emissions include nitric oxide (NO) and nitrogen oxides (NO_2) which together are called **NOx**, sulfur oxides (SO_2), and soot (mostly inert carbon C). Much of the research focusses on eliminating these gases by more efficient combustion systems, their removal from the environment by burial or their conversion into new useful resources. Some of this research includes the following possible solutions:

- **Redesigning of gas turbines** and other combustion engines to work more efficiently with better fuels and combustion processes to produce lower gas emissions with increased power. Whilst these systems put less pollution into the atmosphere, they still rely on the combustion of hydrocarbons producing heat, carbon dioxide and any other waste gases which had not been removed from the gaseous source.

- **Use of hydrogen gas as a fuel** in both combustion engines and in **fuel cells** producing electricity for power in vehicles with the use of fuel cells being expensive but more efficient. So far this has had only a selective application in a limited range of automobiles and buses.

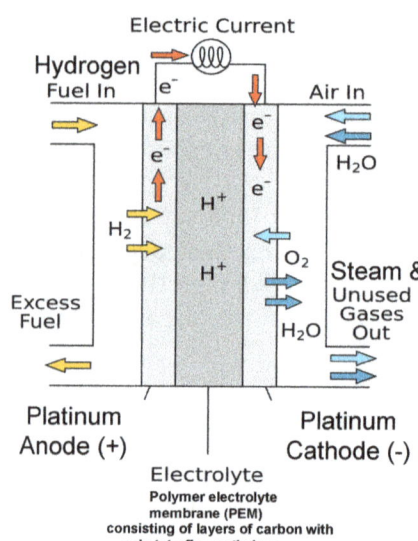

Figure 4.15: Diagram showing the parts of a hydrogen fuel cell (Photo: Public Domain after Sakurambo)

The majority of hydrogen is produced from fossil fuels by partial oxidation of methane and coal gasification and requires energy to do so and some methods also produce carbon dioxide and carbon monoxide in the process. Only a small quantity is produced by renewable methods such as biomass gasification or electrolysis of water using electricity from solar panels or wind farms, but this could improve in the future.

$$\text{pure water} \xrightarrow{\text{Electrolysis}} \text{hydrogen gas} + \text{oxygen gas}$$

$$2H_2O \text{ (liquid)} \longrightarrow 2H_2 \text{ (gas)} + O_2 \text{ (gas)}$$

There is also a problem with storage of pressurized hydrogen and its high volatility because it can easily escape from tanks and its low ignition temperature produces an explosion risk. Currently in the United States, only California seems to have a few refuelling stations.

The big advantage in using hydrogen gas as a fuel is that it produces only steam as its waste gas.

hydrogen gas   **oxygen gas**   **steam**

$2H_2 \text{ (gas)} + O_2 \text{ (gas)} \longrightarrow 2H_2O \text{ (gas)} + \text{energy}$

Hydrogen gas can also be used as a fuel in gas turbine and internal combustion engines either as a stand-alone fuel or mixed with other fuels such as diesel oil. If burnt in air, as would be the most practical case, a small amount of the oxides of nitrogen may also be produced as well as steam. Hydrogen gas can be stored and shipped in tankers but unless seals are efficient, the gas can easily escape. One option to safeguard hydrogen storage is to convert it to ammonia gas by combining hydrogen with nitrogen gas, the most plentiful gas in air. Ammonia storage and transportation has long been used in industry.

See also:

https://www.energy.gov/sites/prod/files/2014/03/f10/fcm02r0.pdf
(Good details about hydrogen as a fuel)

https://afdc.energy.gov/fuels/hydrogen_basics.html
(Use of hydrogen as a fuel in the United States)

https://www.chemistryworld.com/features/hydrogen-still-the-fuel-of-the-future/3009235.article
(Applications and current use of hydrogen fuel)

https://h2tools.org/bestpractices/hydrogen-compared-other-fuels
(Hydrogen compared to other fuels)

https://www.sciencemag.org/news/2018/07/ammonia-renewable-fuel-made-Sun-air-and-water-could-power-globe-without-carbon
(Ammonia gas as a storage/transportation material)

- **Clean coal technologies** using the vast amounts of coal currently being mined and in reserve and using much of the waste products as new resources. These systems may have the advantage of being able to be retro-fitted to existing power stations. The **High Efficiency, Low Emissions (HELE)** coal-fired power plants such as those operating in places such as Japan, Germany and Australia use advanced systems of coal combustion to achieve greater efficiency of power generation with less emissions. They still, however emit flue gases but can be coupled with other systems (mentioned below) to reduce emissions even further.

- **The use of hybrid energy systems** to produce electricity for general power and for transportation. This will probably be the way to the future; instead of relying on a single source, a controlled and integrated combination of renewable electricity generation systems will probably replace reliance on non-renewable systems such as coal-fired steam generated electricity. To a limited extent this has already begun to occur but it will take time for the more efficient but polluting systems to be replaced.

Some other areas of potential which could be used to reduce or even remove flue gas emissions include:

- **Removal of sulfur dioxide** for conversion to useful sulfuric acid with **geo-sequestration** of carbon dioxide by pumping it underground into old saltwater aquifers. This uses current technology on site to remove carbon dioxide and pump it down in a supercritical or liquid state into the old aquifers containing water too saline (salty) for other use. Studies of the geo-sequestration of carbon dioxide from the Norwegian North Sea gas wells have shown this to be feasible. Since 1996, carbon dioxide gas has been pumped down 1000 metres into an old, depleted gas reservoir at the Sleipner West well in the North Sea. Below 800 metres depth, the carbon dioxide remains as a pressurized liquid and becomes trapped within the pore spaces of the rock of the Utsria Formation, a 200-250 metres thick sandstone which is capped with a very non-permeable mudstone layer which prevents any leakage of the carbon dioxide. There has been no reported leakage of any gas from this closely-monitored site.

There has also been some success in the injection of carbon dioxide gas into basalt rock layers which contain fractures and pore spaces. Here the carbon dioxide reacts with the minerals of the rock to form new, solid and non-volatile carbonate minerals.

Sequestration into the deep oceans (say below 3000m) has also been suggested as a way of removing carbon dioxide gas from power stations. At such depths the gas liquefies and forms a pool which stays on the ocean floor as it is denser than sea water. This method may have some problems due to the possibility of disturbance of this pool by ocean-floor currents which may bring the concentrated carbon dioxide to the surface as gas. Currently there are about 17 carbon capture and storage (CCS) sites around the world with many more projects being planned. See:

https://www.globalccsinstitute.com/wp-content/uploads/2018/12/2017-Global-Status-Report.pdf
(A global perspective)

https://www.carbonbrief.org/around-the-world-in-22-carbon-capture-projects
(Interactive map showing CCS locations worldwide)

https://www.tai.org.au/sites/default/files/DP72_8.pdf
(Geo-sequestration in Australia)

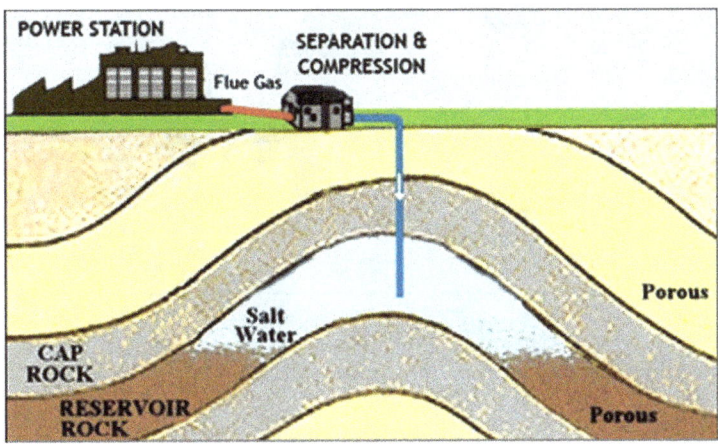

Figure 4.16: Geo-sequestration using saline aquifers

- **Exhausting flue gases** from coal-fired power stations directly into ponds or **photo bioreactors** containing algae which can use the carbon dioxide, even at high concentrations and elevated temperatures, to produce oxygen by photosynthesis. Flue gases typically contain about 82% nitrogen gas (inert), 12% carbon dioxide gas, 5% oxygen gas, 400 parts per million of sulfur dioxide gas, about 120 ppm of gaseous oxides of nitrogen (**NOx**) and some soot. The latter can be precipitated as powder for land fill.

Algae such as *Cyanidium celdanum, Scenedesmus sp., Chlorococcum littorale* and *Synechococcus elongates* can tolerate concentrations of carbon dioxide greater than 60%. There are some problems with this technology yet to be overcome before large-scale application can be made. These algae require large pond systems near the coast and there is considerable expense in land acquisition, construction and piping the gases from inland power stations. Closed systems using racks of pipes containing water with the algae seem to have more potential than pond or open systems which are also subject to massive water evaporation. In addition, the excess algae can then be processed into a hydrocarbon biofuel.

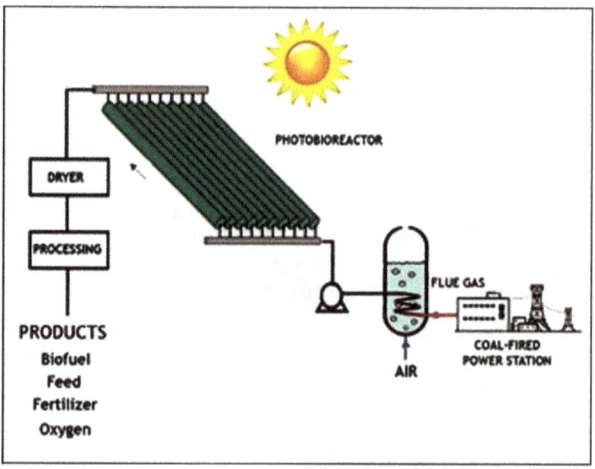

Figure 4.17: A simplified diagram of a photo bioreactor used with flue gas from a coal-fired power station

See also:

http://www.oilgae.com/ref/downloads/Analysis_of_CO2_Capture_Using_Algae.pdf
(Report from India about use of algae)

- **Application of old gasification technology** to new systems which can produce many useable products and fuels with little or no gas emissions. In the early nineteenth century, coal gas (also called town gas) was produced by the direct distillation of coal in confined ovens. This produced an inflammable gas containing hydrogen, carbon monoxide, methane and volatile hydrocarbons together with small quantities of carbon dioxide and nitrogen. The process also produced coke, a soot-free solid used in home fireplaces and in making steel, tar, ammonium products and other organic chemicals.

This 'town gas' was used in the home for lighting and cooking and in industry well into the twentieth century but was gradually replaced by electricity and natural gas. Whilst some of the liquids and solids produced had limited use at that time, most were discarded into the environment as a toxic waste. Modern gasification systems are now under research and also currently coming into use in many countries. These operate with minimal or zero-waste gases so that any organic fuel added to the system will produce gas-synthesis gas or syngas for making synthetic natural gas (SNG), heat for steam for generating electricity and many other useful fuels and products. The raw materials for a gasification plant could include coal, oil or even some organic wastes. Useful products include sulfur, ammonia, hydrogen, methanol and other products already extracted from fossil fuels for the diverse industrial chemical industry.

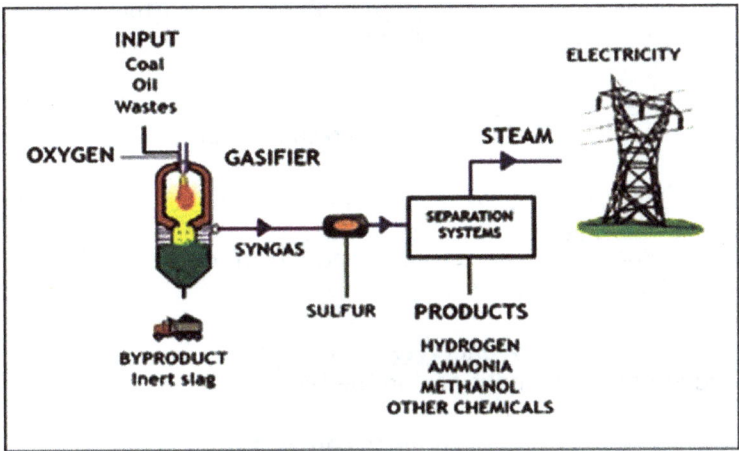

Figure 4.18: A simplified diagram of a gasification system

The modern **integrated gasification combined cycle (IGCC)** technology uses a high-pressure gasifier to turn coal and other carbon-based fuels into pressurized syngas from which impurities such as sulfur, particulates and mercury can be removed. In some cases, waste carbon monoxide can also be converted to carbon dioxide. The carbon dioxide can then be separated, compressed, and stored through geo-sequestration. The heat generated from this total process can then be used to produce steam for power generation.

Whilst the science and technology exist to achieve reduced or even zero emissions from fossil fuel combustion, it is a matter of energy efficiency, cost and above all, motivation for power companies to install such applications. To do so would mean reduced profits and higher energy costs. Governments are also reluctant to take the necessary steps to insist on such actions being taken knowing that such costs will be passed on to a greatly hostile general public. Currently, most large-scale carbon capture and storage (CCS) projects are not yet operational and it should take a few more years before true clean coal technology reduces greenhouse gases.

See also:

https://ourworldindata.org/fossil-fuels
(Data about fossil fuels - production and consumption)

https://www.csiro.au/en/Research/Energy/Fossil-fuel-energy
(Fossil fuels - links from the CSIRO)

http://www.darvill.clara.net/altenerg/fossil.htm
(Many good links for information and some useful data)

https://www.world-nuclear.org/information-library/energy-and-the-environment/clean-coal-technologies.aspx
(Good summary of clean coal and clean oil technologies and their application around the world)

https://www.eia.gov/energyexplained/coal/
(Good information about coal in the USA)

https://reneweconomy.com.au/clean-australias-clean-coal-power-stations-14224/
(Coal-fired power stations in Australia)

https://coal21.com/emissions-reduction/carbon-capture-storage/why-ccs/?doing_wp_cron=1570850369.7304959297180175781250&gclid=EAIaIQobChMIl8nLsLiu5QIVhJCPCh2yigW7EAAYASAAEgLrdfD_BwE
(Scroll down to see some predictions for future energy source parameters)

http://www.oecd.org/environment/cc/34878689.pdf
(Excellent technical paper from the Organisation for Economic Co-operation and Development on international cooperation on energy and global warming)

https://www.energy.gov/sites/prod/files/2017/06/f34/lundquist_algaeccu.pdf
(Information about the use of flue gases in algal ponds)

http://www.oilgae.com/ref/report/The_Comprehensive_Guide_for_Algae-based_Carbon_Capture.pdf
(More on the use of algal ponds from India)

https://iopscience.iop.org/article/10.1088/1755-1315/76/1/012001/pdf
(Future use of clean coal processes in India)

Online Video 4.3: Excellent video showing China's approach to greenhouse gas emissions.

Go to: https://www.youtube.com/watch?v=r9rwMnLzGqU

Final Remarks

1. Fossil fuels are those energy sources (fuels) which have been created due to ancient processes which have converted animal or plant material. They include coals and coal-seam gas from plants and oils and natural gases from marine life.

2. The burning of fossil fuels put large amounts of carbon dioxide and steam into the atmosphere from their carbon and hydrocarbon content as well as sulfur dioxide and oxides of nitrogen from impurities.

3. Combustion in reduced oxygen in furnaces also produces a large amount of particulate matter in the form of unburnt carbon or soot.

4. Extraction of coal and oil also adds to the destruction of the environment due to open cut mining, transportation needs, processing waste products and fracking of sub-surface seams.

5. The burning of fossil fuels, especially coal could be continued with the use of modern technologies which could reduce or remove waste heat, gases and solids. Current technologies which remove sulfur and particles could be augmented with those which remove the carbon dioxide and other components of flue gases.

6. These technologies include burial within old saline traps (geo-sequestration), the use of algal ponds (photobioreactors) and enhanced conversion of useful and less polluting gases and chemical products (gasification).

7. Currently, many coal-fired power stations use High Efficiency, Low Emissions (HELE) systems which give greater efficiency with less reduction. Similarly, the use of gas-fired power stations also gives reduced emissions but these still add to the sum total of greenhouse gas emissions.

8. Considerably more research and development will be required if fossil fuel technologies are to be used in the future and there will be a need to reduce the current about of greenhouse gases already emitted

More Applications

1. Do a fuel audit of your daily life-style noting what type of energy is used over the hours e.g. in the home during 24 hours, driving the car and using public transport.

2. Contact your preferred energy supplier and find out how much of their supply comes from fossil fuels. Shop around for greener supplies and compare costs.

3. Call up coal-mining and environmental activists' websites and review their material for facts as opposed to emotional statements.

4. Find out the localities of fossil fuel extraction sites and fossil fuel power stations nearby. Assess pollution potential.

5. Fossil fuel domination must come to an end. Attempt to research a timeline for the closure of local fossil fuel power stations.

6. Research the dependence of local industry and common, everyday materials on fossil fuels.

Further References

Forinash, Kyle. 2017. Physics and the Environment. Morgan & Claypool Publishers. Online ISBN: 978-1-6817-4493-3.

Jean-Marc Jancovici 2013. Using Coal? But What For? https://jancovici.com/en/energy-transition/coal/using-coal-but-what-for/

Khan, Shahriar (Edit.). 2012. Fossil Fuel and the Environment. InTechOpen. ISBN: 978-953-51-0277-9

Kunstler, James Howard.2006. The Long Emergency: Surviving the Converging Catastrophes of the Twenty-First Century. Grove Press. SBN 0802142494.

Miller, Bruce G. 2016.Clean Coal Engineering Technology. Second Edition. Butterworth-Heinemann. eBook ISBN: 9780128113660

World Coal Institute 2009. The Coal Resource: A Comprehensive Overview of Coal. https://www.worldcoal.org/file_validate.php?file=coal_resource_overview_of_coal_report%252803_06_2009%2529.pdf

Also see:
https://www.palgrave.com/gp/series/14966
(a series of books on the environment)

Chapter 5: Effects of Global Warming

5.1 Impact of Climate Change on the World's Systems

Studies of global warming and climate change have been long and extensive and despite a small but vocal opposition from some ill-informed people, most scientists predict that there will be long-term effects in the near future. The overall consensus is that climate change from global warming is due to the emissions of greenhouse gases, mainly by the use of fossil fuels in power generation and transport. Some of the effects of global warming may be reduced or even reversed with a massive change in thinking and concerted efforts by individuals and governments.

Apart from having a hotter local climate, the worldwide effects of global warming might generally include:

- **Ecosystem changes** in the biosphere will cause destruction of many **ecosystems** as climate changes and habitats become too warm for the optimum temperatures of various organisms. Those which cannot move will die out and those which can move will migrate to more temperate regions e.g. increased sea temperature would mean that coral reefs will move further away from the Equator with the death of older reefs closer to the Equator. Within ecosystems, many plant and animal species will face increasing competition for survival as other migrating species invade occupied ecosystems. In addition to altering plant and animal communities, climate change will also reduce biodiversity and adversely affect the main biosphere cycles involving water, energy, carbon, nitrogen and other natural cycles.

Usually when climate changes are from natural causes and are relatively slow, most plant species are able to relocate over multiple generations because they have evolved to survive and compete in a selective range of climatic and other ecological conditions. For example, with slow uplift of many of the world's mountain chains through orogeny, larger plant species gradually move down slope to remain in their preferred environments below the snowline, leaving the upper slopes for smaller plants which are more adapted to months below snow. With the onset of modern global warming however, plant species and some sedentary animal species are unable to change their habitats and so die out in areas greatly stressed. As well, human activities, such as agriculture and urbanization, have increasingly destroyed many of the Earth's natural habitats, and this encroachment frequently blocks plants and animals from successfully migrating.

Recent studies have found a shift of biomes, or major ecological community types, toward the Earth's poles and higher, cooler areas. These biomes include huge areas of temperate grasslands and boreal forests with the regions which will probably undergo the greatest degree of species turnover include regions in the Himalayas and the Tibetan Plateau, eastern equatorial Africa, Madagascar, the Mediterranean region, southern South America, and North America's Great Lakes and Great Plains areas. The extent of the ecological sensitivity is shown in the following map:

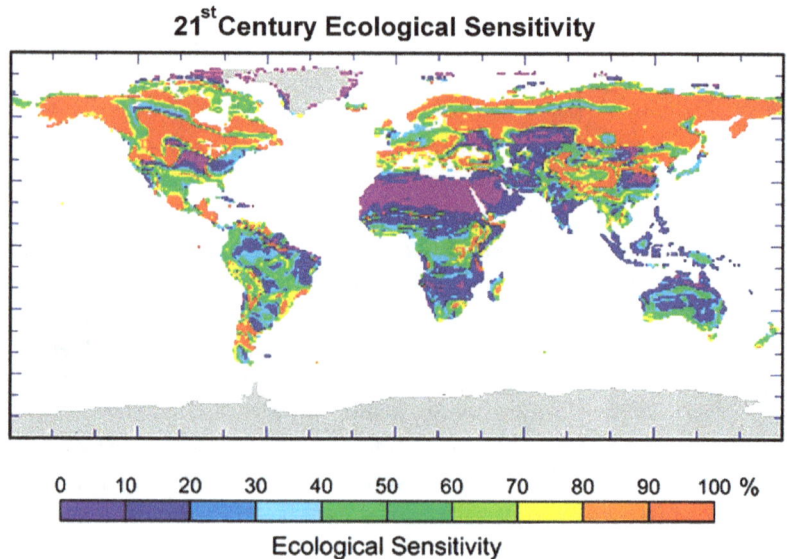

Figure 5.1: Map of ecological sensitivity around the world showing predicted percentage of ecological landscape being driven toward changes in plant species as a result of projected Human-induced climate change by 2100 (Photo: NASA/JPL-Caltech)

In simple terms, drier places such as the American south-west, the interior of Australia, northern Africa and along the west coast of South America will get drier and some places within intense rainfall regions near the tropics will probably get wetter. Change in the regular seasonal pattern has already been seen in some parts of the world where Spring has come early and cold snaps have occurred suddenly in Autumn (Fall). Some biomes and the human activities which rely upon them may be able to adapt, but others may not and either migrate or die out.

See also:

https://theconversation.com/ecosystems-across-australia-are-collapsing-under-climate-change-99367
(Changes to Australian ecology due to global warming)

https://www.wwf.org.au/what-we-do/oceans/great-barrier-reef#gs.d8qM=Ec
(The Great Barrier Reef and global warming)

https://esajournals.onlinelibrary.wiley.com/doi/pdf/10.1890/120282
(Major biosphere changes in the United States)

https://aambonceanservice.blob.core.windows.net/oceanservice-prod/education/pd/climate/teachingclimate/ecological_impacts_of_climate_change.pdf
(More ecological changes in the United States)

https://www.ipcc.ch/site/assets/uploads/sites/2/2019/02/SR15_Chapter3_Low_Res.pdf
(Impacts of 1.5°C of Global Warming on Natural and Human System from the Intergovernmental Panel on Climate Change of the UN)

- **Reduction in the world's ice** which is often referred to as the cryosphere. This includes reduction in both the thickness and area of sea ice, some ice caps especially in the Northern Hemisphere and a corresponding retreat or loss of glaciers. The latter would reduce drinking and irrigation water to communities bordering alpine glacial regions including the Andes, Himalayas and Rocky Mountains. For example, India is drained by 15 major rivers with a total drainage area greater than 20,000 km^2. Much of the north-western part of the country is arid and the population of India is the second largest in the world at about 1300 million people. Most of India's important rivers come directly from meltwater from the glaciers of the Himalaya range. Rising temperatures have caused many of the Himalayan glaciers undergo fast melting, and they could diminish significantly over coming decades with catastrophic results. Water flow in the Ganges is predicted to drop by two-thirds, affecting more than 400 million people who depend on it for drinking water. In the short term however, rapid melting of ice high up in the Himalayas might also cause river swelling and floods, especially those sudden outburst floods which come from swollen glacial lakes. In some dry countries adjacent to mountain ranges, such as Chile and Peru in South America, civilization was only possible in the coastal deserts of these countries by the short rivers which drain off the glaciers of the Andes to the east.

Figure 5.2: A populated valley along a stream fed by a glacier (top left) in the Andes of Chile.

See also:

http://www.agta.asn.au/files/Conferences/conf17/presentations/workshops/Turner_Griffiths/GreenlandGTAVimpactsfinishedversion.pdf
(Global warming and Greenland – a case study)

http://unesdoc.unesco.org/images/0025/002581/258168e.pdf
(The impact of glacial retreat in the Andes)

http://www.alpconv.org/en/newsevents/alpine/alpineSpringfestival/meetingwater/Documents/Spring_Festival_Bottarin_2013.pdf?AspxAutoDetectCookieSupport=1
(Good data about European glaciers)

- **Rise in sea level** with the inundation of some present coastlines and coastal communities will lead to destruction of parts of some great ports, coastal communities and a corresponding mass migration of populations living in low-lying countries such as Bangladesh, Vietnam, most countries with low-lying coastlines and river deltas and estuaries as well as oceanic countries such as Kiribati, Tuvalu, Palau and the Maldives just to name a few. This is not a future event but is happening today. Whilst the average rise in sea level has been estimated at about 3 millimetres per year, most of the islands of Kiribati (pronounced '*Kiribass*') are less than three metres above sea level. Already salt water has infiltrated a rising water table into the central parts of some islands and has destroyed coconut and swamp taro plantations, the main staple food, through salination. The Kiribati government has already had talks with the Fijian government about the possibility of migration of the I-Kiribati people and land has been purchased there to grow food. Apart from simple inundation, an increase in sea level will also increase the effects of local tides and storms with increased effects of storm surges and cyclones. The country is poor but the people have great heart and have taken to building sea walls, planting mangrove trees and relocating villages more inland to protect against increasing storms.

Figure 5.3: Many parts of Kiribati now suffer flooding at higher tides ((Photo: Tioti Timon, Kiribati)

See also:

https://climate.nasa.gov/vital-signs/sea level/
(NASA Sea level change website)

http://www.cmar.csiro.au/sealevel/downloads/SLR_PA.pdf
(Sea level rise and its consequences - from CSIRO)

http://www.ecosmagazine.com/?act=view_file&file_id=EC148p10.pdf
(Sea level rise from ground zero)

http://www.climatehotmap.org/global-warming-locations/republic-of-kiribati.html
(Sea level rise and the plight of Kiribati)

http://siteresources.worldbank.org/INTPACIFICISLANDS/Resources/4-Chapter+4.pdf
(Kiribati and climate change - World Bank report)

https://unfccc.int/resource/docs/napa/kir01.pdf
(Kiribati Adaptation Program of Action)

https://www.researchgate.net/figure/A-map-of-sea-surface-height-anomaly-during-a-King-Tide-event-The-blue-star-indicates-the_fig6_263056778
(Great map showing the Pacific sea level rise near Tuvalu and a set of good scientific papers to download)

https://coastadapt.com.au/climate-change-and-sea   level-rise-australian-region
(Climate change and sea level rise around Australia)

https://www.ilo.org/dyn/migpractice/docs/261/Pacific.pdf
(Sea level rise across the Pacific nations and migration)

https://www.weforum.org/agenda/2019/01/the-world-s-coastal-cities-are-going-under-here-is-how-some-are-fighting-back/
(Good site on how some coastal regions are combating sea level rise)

https://calculatedEarth.com/
(Interactive world map showing sea level rise effects)

- **Increase in permafrost thawing** will dramatically add to the greenhouse gases in the atmosphere by the liberation of the vast amounts of methane stored in the **permafrost** by the decomposition of its organic component. In addition to the release of methane, permafrost is really frozen marsh and bog country and so with its thawing, there will be great difficulties in using and moving across huge areas of Russia, Canada and Alaska. There will be an increase in the numbers of insect pests, such as mosquitos and some water-borne parasites in these countries.

See also:

http://whrc.org/wpcontent/uploads/2017/11/PB_Permafrost.pdf
(Good summary of the problems with permafrost)

- **Decreased water resources** in semi-arid regions where temperatures will become higher adding more heat stress to countries which have been climates which often have limited water and periods of extreme drought. Countries such as Australia, parts of the Middle East, northern Africa, Central Asia and the south-western parts of the United States would become desert environments with continued global warming. Biota, including humans have become adapted to these areas and are able to cope with normal climatic conditions of high temperatures and sunlight intensity and low humidity and surface water. However additional heat stress will probably cause a tipping point beyond which many plant and animal species will be beyond their ability to exist. Difficulties associated with human habitation, farming and grazing in these areas will become more extreme and many of the less-developed countries with low adaptive capacity – the ability to adjust to global warming – will find it difficult to maintain their way of life and economies. Water resources and their ownership will become a very big political and social issue as water becomes limited and may lead to conflict or mass migration if many of the world's less-developed nations cannot obtain enough water for their communities. For example, the Nile River runs through several countries such as Ethiopia and South Sudan and Sudan before providing the only true source of water and electrical power in Egypt. Recent indications that Sudan may build a dam on the river before it flows out of their country has given considerable concern to Egypt and some clashes between the two countries.

See also

http://www.garnautreview.org.au/pdf/Garnaut_Chapter5.pdf
(Climate change and Australia with special reference to drought)

http://www.bom.gov.au/climate/droughtec/ec-report.pdf
(Good detail from the Australia Bureau of Meteorology on drought)

http://www.cawcr.gov.au/projects/Climatechange/wp-content/uploads/2016/11/rainfall-paper.pdf
(Good summary about rainfall in Australia)

- **Spread of diseases** due to increased temperature has become a major concern with many world health organisations. The spread of the effects of climate change such as increased temperature and rainfall in some places will have enormous implications for human health, especially for the burden of vector-borne and water-borne infectious diseases. **Vector borne diseases** are those transmitted by insects and other animals which carry parasites which cause the disease e.g. some mosquitoes carry the malaria parasite. Already some bacterial diseases such as eye-infections in some species of penguins in Antarctica are thought to be a result of these organisms moving south as the water becomes warmer. However, some reports suggest that these infections have been carried there through increased tourism and that the warmer conditions are now better suited for the bacteria. Tropical diseases such as malaria could also spread into more temperate climates as their climates become hotter and wetter. There are also disease hazards during times of drought, when water scarcity and poor sanitation result in the population being exposed to potentially contaminated water. For example, epidemics of cholera can occur in less developed nations such as in Africa in the wake of a severe droughts. At the opposite end of extremes, flooding can also contribute to epidemics of waterborne infectious diseases due to poor sanitation when drinking water supplies are contaminated by sewerage from overwhelmed utilities or from contamination of water by livestock. For example, in 1993 an epidemic of diarrheal disease was caused by contamination of water in the Milwaukee River after heavy Spring rains.

  See also:
  http://www.who.int/globalchange/climate/en/chapter6.pdf
  (Summary of health problems and climate change from the World Heath Organisation)

https://www.who.int/globalchange/summary/en/index4.html
(A good summary of climate change and health hazard potential)

- **Reduction in crop production** as countries which have significant agriculture adapted to their current climate will need to change their practices if the climate changes. Climates in large, agricultural regions may become hotter and drier with a corresponding removal of topsoil or they may become wetter with severe storms which can destroy crops. In the short-term farmers and graziers will have to cope with increases in production costs including those for irrigation, pest control, drought resistance strategies and transportation of their products to markets. There could also be positive aspects of climate change as areas become more suitable to agriculture because of increased temperature, rainfall and carbon dioxide levels but whether or not these increases will outweigh the negative impact of climate change is yet to be seen. Never-the-less, agriculture is a resilient enterprise with new methods and technology being introduced to enhance current methods and productivity. The following table gives some indication the world crop production and the effects of climate change:

Crop	Global Production (1998-2002) (millions of tonnes)	Global Yield Impact due to Temp. Trends %	Global Yield Impact due to Precipitation Trends %	Global Yield Impact due to CO2 Trends %	Total %
Maize	607	-3.1	-0.7	0.0	-3.8
Rice	591	0.1	-0.2	3.0	2.9
Wheat	581	-4.9	-0.6	3.0	-2.5
Soybeans	168	-0.8	-0.9	3.0	1.3

Data from the Agricultural Marketing Resource Center, Iowa State University
https://www.agmrc.org/renewable-energy/climate-change-and-agriculture/climate-change-beginning-to-impact-global-crop-production

Table 5.1: Estimates of Global Impacts of Temperature and Precipitation Trends on Yields of Four Major Crops, 1980-2008.

See also:

https://sustainable.unimelb.edu.au/__data/assets/pdf_file/0009/2752272/MSSI_AppetiteForChange_Report_2015.pdf
(Good summary of the future of farming and climate change)

https://www.imf.org/external/pubs/ft/fandd/2008/03/pdf/cline.pdf
(Climate change and the world's agriculture)

https://www.usda.gov/oce/climate_change/effects_2012/CC%20and%20Agriculture%20Report%20(02-04-2013)b.pdf
(Good summary of the problems of agriculture with climate change)

- **Mass migration** of both human populations and flora and fauna due to sudden changes in the local environment which prevent adaptation. Many of the reasons for these migrations have been given above but this will be a major social and geopolitical effect of global warming as people relocate and there is competition for land. There has already been consideration for some Pacific nations to migrate due to sea level rises flooding their islands. This is also true of marginal lands in semi-arid regions where the heat and the lack of water will require migration.

- **Changes in the geosphere** usually cause climate change and global warming, but there may also be some changes to the geosphere or rocks and rock structures itself as temperatures increase in the atmosphere and the Earth's waters or hydrosphere. Any impact on the geosphere would probably be localised and specific to those parts which are affected by sudden changes. There would be major reductions in the moisture content of soils of many of the world's soils, especially those which have been exposed by deforestation or vegetation failure due to heat stress and drought. There might also be some geological changes under extreme conditions such as uplift and doming and in places such as Greenland and Antarctica as the massive weight of the ice sheets is partially removed through melting. This may cause some movement along local fault lines with subsequent local Earth tremors and possibly some landslides off the coastal margins with associated tsunamis. There have also been some suggestions that such uplift may also influence volcanic magma chambers of volcanoes which lie under massive amounts of ice such as Eyjafjallajökull volcano in Iceland which erupted in 2010. Considering the relative energies and forces involved between volcanism and unloading of massive weights of ice, the possibility of large-scaled earthquakes and volcanic activity being a direct result of climate change is a rather tenuous hypothesis.

See also:

http://rsta.royalsocietypublishing.org/content/roypta/368/1919/2317.full.pd
(Scientific paper on the potential geosphere hazards due to global warming)

- **Increase in storm activity** and associated heavy precipitation will be a consequence of addition heat being put into the atmosphere and the hydrosphere. This will increase the threat of more intense tropical cyclones (hurricanes) and perhaps their movement into more temperate regions. This has been discussed elsewhere in this book but it is important to emphasise the effects of having greater intensity of storms and their associated heavy rainfall, flooding, coastal storm surges and high winds. Many places on the fringes of tropical cyclone and hurricane regions will now experience these extreme forms of weather and such places as the Atlantic coast of the United States and the southern eastern coast of Australia with their major centres of population such as New York, Brisbane and Sydney may experience the effects of these extreme storms for the first time. More on this topic later.

- **Increase in wildfires** as vegetation becomes drier and storms with lightning strikes to ignite it become more frequent. Some places in the world such as within the lush vegetative regions of the tropics may experience wildfires for the first time. More on this later.

For some excellent reference material for further study on the effects of climate change in general, see also:

https://unfccc.int/resource/docs/publications/impacts.pdf
(Climate change impact on developing countries)

http://www.ccma.vic.gov.au/soilhealth/climate_change_literature_review/documents/organisations/ago/science-guide.pdf
(Climate change impact on Australia)

http://www.south-westnrm.org.au/sites/default/files/uploads/ihub/hughes-l-2003climate-change-and-australia-trends-projection.pdf
(Scientific paper on climate change and the future in Australia)

http://www.environment.gov.au/climate change/climate-science-data/climate-science/impacts
(General impact of Global warming on Australia)

https://www.schroders.com/nl/sysglobalassets/digital/us/pdfs/the-impact-of-climate-change.pdf
(Impact of climate change on the global economy)

http://www.ase.tufts.edu/gdae/education_materials/modules/The_Economics_of_Global_Climate_Change.pdf
(Good article on the economics of climate change)

https://climatechange.environment.nsw.gov.au/About-climate-change-in-NSW/Causes-of-climate-change
(Causes of climate change - a general view with several good links)

https://www.geolsoc.org.uk/~/media/shared/documents/policy/Statements/Climate%20Change%20Statement%20final%20%20%20new%20format.pdf?la=en
(Climate change from the Geological Society of London)

https://www.amnh.org/explore/science-bulletins/Earth/documentaries/archived-in-ice-rescuing-the-climate-record/the-coming-and-going-of-an-ice-age
(Good site about the Milankovitch Cycles etc.)

https://www.science.org.au/files/userfiles/learning/documents/climate-change-r.pdf
(Good climate change details from the Australian Academy of Science)

https://www.ncdc.noaa.gov/data-access/paleoclimatology-data
(Good links to palaeoclimatology data sites from NOAA)

https://cpo.noaa.gov/warmingworld/air_temperature_over_land.html
(Good interactive site from NOAA about climate change)

http://www.ipcc.ch/
(Website of the Intergovernmental Panel on Climate Change - IPCC)

https://www.noaa.gov/infographic/infographic-how-does-climate-change-affect-coral-reefs
(Climate change and coral reefs)

https://Earthobservatory.nasa.gov/features/GlobalWarming
(Good summary of climate change and causes)

https://www.odi.org/sites/odi.org.uk/files/resource-documents/11874.pdf
(Good information on forced migration due to climate change from the United Nations)

https://siteresources.worldbank.org/EXTSOCIALDEVELOPMENT/Resources/SDCCWorkingPaper_MigrationandConflict.pdf
(The social dimensions of climate change from the World Bank)

Some of the regional impacts of global change forecast by the Intergovernmental Panel on Climate Change include:

REGION	FUTURE IMPACT
Australia and the Pacific	Decreasing rain in the interior and more intense storms and tropical cyclones in the north and east. Severe droughts and reduced crops. Inundation of some Pacific islands.
North America	Decreasing snowpack in the western mountains; 5-20 percent increase in yields of rain-fed agriculture in some regions; increased frequency, intensity and duration of heat waves in cities that currently experience them.
Latin America	Tropical forest replaced by savannah in eastern Amazonia; loss of significant biodiversity and species extinction in many tropical areas; significant changes in glacial melt with reduced water availability for Human consumption, agriculture and energy generation.
Europe	Increased in inland flash floods; more frequent coastal flooding and increased erosion from storms and sea level rise; glacial retreat in mountainous areas; reduced snow cover and Winter tourism; extensive species losses; reductions of crop productivity in southern Europe.
Africa	Large populations facing increased water stress; reduced yields from rain-fed agriculture; access to food becoming more difficult; droughts.
Asia	Freshwater decrease in Central, South, East and South-east Asia; coastal areas will be at risk due to increased flooding; death rate from disease associated with floods and droughts expected to rise in some regions.

Table 5.2: Summary of some of the problems which will be encountered by different continents

Final Remarks

1. All countries and populations in the world are affected by an increase in temperature of the air and sea.

2. Current events and scientific projections suggest that global warming will cause changes in the delicate and interrelated ecosystems which make up the Earth's Biosphere. This will mean loss of species of migration into new regions with competition with indigenous species.

3. It will also cause loss of the world's ice or cryosphere with flooding of valleys above glaciers and eventually lack of water. Melting ice caps will also lower the temperatures of nearby oceans with a corresponding reduction in evaporation and therefore rain on nearby continents.

4. In addition, large areas of the permafrost – that frozen soil of northern latitudes – will thaw, giving up considerable methane gas stored in it as well as causing insect-ridden marshlands.

5. With an increase in atmospheric temperatures, many valuable semi-arid or marginal lands such as most of inland Australia and the American south-west will become drier as water sources dry up. Competition for diminishing water resources will cause conflict between countries and mass migration.

6. The increase in temperature could also allow for the spread of diseases such as tropic diseases spreading into previously more temperate zones.

7. Warming may also reduce crops, especially those which are more dependent on specific temperatures and amount of water. Some areas may become totally unable to support crops and livestock.

8. There may be more long-term changes as global warming effects cause changes to the land surface trough soil, erosional and structural change.

Final Remarks continued

9. There may be an increase in storm activity as the ocean becomes warmer further away from the equator. Places which were outside of cyclone (hurricane) belts may now experience such severe storms. Warmer oceans may also mean an increase in intensity as well as frequency of major storms.

10. The drying of vegetation will also mean the increased risks of wildfires (bushfires) and already there have been major catastrophic events in Europe, west coast USA and Australia which have been considered unprecedented.

More Applications

1. Do a risk analysis of the local area considering the problems which might be encountered due to increased warming. Especially consider the availability of water and food production as well as the dangers due to storms, fires and floods.

2. In hot weather, open all windows at late afternoon for the cool air and close up doors, windows and curtains/blinds at sunrise to keep the cool air in. Shade westerly-facing windows and doors.

3. Research where local food supplies are sourced and try to determine what would happen if these sources became drier and hotter.

4. Review media news items about populations such as those in the Pacific, such as Kiribati, Tuvalu and others who are now beginning to feel the effects of sea level rise. What are these countries doing about these effects and how should the industrialised nations assist?

5. Debate with friends or in community groups the moral aspects of inequality of global warming effects. Should those countries which have caused global warming through the mass use of fossil fuels assist those innocent rural or island communities which are now beginning to suffer? How can you help through aid programs or lobbying government members?

Further References

McMichael, A.J. *et al.* (Edits). 2003. Climate Change and Human Health: Risks and Responses. Geneva, World Health Organisation.
https://www.who.int/globalchange/publications/climchange.pdf
(Good summary of potential health hazards)

Mal, Suraj, Singh, R.B., Huggel, C. (Eds.) 2017. Climate Change, Extreme Events and Disaster Risk Reduction. New York, Springer Publishing. ISBN 978-3-319-56469-2

Union of Concerned Scientists. 2011. Climate Hot Map.
http://www.climatehotmap.org/
(Interactive map of some of the World's problem areas with solutions)

UNFCCC (United Nations Framework Convention on Climate Change), 2007. Impacts, Vulnerabilities and Adaptation in Developing Countries.
https://unfccc.int/resource/docs/publications/impacts.pdf
(Good information about developing countries including the Pacific region)

United Nations Intergovernmental Panel on Climate Change. 2019. (Chapter 3) Impacts of 1.5°C of Global Warming on Natural and Human Systems.
https://www.ipcc.ch/site/assets/uploads/sites/2/2019/02/SR15_Chapter3_Low_Res.pdf
(Extensive technical paper on the impacts of climate change)

University of Colorado National Snow & Ice Data Center 2019: SOTC (State of the Cryosphere) Overview https://nsidc.org/cryosphere/sotc
(An extensive site with many links about snow and ice)

Wang, J and Chameides, B. 2005. Global Warming's IncreasinglyVisible Impacts. New York. Environmental Defense.
https://www.edf.org/sites/default/files/4891_GlobalWarmingImpacts.pdf
(Compact article on the effects of climate change)

Chapter 6: Rain, Wind and Fire

6.1 Introduction

There are many hazards associated with the geosphere such as earthquakes, volcanoes, landslides just to name a few. There are also hazards associated with the atmosphere and the hydrosphere and some of them, such as hazardous weather patterns and floods have been mentioned briefly in previous chapters. Fire, too is a constant hazard affecting, and in part caused by the biosphere and weather conditions. This chapter will examine these other hazards more closely because they affect most countries around the world.

6.2 Extremes of Weather

Every country in the world experiences hazardous weather patterns which can cause damage to ecosystems, property and infrastructure, soil erosion as well as damage to crops and loss of livestock and sometimes causing great loss of life in extreme situations. Extreme weather conditions also cause flooding, drought and fires.

For example, Australia has a wide range of climates, ranging from a wet monsoonal tropical climate in the north to a temperate climate having four distinct seasons in the south. Rainfall is usually low, with more than 80% of the continent having an annual rainfall of less than 600 mm (24 inches) per year. Climates are variable and many regions are often prone to droughts, heatwaves, fires, intense rainfall and floods.

See also:
https://www.climatechangeinaustralia.gov.au/en/climate-campus/climate-system/australian-climate-influences/
(Australian climate and the monsoon season)

The main variations to the world's normal weather patterns are due to irregular changes in **sea surface temperature gradients (SSTG)** across the world's great oceans. These include the El Niño-Southern-Oscillation (ENSO) in the central Pacific, the Pacific Decadal Oscillation (PDO) in the Northern Pacific, the Indian Ocean Dipole (IOD), the Tropical Atlantic Dipole (TAS) and the Southern Annular Mode (SAM), which is also called the Antarctic Oscillation. Whilst these variable anomalies may have little impact on increasing global warming, they do impact the weather of the areas affected by these changes and with increased global warming these patterns may alter or become more severe.

The **El Niño-Southern Oscillation** is a periodic variation in winds and ocean surface temperatures over the Equatorial Pacific Ocean which sometimes produces a warming of the ocean in the east known as El Niño and at other times cooling, known as La Niña. An El Niño effect in the eastern and central Pacific in the tropics causes a higher degree of rainfall along the coastline off South America but also a cooling effect and drought in the western Pacific near Australia. Conversely, a La Niña effect off the South American coast brings drought in the Americas but heavy rain in Australia. The Southern Oscillation concerns the atmospheric changes associated with these changes in sea temperature. The strength of the Southern Oscillation is measured by the **Southern Oscillation Index** (**SOI**) which is computed from fluctuations in the surface air pressure difference between Tahiti in the Pacific and Darwin on the Indian Ocean. Sustained negative values of the SOI indicate an El Niño event whilst sustained positive values of the SOI indicate a La Niña event.

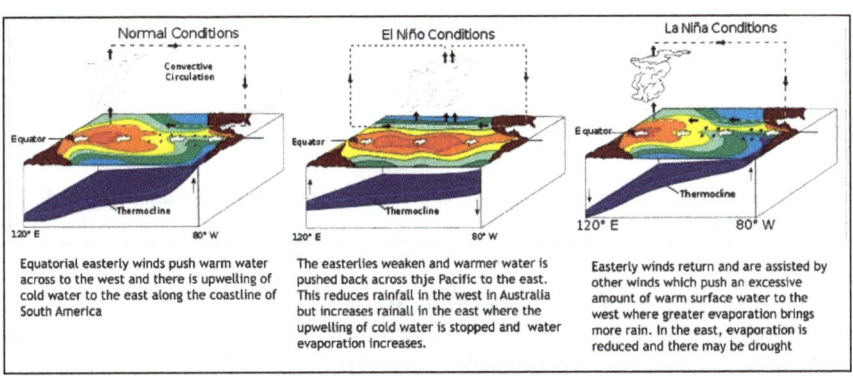

Figure 6.1: A diagram showing the El Niño – La Niña effect (Photo: NOAA)

There is also a long-term recurring variation of sea surface temperature (SST) in the northern Pacific Ocean. This is the **Pacific Decadal Oscillation** (**PDO**) and like the El Niño – Southern Oscillation (ENSO) it has a warm or positive phase and a cool or negative phase. However, unlike the relatively simple ENSO, the PDO is a mixture of mostly independent processes which influence a broader area. One influence in this system is the variation in the strength of a semi-permanent area of low atmospheric pressure off the Aleutian Islands in the north eastern Pacific, called the **Aleutian Low**. Another is a persistence of a previous **anomaly**, or variation from normal, because it had been submerged deeper into the ocean depths. It may later re-emerge to add to the surface conditions. A third factor in the PDO is the variation in strength of the Kuroshio Current, a north-flowing ocean current on the western side of the North Pacific Ocean, similar to the Gulf Stream in the North Atlantic. The PDO largely represents an ocean response to atmospheric changes but it can influence local climate in countries bordering the North Pacific.

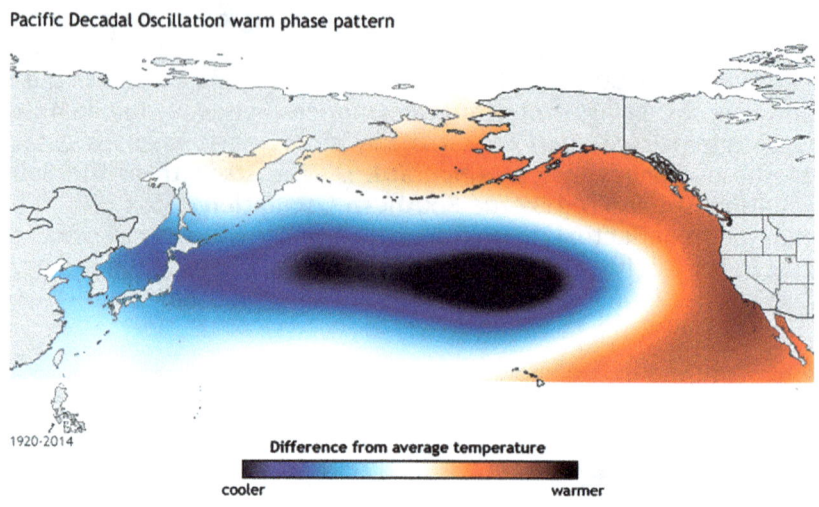

Figure 6.2: A diagram showing the effects of the Pacific Decadal Oscillation PDO (Photo: NOAA)

For the purposes of measurement and evaluation, the **Pacific Decadal Oscillation index** is used. This is a measure of the variations of monthly sea surface temperature anomalies over the North Pacific after the global average sea surface temperature has been removed. For example, on the west coast of North America, a positive PDO can lead to changes in the weather through the direct effects of warmer-than-average ocean waters giving more rainfall. Similarly, a negative value would have colder sea surface temperatures on the American coast with less rain. It should be remembered however, that these are long-term oscillations probably occurring about every 20 to 30 years and lasting only about 12 months so they may have little effect on the overall global warming of the planet.

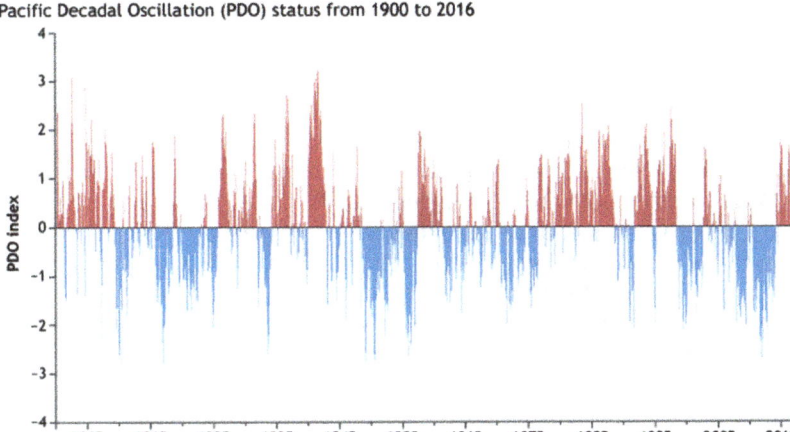

Figure 6.3: A graph showing variations Pacific Decadal Oscillation PDO over time (Photo: NOAA)

The **Indian Ocean Dipole (IOD)** is also an irregular oscillation of sea-surface temperatures in which the western Indian Ocean becomes alternately warmer, called its positive phase, and then colder or its negative phase, compared to the eastern part of the Indian Ocean. The intensity of the IOD is due to a sea surface temperature (SST) difference or gradient between the western equatorial Indian Ocean and the south eastern equatorial Indian Ocean. This gradient is named the **Dipole Mode Index (DMI)**. When the DMI is positive, so is the IOD and when it is negative, there is a negative IOD. These variations usually occur every three to five years with a neutral phase in between. These events usually start around May or June, peak between August and October and then rapidly decay when the monsoon arrives in the Southern Hemisphere around the end of the Southern Hemisphere Spring. A positive phase causes higher sea-surface temperatures, more evaporation and therefore greater precipitation in the western Indian Ocean region with higher rains in east Africa and western India, enhancing the Indian Monsoon or wet season. In the eastern Indian Ocean at this time there is a corresponding cooling of waters which tends to cause droughts in the adjacent land areas of Indonesia and Australia due to the delay in the Monsoon in these regions. The negative phase of the IOD brings about the opposite conditions, with warmer water and greater precipitation in the eastern Indian Ocean, and cooler and drier conditions in the west.

See also:

http://www.bom.gov.au/climate/iod/
(Indian Ocean influences on the Australian climate)

The **Tropical Atlantic Dipole** is a complex, long-term sea surface temperature dipole in the Atlantic Ocean which has a period of about 12 years. This brings changing conditions similar to the other dipoles and occurs between about 10-15 degrees of latitude north and south of the Equator in the Atlantic and has effects in north eastern Brazil on one side of the ocean and the Sahel or north western Africa on the other. The **Tropical Atlantic Dipole Index (TASI)** is a measure of the surface temperature gradient across this zone and is calculated as the difference between the variations in the North and South Atlantic Oceans.

See also:

https://www.metoffice.gov.uk/research/climate/seasonal-to-decadal/gpc-outlooks/atlantic-indian-ocean
(An interactive site for the Indian and Atlantic Oceans)

https://www.cpc.ncep.noaa.gov/products/GODAS/ocean_briefing_gif/global_ocean_monitoring_current.pdf
(A detailed site from NOAA on these ocean dipoles and predictions for the future)

https://www.esrl.noaa.gov/psd/data/climateindices/list/
(List of various oceanic indices from NOAA)

The **Southern Annular Mode (SAM)** is also called the **Antarctic Oscillation (AAO)** and involves the regular low-pressure westerly winds which flow around Antarctica. They sometimes vary in their north-south position bringing cooler temperatures to southern Australia as they move north. They also influence the strength and position of cold fronts and mid-latitude storm systems which are important drivers of rainfall variability in southern Australia. In a positive SAM event, the belt of strong westerly winds contracts towards Antarctica resulting in weaker than normal westerly winds and higher pressures over southern Australia. This restricts the penetration of cold fronts inland and generally causes fine weather. A negative SAM event causes an expansion of the strong westerly winds further north which results in low pressure systems and more storms.

These extremes of climate can have a significant impact on communities in the countries they affect; their economies and the varied natural environments of the country. An understanding of these extremes is often more important than a simple understanding of average climates often quote in weather predictions and literature.

6.3 Severe Thunderstorms and Tornadoes

Thunderstorms are characterized by strong winds, torrential rain and lightning which produces thunder. They usually come from the rapid build-up of clouds known as **cumulonimbus** or thunderheads, which build up a bulbous head at its top resembling an old-fashioned blacksmith's anvil.

Figure 6.4: Thunderheads building up over Port Denarau, Fiji during a hot afternoon.

These clouds are formed from the rapid upward movement of warm, moist air during hot weather. They can occur as an isolated system or as part of an incoming **weather front** when a wall of air of one temperature meets a mass of air at another temperature e.g. a warm front when warm, moist air moves into cooler air. The warm moist air, often taken up from the sea, rapidly moves upwards because it is lighter than the surrounding cooler air. With increased height, the moisture content condenses until it reaches its dew point when it forms droplets of water seen as the cumulonimbus cloud. Sometimes these thunderheads may reach heights of over 20 kilometres. This condensation further reduces the air pressure within the thunderstorm cell bringing up more water vapour. If the rise of the water vapour is rapid and it goes quickly to higher altitudes it will condense quickly and then form into ice and fall as hail. The presence of any particulate material such as dust assists in this process, as the ice forms around the dust particles which act as a nucleus and a hailstone will show a concentric structure around it. Liquid water droplets which form by combining with others will fall from this great height, pulling a downdraft of cold air to the surface along with the precipitation. This cold air spreads out at the Earth's surface, causing the strong winds which are associated with such thunderstorms. Sudden concentrated pockets of downdraft are called **microbursts**.

Storm cells dissipate when the precipitation and subsequent downdraft prevents further uplift of warm air and moisture, although new storms can develop close by if there is sufficient moisture in the air. Thunderstorms usually move through the **troposphere** along with the general wind direction. Sometimes there may be a collection of storm cells formed as a long line, sometimes several hundred kilometres in length, which share common precipitation cores or cloud masses. These are called **squall lines** and they can last for hours or even days, with new storm cells continually forming along the leading edge of the line. They form when there is moist air near the ground and when there are larger vertical updrafts, with the winds near the surface being very different from the winds higher up. These storms are more common in the northern parts of Australia with the approach of the squall line being felt as a sudden onset of gusty winds followed by the arrival of an ominous dark bank of low cloud called a **shelf cloud**.

Lightning is caused by the sudden discharge of the static electricity which builds up by the friction of the ice particles and other droplets in the clouds of the central part of the thunderheads. This discharge can occur between the negative and positive sections of the same cloud as **intra-**

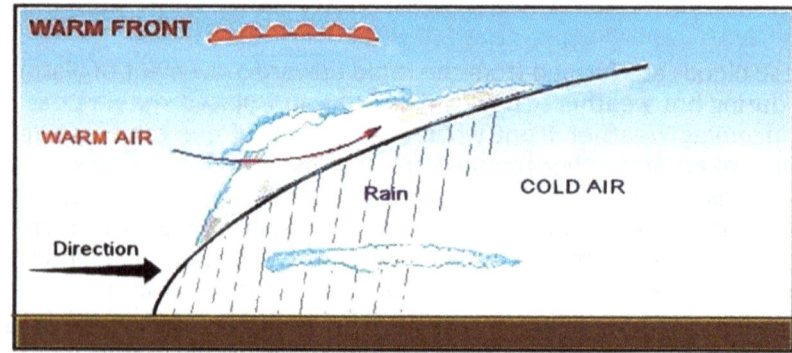

Figure 6.5: Thunderheads building up during a warm front.

cloud lightning (IC), between two clouds as cloud-cloud lightning (CC) or between the cloud and the ground (or CG lightning).

Figure 6.6: Lightning discharge striking a tree (Photo: NOAA)

As the charged cumulonimbus cloud moves across the land surface, an opposite charge is induced upon the ground. When the charge in the cloud has enough electrical potential, and if the cloud is sufficiently close to the ground, a discharge will occur in the form of a lightning strike. This will take the best pathway through air of least electrical resistance and so the path of the hot plasma which forms the discharge will be seen as a bright, jagged flash. Lightning will naturally conduct itself through any tall object which is closer to the cloud, such as a tall tree or spire atop a building. In open country it is recommended that if one cannot quickly get indoors when caught in a lightning storm that one should lie flat to the ground. Seeking shelter below a lone tree is not recommended. One can seek shelter in a vehicle as the metal body takes the charge on the outside and so acts as a shield. The sudden increase in temperature from the lightning discharge rapidly expands the air which creates a sonic shock wave which is heard as a peal of thunder. At sea, tall-masted ships would have a lightning conductor strip running down the mast and to the sea through the iron hull or another conducting strip. Today, yachtsmen may attach a length of chain to their aluminium mast to dissipate any charge which may strike it. Lightning bolts striking the sea spread out in a decreasing flower-like pattern.

An extreme form of thunderstorm often comes as a broad front as part of a **supercell**. These storms have a rapid and rotating vortex of warm, moist air which rotates in the usual direction of a low-pressure area i.e. clockwise in the Southern Hemisphere and anti-clockwise in the Northern Hemisphere. They usually are relatively isolated from other storm cells, move rapidly across the sky with a rotating motion as a long wall of dark cloud. Supercells occur all over the world but some places such as the Great Plains area of the United States and the central, eastern area of South America near the boundaries of Argentina, Uruguay and Brazil seem to have them frequently. The eastern seacoast of New South Wales and Queensland also experience supercell activity. On November 27, 2014, a supercell with winds over 100 km/hour hit the inner suburbs of Brisbane dropping hailstones over 10 cm in diameter, cut the power for a short time and created considerable damage to cars, buildings and vegetation as well as injuring 39 people.

Figure 6.7: A supercell cloud formation (Photo: NOAA)

Figure 6.8 and 6.9: A tornado moving across central USA at left and a waterspout on one of the Great Lakes of America at right (Photos: NOAA)

Supercells can sometime give rise to more concentrated rotating systems called **tornadoes**. These occur when the bottom of the rotating current, or **vortex**, within a cumulonimbus cloud drops down below the base of the cloud and connects with the ground. Here it will move along the surface in the same direction of the cloud above, creating a narrow funnel with an intense updraft which lifts and destroys anything with which it makes contact. At sea, the same processes can form a **waterspout**.

In the United States, the weather varies greatly across the continental part of the country, in Alaska and Hawaii. In Alaska, the weather can be unpredictable with a short but moderately warm summer from June to August and cold winters. The climate of Hawaii is moderated by its oceanic and tropical location and so temperatures are generally warm with good rainfall. In the continental United States, summers are generally hot and humid in the plains and southern states, while the south-west is very hot and dry. The climate in California and Florida are often good all year round and other states in the Gulf of Mexico. In the Pacific northwest and New England states, summers are warm with cool mornings but winters can be cold and harsh with regular snow and severe storms. Severe thunderstorms and devastating tornados usually occur in spring and summer, especially on the central Great Plains. Hurricanes occasionally strike the eastern coastline and Gulf of Mexico states from June through October.

Some useful sites can be found at:

http://www.bom.gov.au/
(Main site of the Australian Bureau of Meteorology with many links and RADAR images)

http://www.bom.gov.au/weather-services/severe-weather-knowledge-centre/severethunder.shtml
(A brief summary from the Australian Bureau of Meteorology)

http://worldweather.wmo.int/en/home.html
(World's weather with a linked list of any countries from the World Meteorological Organisation)

https://www.weather.gov/media/owlie/ttl6-10.pdf
(Good summary with links from the US Department of Commerce)

http://nationalaglawcenter.org/wp-content/uploads/assets/crs/R40097.pdf
(Technical paper about predicting storms and tornadoes in the United States)

https://www.dhs.wisconsin.gov/publications/p01037.pdf
(Good handbook for tornadoes with much of the safety advice applicable to all countries and storms in general)

https://www.weather.gov/afg/roundup.html
(Weather in the United States with RADAR and predictions from the National Weather Service)

6.4 Tropical Cyclones, Typhoons and Hurricanes

It is in the tropics, between 30^0 S and 30^0 N and the adjacent sub-tropical regions that longer periods of extreme storms often develop. The most severe are the intense low air pressure systems which develop in the hotter months and which are known by several names: **tropical cyclones** in Australia and New Zealand; **typhoons** in South-east Asia; and **hurricanes** in the Americas. They occur in regular time periods just after the advent of the hotter months. For example, in northern Australia and the southern Pacific they occur from November to April, and in the Northern Atlantic they occur from June to November. At these times, there is a greater difference in temperature between the sea and the air above. The warmer water experiences rapid evaporation with the water vapour rising in a circular motion as described by the **Coriolis Effect** which is due to the Earth's rotation. These currents rotate clockwise as seen from above in the Southern Hemisphere and anti-clockwise in the Northern Hemisphere. The rapid rise causes an intense low pressure at the surface and a massive build-up of storm clouds with high winds, often over 100 kilometres per hour. Torrential rain is common and **storm surges** sometimes occur when the high winds push the surface of the sea outwards from the cyclone's centre. These storm surges often reach land and carry flooding seawater

Figure 6.10: A sub-tropical cyclone moving across the eastern seaboard of the United States. Note the rotation is anti-clockwise in the Northern Hemisphere and around the calm central eye (Photo: NOAA)

inland, demolishing crops and buildings. These intense low-pressure zones move in from the sea bringing destructive winds, waves, local flooding and general destruction. These events occur at their edges, around the central part of the tropical cyclone which experiences relative calm.

In the United States, the **Saffir-Simpson scale** is used to classify hurricanes in the Atlantic and Northern Pacific regions. It uses a five-point scale based on the maximum velocity of wind recorded in any one-minute time span. A summary of this system is shown in the following table:

CATEGORY	STRONGEST GUST	TYPICAL EFFECTS	EXAMPLES
ONE	119-153 km/h	Very dangerous winds will produce some damage	Diane 1955, Agnes 1972,
TWO	154-177 km/h	Extremely dangerous winds will cause extensive damage	Unnamed (North Carolina) 1883, Unnamed (Florida) 1906
THREE	178-208 km/h	Devastating damage will occur	Betsy 1965, Katrina 2003,
FOUR	209-251 km/h	Catastrophic damage will occur	Hazel 1954, Audrey 1957, Floyd 1999
FIVE	≥ 252 km/h	Catastrophic damage will occur	Camille 1969

Table 6.1 The Saffir-Simpson classification of hurricanes in the USA

See also:

https://en.wikipedia.org/wiki/List_of_United_States_hurricanes
(Interactive list of US hurricanes by State)

Tropical cyclones, typhoons and hurricanes, which are the same type of extreme low-pressure system, are given personal first names for ease of reference and communication as in any yearly season there are likely to be several of these storms in the world's cyclone regions. The first use of names for cyclonic systems is generally attributed to the Queensland Government Meteorologist Clement Wragge (English: 1852 – 1922), who gave female names to weather systems between 1887 and 1907. This practice fell into disuse for several years but was revived after 1953 when cyclones were again given names from lists maintained and updated by an international committee of the World Meteorological Organization.

However, naming rights are in the hands of the regional member associations and formal naming schemes have subsequently been introduced for the North Atlantic, Eastern, Central, Western and Southern Pacific basins as well as the Australian region and Indian Ocean. The only time that there is a change in the list is when a storm is so deadly or costly that the future use of its name on a different storm would be inappropriate for reasons of sensitivity. When this happens, the names are permanently retired from all lists. Usually there are 21 names in each list, if there are more cyclones in any year than there are allocated names, the names from the Greek alphabet are used e.g. alpha, beta, gamma, delta and so on.

In the Australian region, the list of names given to tropical cyclones is the responsibility of the Bureau of Meteorology Tropical Cyclone Warning Centres (TCWC). This system of a single list replaced several lists used prior to the start of the 2008/09 season and the names are allocated in alphabetical order. For example, the 2008/09 season listed potential names in order as:

Australian Regional Names for tropical Cyclones				
Anika	Anthony	Alessia	Alfred	Ann
Billy	Bianca	Bruce	Blanche	Blake
Charlotte	Courtney	Catherine	Caleb	Claudia
Dominic	Dianne	Dylan	Debbie	Damien
Ellie	Errol	Edna	Ernie	Esther
Freddy	Fina	Fletcher	Frances	Ferdinand
Gabrielle	Grant	Gillian	Greg	Gretel
Herman	Hayley	Hadi	Hilda	Harold
Ilsa	Iggy	Ivana	Irving	Imogen
Jasper	Jenna	Jack	Joyce	Joshua
Kirrily	Koji	Kate	Kelvin	Kimi
Lincoln	Luana	Laszlo	Linda	Lucas
Megan	Mitchell	Mingzhu	Marcus	Marian
Neville	Narelle	Nathan	Nora	Noah
Olga	Oran	Olwyn	Owen	Odette
Paul	Peta	Quincey	Penny	Paddy
Robyn	Riordan	Raquel	Riley	Ruby
Sean	Sandra	Stan	Savannah	Seth
Tasha	Tim	Tatiana	Trevor	Tiffany
Vince	Victoria	Uriah	Veronica	Vernon
Zelia	Zane	Yvette	Wallace	Ann
Anika	Anthony	Alessia	Alfred	Blake

Table 6.2 Regional names for tropical cyclones in Australia

The name of a new tropical cyclone is usually selected from this list of names. If a cyclone named in another country's region moves into the Australian region, then the name assigned by that other country is retained. For example, on the 30th of January, 2011, a tropical cyclone developed near Fiji and was name Yasi by the Fiji Meteorological Service. It moved westward across the Pacific, intensified as it picked up warm water and then struck the north Queensland coast on the 3rd February as a Category 5 Tropical Cyclone – one of the most damaging events in recent times.

See also:
http://www.aoml.noaa.gov/hrd/tcfaq/tcfaqB.html
(Naming systems in general and in other regions from NOAA)

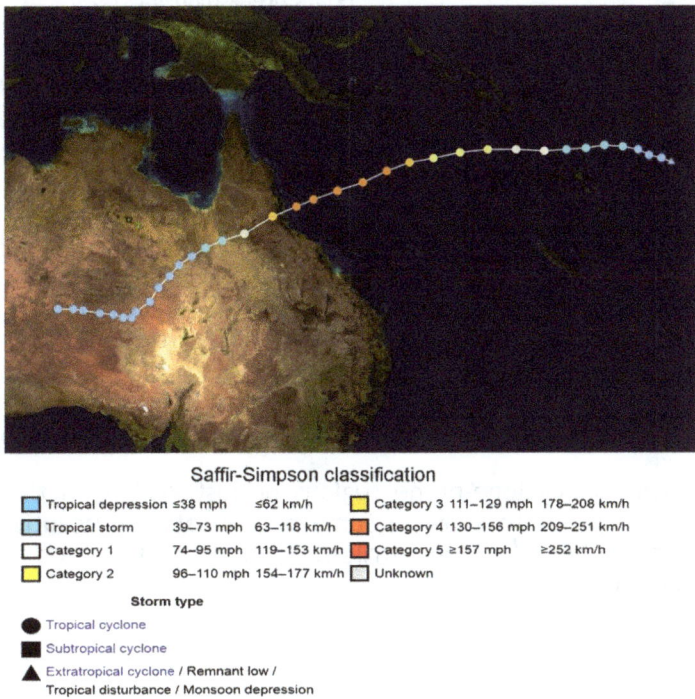

Figure 6.11: Track of tropical Cyclone Yasi (Photo: Public Domain)

Some of the most destructive tropical cyclones in Australia are given in the following table :

CATEGORY	STRONGEST GUST (Km/hr)	TYPICAL EFFECTS	EXAMPLES
1. Tropical Cyclone	Gales - winds less than 125 km/h	Minimal house damage, some damage to crops, trees and caravans. Boats may drag moorings.	Olga 2010
2. Tropical Cyclone	Destructive winds 125-164 km/h	Minor house damage and significant damage to signs, trees and caravans. Heavy damage to some crops, risk of power failure and small boats break moorings.	Anthony 2011
3. Severe Tropical Cyclone	Very destructive winds 165-224 km/h	Some roof and structural damage and caravans destroyed. Power failure likely.	Magda 2010
4. Severe Tropical Cyclone	Very destructive winds 225-279 km/h	Significant roof and structural damage and caravans will be destroyed or blown away. Dangerous flying debris and widespread power failure	Tracy 1974 & Larry 2006
5. Severe Tropical Cyclone	Extremely destructive winds more than 280 km/h	Extremely dangerous with widespread destruction	Yasi 2011

Table 6.3 Some of the most destructive tropical cyclones in Australia

Whether living in cyclone-prone regions or just visiting in the cyclone season, it is prudent to know of some of the safety precautions which would needed when a cyclone strikes. Some suggestions for action are given in the following table:

WHEN A CYCLONE (HURRICANE) STRIKES
• Disconnect all electrical appliances. Listen to your battery radio for updates. • Stay inside and shelter well clear of windows in the strongest part of the building i.e. bathroom, internal hallway or cellar. Keep evacuation and emergency kits (food, water for two days, battery radio + spare batteries, torch, First Aid kit) with you. • If the building starts to break up, protect yourself with mattresses, rugs or blankets under a strong table or bench or hold onto a solid fixture, e.g. a water pipe. • Beware the calm eye of the storm if the wind drops. Don't assume that the cyclone is over, violent winds will soon resume from another direction. Wait for the official all clear. • If driving, stop put the handbreak on and the car in gear. Stay well away from trees, power lines, streams and beaches. Stay in the vehicle unless there is a sturdy shelter nearby.

Figure 6.12: Emergency action in a cyclone

Weather details are often shown as **synoptic charts** or weather maps, as satellite infra-red images which show the temperature within the body of the cyclone and as radar images showing cloud mass and rainfall. The synoptic chart shown below is a good example of the type of chart usually given during weather reports on the media. It shows an unnamed tropical cyclone off the north coast of Queensland. These cyclones usually develop out to sea to the east above the warm waters of the Pacific during the cyclone season which usually occurs from November to April. On this synoptic chart, the tropical cyclone is shown by the closely-packed concentric circles of isobars.

As **isobars** are lines joining places of equal air pressure, having a tightly packed set of circles means that there is a large pressure drop across a relatively small area. This means that winds here will be strong. The air pressure values in hectopascals (hPa) also indicate a very low pressure associated with such cyclones which often will go as low as 980 hPa (normal air pressure is 1013 hPa) and the wind symbol shows a local velocity of about 200 km/hour.

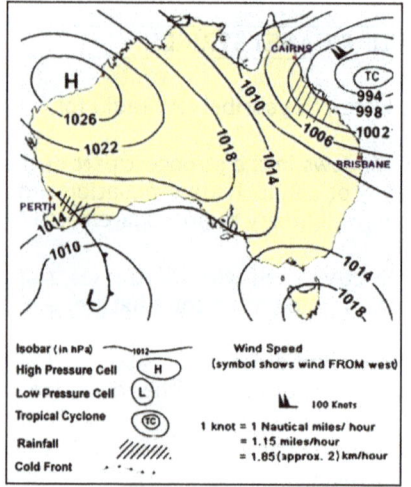

Figure 6.13: A synoptic chart showing a tropical cyclone off the north eastern coast of Australia

Satellite images are excellent for showing the development of a tropical cyclone because they can take infra-red images which clearly show the heat within parts of the growing system. These images taken over time can be used to make predictions about the nature and possible severity of the developing storm. The darker red or black colours indicate the lower temperatures. The image below was taken by the **Moderate Resolution Imaging Spectroradiometer (MODIS)** instrument aboard NASA's Aqua Modis satellite and shows cold cloud tops extending over most of the cloud cover which had temperatures near -63 degrees Fahrenheit (i.e. -53^{0}C). Strongest storms with the coldest cloud top temperatures were just east of Florida and are shown in red.

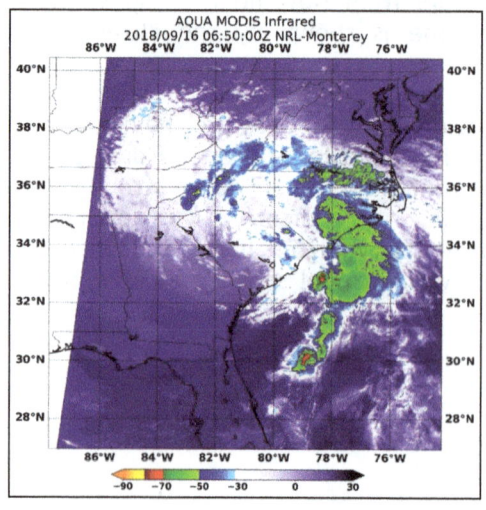

Figure 6.14: An infra-red image showing a tropical cyclone and temperature data shown below in temperature units of Fahrenheit degrees (Photo: NASA/NRL)

RADAR, which is an acronym for **RA**dio **D**etecting **A**nd **R**anging, is used locally and nationally for detecting the rain content, speed and direction of storms, including tropical cyclones. Ground stations send out electromagnetic pulses which strike any water-bearing clouds and return the reflections to the receiver part of the radar dish. These signals are then converted to false colour images by a computer and then published on the meteorological services' website.

See also:

http://www.bom.gov.au/australia/radar/
(Australia's radar locations from the Bureau of Meteorology - click on the appropriate station location).

As well as showing any rain potential as distinct colours with the reds and black colours being the potential heaviest rain, these radar images also make use of the **Doppler Effect** to show the direction and speed of the storm. This effect was discovered by Christian Doppler (Austrian: 1803 - 1853) when investigating sound waves. He found that if the source of the sound was moving towards an observer, the frequency of the sound shortened and the notes became higher in pitch as the waves became bunched up. If the sound source was moving away from the observer, the frequency and the pitch became longer. The same effect can be observed with radar electromagnetic waves. Reflected waves from the storm will show a shortening of frequency when the storm is approaching the radar station and an increase when receding from it. These are shown on the Doppler image for that station as blue and red patterns respectively.

The strength of the pulse returned to the radar depends on the size of the particles, how many particles there are, what state they are in such as solid-hail, liquid-rain etc. and also their shape. After making assumptions about these factors and others, the approximate rain rate at the ground can be estimated. In fact, the most reflective precipitation particles in the atmosphere are large and usually have a liquid surface as water-coated hailstones.

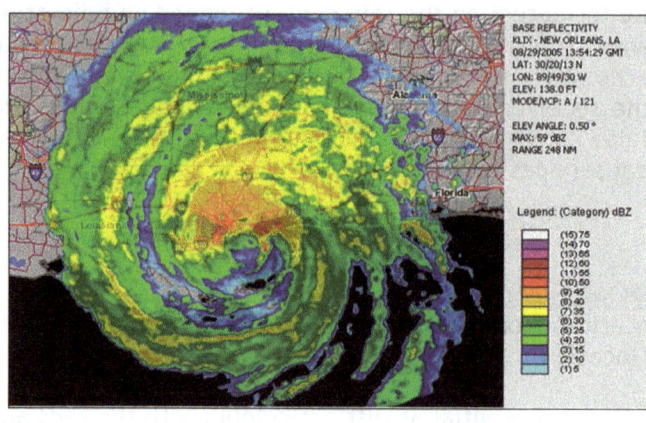

Figure 6.15: A radar image for New Orleans during Hurricane Katrina. Notice the intense red colour showing maximum rainfall over the city centre (Photo: NOAA)

See also:

http://www.bom.gov.au/cyclone/about/
(Australian Bureau of meteorology about Tropical Cyclones)

https://www.meteo.physik.uni-muenchen.de/~roger/Mtheory/Ch10_Tropical_Cyclones.pdf
(Excellent information with much detail, maps and charts about Tropical Cyclones)

https://www.nhc.noaa.gov/pastdeadly.shtml
(Information about deadly Hurricanes in the United States and other useful links from NOAA)

http://www.australiasevereweather.com/cyclones/
(Good data for Tropical Cyclones for Australia and other Southern Hemisphere countries)

https://coastadapt.com.au/cyclones-and-east-coast-lows
(Good data including T.C. tracks for eastern Australia)

https://iumi.com/images/gillian/HongKong2014/Tuesday/IFY/23_1505_Ward.pdf
(Case Studies: T.C. Yasi and T.C. Oswald)

https://www.nccarf.edu.au/sites/default/files/attached_files/Synthesis_Summary_Cyclones_web.pdf
(Good risk analysis of storms and cyclones)

https://www.climatecouncil.org.au/uploads/3cf983377b8043ff1ecf15709eebf298.pdf
(Short but clear information about cyclones and storm surges)

http://portal.gdacs.org/GDACSDocuments/013-TC_Seasonal_Forecast_CaribbeanRegion.pdf
(Hurricanes in the Caribbean area)

https://www.aao.gov.au/science/observing/useful-australian-weather-resources
(Useful Australian weather links)

https://www.ipcc.ch/pdf/special-reports/srex/SREX-Chap9_FINAL.pdf
(Case studies of natural disasters and their management)

https://en.wikipedia.org/wiki/Disaster_risk_reduction
(Disaster risk management as benefical cyclones)

https://practicalaction.org/disaster-risk-reduction
(More risk management)

https://www.researchgate.net/figure/The-Benefits-of-Cyclones-from-Literature-and-Case-Studies-Linked-to-Ecosystem-Services_fig1_303858571
(More on the usefulness of cyclones)

https://www.researchgate.net/publication/303858571_The_Benefits_of_Cyclones_as_Ecosystem_Services
(More on this)

http://www.ifrc.org/Global/Publications/disasters/reducing_risks/194300-Empowering-communities-to-prepare-for-cyclones.pdf
(More with local communities)

6.5 Floods

Floods are a natural part of any river system. They can occur at anytime but in many countries they are more commonly associated with seasonal climate such as the monsoon season in Asia, northern Australia, central America and Mexico and the southweatern part of the United States. There are also occasional climatic events such as El Niño and La Niña events which also cause flooding in their rain cycles. Flooding occurs when there is more water than the soil can absorb or can be contained in storage capacities in the soil, rivers, lakes and reservoirs. There are several types of flooding events including:

- **River flooding** which is usually seasonal such as the annual flooding of the Nile and other great rivers.

- **Flash flooding** which occurs unexpectedly and suddenly due to a severe storm upstream and so can occur at some distance from the areas most affected. These are especially dangerous in mountainous areas where flash flooding can sometimes quickly fill canyons and cave systems without warning. Flash flooding can also occur in hilly urban areas after sudden rainfall, the situation is made worse by the large amounts of concrete and asphalt surfaces that do not allow water to penetrate into the soil easily. These are especially a problem in urban developments, where there have been inadequate drainage pipes or channels to remove the water rapidly. For example, parts of Brisbane, Queensland regularly go under water during brief storms, especially in the summer months.

- **Storm surges** and subsequent coastal inundation are caused when high winds from severe weather systems such as tropical cyclones drive the sea inland onto the coast and up rivers. A storm surge brings water which is above the usual tide level and its maximum height at any given location depends upon the angle at which the storm strikes the coastline, the seafloor shape and its coastline angle. It also depends upon the intensity, size, and speed of the storm and an intense low-pressure zone is able to lift up a relatively large body of water in a small local area. Storm surges are also dangerous when they are in conjunction with high rainfall associated with intense low-pressure systems as they also prevent the escape of the excess water flow of rivers at their estuaries and so cause much flooding near the river mouth. In places where the risk of coastal flooding is particularly high, such as in the Netherlands, Spain, the United States and Great Britain, specific storm surge warnings are often given. To counter the effects of storm surges in urban areas near river mouths, the construction storm-surge barriers can be built. These can be closed when the land is under threat of a storm surge. Major storm surge barriers have been built at Oosterscheldekering and Maeslantkering in the Netherlands, the Thames Barrier near London and the Saint Petersburg Dam in Russia.

Figure 6.16: Storm surge damage on the Bolivar Peninsula of Texas following Hurricane Ike in 2008 (Photo: US National Weather Service)

See also:

https://www.weather.gov/media/owlie/surge_intro.pdf
(A brief introduction to storm surges)

- **Snowmelt flooding** occurs when the major source of water is melting snow. Areas adjacent to high mountains with permanent snowfields are subject to this type of flooding, especially in the European Alps and other ranges of Europe, along the northern Rocky Mountain states of the United States and along the narrow coastlines of western South American countries where the rivers are short but drain off high peaks. Unlike rainfall which can cause immediate flooding in rivers, snow melt may be delayed for quite some time after heavy snowfall because the snow packs down and will only melt when there is a sudden significant rise in temperature. Once it begins to melt, water from snowmelt behaves much as it would if it had come from rain instead of snow by either infiltrating into the soil, running off, or both. Flooding can be more severe if there has been enough moisture in the soil to saturate it, either from local rainfall or generally humid conditions at lower altitudes. Also, deep, hard ground frost prevents snowmelt from infiltrating into the soil when there is a period of cold weather prior to heavy snowfalls of the winter season. Usually snowmelt runoff is slow and regular with little or no flooding. However extensive flooding can occur, especially during unusually warm periods when there are high dew point temperatures, and when night temperatures remain above freezing. Another type of flooding due to rapid snowmelt is the **lahar**, or volcanic mud flow, which can occur due to the sudden melting of snow on the top of a volcano which suddenly becomes more active. In these lahars, the warmed water from the melted snow mixes with fine volcanic ash and then rapidly flows for very long distances down any natural water channel without warning. Parts of the Andes Mountains can produce these deadly meltwater flows with only a small amount of volcanic activity required to set off the mudflow.

- **Collapse of confining structures** such as natural **levee banks** hold in rivers above the lower flood plains, man-made dams and reservoirs and large mine tailings ponds. These breaches are often sudden as the structure gives way with little warning. Many river systems have complex systems of natural levee banks which contain the river and build up above the level of the surrounding floodplain as the river both deposits sediment and then cuts down through it. The Mississippi River around the New Orleans region has flooded many times due to levee failure.

Figure 6.17: Levee breach of the Sacramento River (Photo: US Army Corps of engineers)

During periods of extreme rainfall, the flood gates of dams upstream are often opened to prevent additional stress on the dam wall. This sudden release, if not done in a controlled manner can cause flooding downstream. The Brisbane flood in Australia of the 11th of January 2011 was caused by tropical cyclone Oswald and by the release of water from the Wivenhoe Dam upstream. Thirty-five people were killed in the flood waters. There also have been several catastrophic flooding events due to the sudden collapse of dam walls, especially huge dams designed to hold mine tailings such as the Marianas and the Córrego do Feijão tailings dams in Minas Gerais, Brazil. Both dams caused a large amount of damage to the local populace and the environment when they suddenly released huge amounts of toxic water across the countryside. Nineteen people were killed in the 2015 Marianas collapse and over 100 were killed in the Córrego do Feijão collapse in 2019.

See also:

https://www.wise-uranium.org/mdaf.html
(A list of major mine dam failures)

Flooding due to climatic conditions is part of the natural cycle of the environment and it can have both a good and bad impact on the environment. The benefits of natural floods generally outweigh their negative aspects. The negative impact of flooding is usually seen when it occurs in places of human habitation but flooding also damages the natural environment. In general, the negative impact of flooding includes:

- Disruption of drainage and sewerage systems in cities causes the mixing of raw or partially raw sewage into the flooded area. This polluted water is ideal for the spreading of diseases such typhoid fever, cholera, leptospirosis and hepatitis A. Flooding also assists in the transmission of diseases by parasites, or **vector-borne diseases** which are caused by microorganisms which are transmitted by blood-sucking mosquitoes, ticks and fleas known as vectors. Vector-borne diseases include malaria, dengue and dengue haemorrhagic fever and yellow fever which can cause serious illness and even death.

- Destruction of infrastructure such as homes and other buildings, roads, bridges and other communication links. On a personal scale this can be very damaging to the population both physically and psychologically. Floods disrupt many people's lives each year with personal tragedies especially if flooding occurs frequently.

- Spreading of toxic materials such as pesticides, agricultural chemicals, paints, oils, solvents and other substances which can cause major damage to the local ecology. In 2011, coastal flooding in Japan caused radiation from the destroyed Fukushima nuclear plant to be spread over a wide area.

- Dislocation of animal species and destruction of local habitats also can occur with massive flooding. If the flooding is gradual, many of the more mobile animal species will leave the area but smaller animals, plants and microorganisms will be severely impacted. Habitat destruction also disrupts the natural food chain and so the impact of the flooding often lasts well after the waters have subsided. These longer-term effects impact on all species which have survived, especially on those species considered endangered.

- Impact on marine environments due to sedimentation and turbidity, litter, human waste and toxins deposited from the land. For example, flood effluent from rivers can affect the health of mangrove communities, seagrass and coral communities along the coast and those species which depend upon them.

- River bank destruction and soil erosion which also removes natural habitats along the river as well as lands for agricultural and urban use. Severe surface runoff also removes valuable topsoil necessary for good plant growth.

- Financial loss to the local economy as well as that of state and national governments. In Australia, floods are the most expensive type of natural disaster for example, in 1974, floods affected New South Wales, Victoria and Queensland and resulted in a total cost of $2.9 billion. The 2011 floods in Queensland were even more severe with the total damage to public infrastructure across the state at between $5 and $6 billion.

In the long-term however, flooding can have some positive benefits to the environment. These include the:

- Distribution of large amounts of water over vast areas with subsequent infiltration. The water provides moisture to soils which may have otherwise been dry. In countries such as Egypt, farmers along the Nile once relied upon the annual flooding for their crops, although today this flooding is controlled by releases from the Aswan Dam.

Figure 6.18: Crops along the banks of the Nile rely on the water from the river as there is little or no rainfall. The dry land can be seen in the distance.

- Spread of sediment which helps replenish valuable topsoil components to agricultural lands. In some low-lying coastal communities, such as around the Mississippi Delta and in the Netherlands, this new soil also helps to can keep the elevation of a land mass above sea level and slows coastal erosion. In the Mississippi Delta region, the associated rivers would frequently spill over their banks and increase the height of the surrounding land as the area is gradually subsiding. Unfortunately, massive flood-prevention projects have stopped this natural process of flood sediment replenishment and much of the surrounding land has dropped to elevations below natural sea level.

- Provision of natural nutrients from the land and into coastal wetland and marine habitats for the well-being of bird life, fish and other forms of marine life.

- Flushing of the land, river and adjacent wetlands with the removal of accumulated agricultural and urban wastes and toxins.

- Maintenance of local climatic conditions by moderating local temperatures and atmospheric humidity. For example, in the large Bay of Bengal there is an enormous amount of less-dense and low-saline layer of water brought down by the flooding of the major rivers which flow into the bay such as the Ganges, the three major rivers of Bangladesh the Padma, the Jamuna and Meghna and the Irrawaddy River of Myanmar. The presence of this low-salinity layer helps in the maintenance of high sea-surface temperatures which are usually greater than 28°C. It is thought that this surface water temperature is responsible for the intensification of summer monsoon in the region.

The scientific study of floods comes under the branch of surface **hydrology**, the study of water or more specifically, **hydrodynamics** or the study of moving water. Usually waterways such as river systems will have a network of stream gauges which measure the water discharge or volume per time measured in cubic metres per second (cms^{-1}) and its **stage**, that is its height above a specific datum mark such as mean sea level. As discharge increases, stage usually increases before reaching its **peak** or crest which is the highest water level. However, the relationship between rainfall and the time for its water to flow downstream, or its **lag time**, is not a linear relationship. That is to say, that if a certain amount of rain falls then it will not take a specific corrsponding time for it to reach a certain point downstream. This is because there are many factors including the intermittent nature and amount of the rainfall, the effect of infiltration as water seeps into the ground, the degree of saturation of the ground, the varying width of the stream channel and the amount of loose sediment and debris within the channel. Following an event of heavy rainfall in the **catchment area**, the land surface which collects the rain, hydrologists can use the stream gauges to determine if the water level will reach a **flood stage** i.e. a height of water flowing over the river's banks which are of sufficient magnitude to cause widespread inundation of land or significantly threaten life and property. This should not be confused with **overbank flow** which is a localised flow of water over the river's banks and into another waterway. Never-the-less, hydrologists can make relatively good assumptions about the **lag time**, or the time taken for floodwaters to reach a point after rain upstream. This enables them to give flood warnings to communities further downstream with an idea of the **peak** or the extent of the potential maximum height of the flood and its estimated time of arrival. Data for the characteristics of a particular river can be ploted and given as a hydrograph which shows the peak or maximum time of rainfall, the peak discharge and therefore the difference in time between them as the lag time.

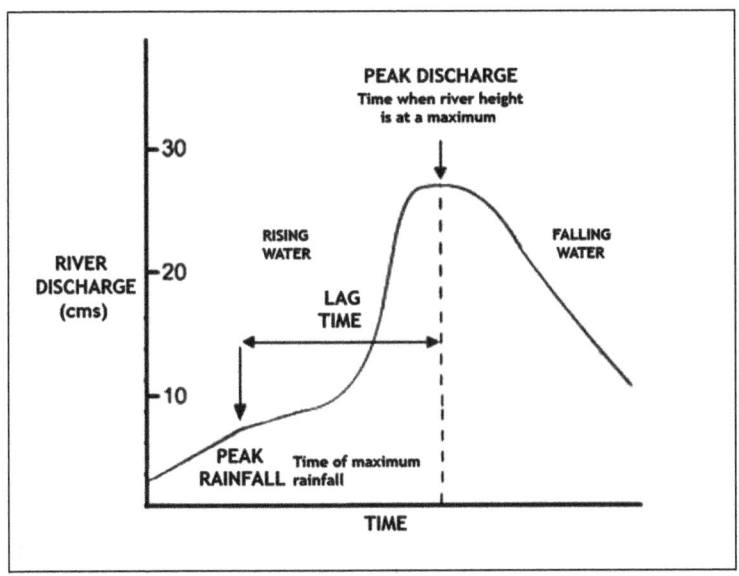

Figure 6.19: Graph showing river discharge against time

Flood predictions are determined by examining rainfall and river water level data from past flooding events, and using statistics based upon historical recurrence intervals and then extrapolating the data to predict future probabilities. There are some statistical errors in this method, notably which of the past occurrences of flood events and the intervals between them will be the same for future occurrences. Because drainage basins are changed by human activities and other events, and rainfall may be changing due to local or global climate variations, the extrapolation of past events to the future may be invalid. The modelling also assumes that there is the same direct relationship between rainfall and the flooding event and that each flood event is independent of the others.

The extrapolations of recurrence intervals are then used to forecast the future probability of a flood of a given discharge. The probability (**P**) of a flood with recurrence interval **T** is given as:

$$P = 1/T$$

Often the term, a one-hundred-year flood, is used to express the probability of a particular flood event. This does not actually mean that such a flood only occurs every one hundred years but that such an event has a 1% chance of occurring in any given year i.e.

$$P = 1/100 \text{ or } 1\%$$

The stage of such a flood can be back-calculated using the rating curve for the river. In addition, based upon the areas which have been flooded in the past, flood maps showing the probability of future flooding can be drawn. These should also involve any more recent changes to the draining of the region and changes in rainfall patterns.

See also:

http://www.bom.gov.au/qld/flood/rain_river.shtml
(Very detailed information and data for Australian rivers)

http://www.bom.gov.au/water/floods/floodWarningServices.shtml
(Australian flood warning services with good links for each state)

http://www.bom.gov.au/qld/flood/
(Interactive map of the conditions in Queensland rivers)

http://www.bom.gov.au/qld/flood/fld_history/brisbane_history.shtml
(Flooding of the Brisbane and Bremer Rivers - Australian Government data)

http://www.bom.gov.au/australia/flood/EMA_Floods_warning_preparedness_safety.pdf
(Flood risk management pamphlet from the Australian Bureau of Meteorology)

https://mdx.mrooms.net/pluginfile.php/221639/mod_resource/content/2/PRS4526%20Week%2014%20Environmental%20Impacts%20of%20floods%20and%20flood%20mitigation.pdf
(Very good summary about flooding and its control)

https://www.dnrm.qld.gov.au/__data/assets/pdf_file/0009/230778/flood-mapping-kit.pdf
(Detailed notes on floods in Queensland)

https://www.brisbane.qld.gov.au/sites/default/files/20161216-flooding-in-brisbane-guide-for-residents.pdf
(Good guide to flood emergency)

https://knowledge.aidr.org.au/media/3518/adr-guideline-7-3.pdf
(Flood hazard in Australia)

http://dnrm-floodcheck.esriaustraliaonline.com.au/floodcheck/
(Interactive flood map for Australia)

https://www.weather.gov/safety/flood-map
(Interactive map for flooding data in the separate states of the United States)

https://waterdata.usgs.gov/nwis/rt
(Excellent interactive map for river conditions in all states of the United States with data)

6.6 Wild Fires

Wildfires are any natural, large-scale burning of of the countryside. In Australia they are called **bushfires** as a general term for the burning of any vegetative environment, in South Africa the term veld fires is often used, particularly in the open savanna and in many countries forest fires burn out large areas of taller trees. In most cases these fires are part of the natural cycle of regrowth and are usually started by lightning strikes during thunderstorms in periods of prolonged hot, dry weather. From historical times and continuing today, many wildfires are started by the activity of humans such as careless use of smaller fires in the home and farm, sparks from machinery, electrical discharges from fallen power lines and on some occassions by pyromaniacs, deranged individuals who like to watch fires regardless of the consequences.

For any fire to start, there must be three essential elements - Fuel to burn, heat to get the fuel to its **ignition temperature** and oxygen to support combustion:

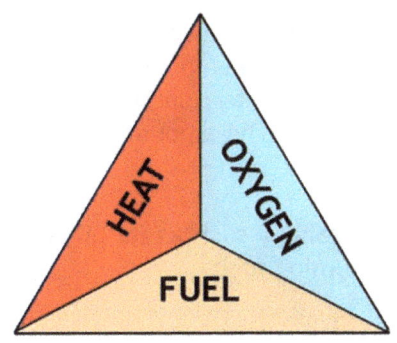

Wildfires can occur in any region where these conditions exist. After a long, season of hot weather, with considerable dry undergrowth and high winds, wildfires can easily be started, especially near urban areas where the chances of an ignition spark or a careless ember can destroy a large area of countryside, homes, wildlife, crops, forests and sometimes human life.

The three primary classes of wild land fires are surface fires, crown fires, and ground fires. Each classification depends on the quantity and types of fuels involved and their moisture content. These conditions have an effect on fire intensity and will determine how fast the fire will spread. There may be several separate parts to a wildfire including:

- **Surface fires** which occur along the land surface through grass, forest leaf litter, low scrub and other low-lying vegetation. These fires burn at a relatively low temperature, usually less than 400 °C and often spread at a slow rate, although they tend to go at much faster speeds up steeper slopes if there is a strong wind behind them. These are easier to control if caught in time and often pose less threat to larger, mature trees. The clearing of dead undergrowth around buildings and roads can reduce the threat of ground fires and they can also be prevented by limited controlled burn-off before the main hot season.

- **Ground fires** are often overlooked after surface fires have been extinguished because they burn by the slow smouldering of subsurface roots and the lower, internal parts of larger trees. They can burn slowly for days to months, such as in some peat fires. Eventually, ground fires may ignite the surface vegetation and produce large and intense blazes which may evolve into a **fire storm** or localised **fire tornado** when the sudden updraft of hot air brings in strong, swirling winds at ground level, further increasing the fire's intensity.

- **Crown** or **canopy fires** burn higher up in the forest canopy where they can be fuelled by natural oils in the foliage of the trees. These fires often spread very rapidly by jumping from treetop to treetop in a motion called crowning. The amount and speed of crowning depends upon the:

    - material of the canopy such as its oil content and leaf size which creates a fast-burning fuel;

- vegetation moisture content of the canopy which is also dependent upon the type of vegetation and the time it has to dry out due to the weather conditions;

- canopy height which allows for an intense under-draught of air to rapidly rise up from ground level and further spread the flames;

- density of the canopy both in terms of the size of the canopy of each tree as well as its proximity to other canopies; and

- weather conditions during the blaze such as general surface winds which can blow across the canopy spreading the flames.

Wildfires can occur all year-round, although there are identifiable seasons which vary by state and by region. For example, in the Australian tropical state of Queensland, the main threat usually comes in spring in the populated south-east. This threat then extends into summer in the state's west and south-west. In the dry interior of the tropical north, past the inland city of Mt. Isa, the fire season occurs in winter and spring.

Figure 6.20: Wildfire racing up a hilltop in California, USA. Note the crowning at top right and the ground fire at bottom left (Photo: US Forest Service

See also:

http://www.bom.gov.au/weather-services/bushfire/about-bushfire-weather.shtml
(Australian map showing the bushfire seasons in various states)

https://climate.nasa.gov/blog/2830/six-trends-to-know-about-fire-season-in-the-western-us/
(Data of fire risk in the United States with predictive graphs and maps)

https://ec.europa.eu/environment/forests/pdf/InTech.pdf
(Monitoring wildfires in Europe)

https://fires.globalforestwatch.org/map/#activeLayers=viirsFires%2CactiveFires&activeBasemap=topo&activeImagery=&planetCategory=PLANET-MONTHLY&planetPeriod=Jan%200000&x=98.597729&y=25.143718&z=2
(Very good interactive map of global wildfires and their effects)

In south-eastern Australia in general, where most of the population is centred, bushfires are most severe during summer and autumn (fall), especially during drought years. These are often caused by a prolonged El Niño event. In northern Australia, bushfires usually are more severe during the dry season around April to September. In the south-west, bushfires occur in the summer dry season and depends upon seasonal vegetation growth. Over the last thirty years, Australia's climate has shown a trend towards hotter, dryer weather with an associated increase in the number and severity of bushfires. It is thought that this is a result of global climate change.

The hazards of being in a wildfire area include:

- Unpredictability of the fire front movement which could be very rapid and the disorientation caused by smoke, noise, heat and general panic.

- Destructive flames at a high temperature capable of burning most structures. Flames may also crown across the tops of trees or start elsewhere as **spot fires**.

- Thick clouds of embers, burning debris moving through the air as a dense cloud. These can penetrate closed structures such as houses through roof openings below tiles and set fire to the interior.

- Smoke inhalation often is the main cause of death. The smoke as fine particulate matter which will penetrate vehicles and buildings. Smoke haze can drift over large distances and be a bad hazard due to respiratory problems and visibility a long way from the fire source. Ash and dust, particularly from burned buildings, may contain toxic and cancer-causing chemicals, including asbestos, arsenic, and lead.

- Toxic gases especially carbon dioxide and carbon monoxide are both odourless and colourless. In addition, there may also be oxygen starvation due to the intensity of the fire.

- Direct radiation from the heat can cause heat stroke at a distance.

- Falling timber as tree limbs and whole trunks as well as falling man-made structures which have been weakened by burning.

- Electrical hazards due to fallen power cables and exposed wiring within burned structures may still carry current.

- Traffic hazards due to fire-fighting vehicles and of those evacuating which cannot be seen in the dense smoke and may be in a hurry.

- Interaction with wildlife escaping from the fire such as snakes.

See also:

https://mobil.wwf.de/fileadmin/fm-wwf/Publikationen-PDF/WWF-Study-Forests-Ablaze.pdf
(causes and effects of forest fires in the world)

https://www.climatecouncil.org.au/wp-content/uploads/2019/11/CC-nov-Bushfire-briefing-paper.pdf
(climate change and bushfires in Australia)

https://pubs.usgs.gov/fs/2006/3015/2006-3015.pdf
(wildfire threats in the US from the USGS)

Some of the world's deadliest wildfires are given in the following table:

WILDFIRE EVENT	LOCATION	DATE	NOTES
Peshtigo Fire	Wisconsin USA	October 8, 1871	1,200-2,500 deaths destroyed over 4900 km^2
Kursha-2 Fire	Soviet Union	August 3, 1936	1,200 deaths
Cloquet Fire	Minnesota USA	October 12, 1918	453 deaths, 1000 km^2
Great Hinckley Fire	Minnesota USA	September 1, 1894	418+ deaths, destroyed 810 km^2 of pine and the town of Hinckley

Name	Location	Date	Impact
Thumb Fire	Michigan USA	September 5, 1881	282 deaths and 4,000 km² burnt out
1997 Indonesian forest fires	Sumatra and Kalimantan, Indonesia	September 1997	240 deaths and smoke covered all of Indonesian, Malaysia and much of the nearby countries.
Matheson Fire	Ontario, Canada	July 29, 1916	223 deaths and 2,000 km²
Black Dragon Fire	China and Soviet Union	May 1, 1987	191 deaths and over 73000 km²
Black Saturday bushfires	Australia	February 7, 2009	180 deaths, many injured, 4500 km²
Miramichi Fire	Canada	October 7, 1825	160-300 deaths, 16,000 km²
Attica wildfires	Greece	July 23, 2018	99 deaths
Camp Fire	California USA	November 28, 2018	88 deaths with 1,276 still missing. Destroyed the town of Paradise. 620 km²
Great Fire of 1910	Montana and Idaho USA	August 20, 1910	87 deaths, 12,100 km²
2007 Greek forest fires	Greece	June 28, 2007	84 deaths, 2,700 km²
Landes forest fire	France	August 19, 1949	82 deaths, 500 km²
Ash Wednesday bushfires	Victoria & South Australia	February 16, 1983	75 deaths across two states. Over 2,545 homes lost
Great Porcupine Fire	Canada	July 11, 1911	73-200 deaths, 2000 km²
Black Friday bushfires	Australia		71 deaths, 20,000 km²
2017 Portugal wildfires	Portugal	June 17, 2017	66 deaths
Yacolt Burn	Washington and Oregon USA	September 8, 1902	65+ deaths, over 2000 km²
1967 Tasmanian fires	Australia	February 7, 1967	62 deaths, 2,640 km²
1926 Victorian bushfires	Australia	January 26, 1926	60 deaths
1991 Indonesian forest fires	Indonesia	August 1991	57 deaths
1992 Nepal wildfires	Nepal	March 1992	56 deaths

Further information about the world's wildfires can be found at: https://en.wikipedia.org/wiki/List_of_wildfires

Table 6.4: Some of the world's worst wildfires in order of number of lives lost

On a personal note, it is now late spring in Brisbane on the east coast of Australia. It is 40^{0C} (104^{0F}) and the end of our short street is obscured by a thick, brown haze. There has been a government warning to stay indoors as the smoke haze in this state and in our neighbouring state to the south (New South Wales), is 'unprecedented' (fire authority usage) due to extensive wildfires - the nearest being about 100 km. away. The concern here is that many of the fires in the state of Queensland are in what normally would be lush, tropical rainforests.

Many environmental Earth scientists are interested in how wildfires affect the world's ecology. **Fire ecology** focuses on the origins of wildfires fire and their relationship to the environment, both living and non-living. Fire ecologists recognize that fire is a natural process, and that it often operates as an integral part of the ecosystem in which it occurs. Some of the benefits of wildfires include the:

- Role of small wildfires burning off accumulated undergrowth which may be built up to be an excellent fuel supply at a later date for larger, more disastrous wildfires. This is often the reason why fire authorities carry out regular, controlled spot fires in potentially dangerous areas, especially those on the fringes of urban habitation.

- Regeneration of native species, many of which will only germinate or grow after a fire. **Pyrophytes** (Greek: *pyros* for fire and *phytos* for plant) are plants which have adapted to tolerate fire because it has become part of their natural environment. These plants require fire so that their reproductive cycle will continue and they usually have features which resist destruction by fire including thick bark, tissues with high moisture content and underground storage systems. Many pyrophytes actively encourage wildfires by having volatile oils within their leaves. Many species of Australian eucalypts are active pyrophytes and in other countries, many species of pine, oaks and redwoods are relatively fire-resistant. Many of these trees such as the eucalypts and banksias and some pines often have their seeds encased in hard, protective resins which maintain the seeds through the many drought periods of their environment but require a major fire to open the seed cases and allow them to germinate. In addition, some species of trees have tall crowns which would remain above the level of smaller wildfires, some have fire resistant buds within protective bark and have abundant flowering systems.

- Removal of alien plants which have been introduced to the environment which compete with native species for nutrients and space. Because they have not evolved within fire-dependent environments, these introduced species do not have the ability to resist major fires.

- Removal of compact undergrowth allows sunlight to reach the forest floor, supporting the growth of native species.

- Provision of nutrients from the ashes which remain after the fire has passed. When burnt trees decay, they return even more nutrients to the soil.

- Control of insect pests by killing off the older or diseased trees which are often contaminated with pests and disease leaving the younger, healthier trees which are more able to resist the fire and regrow.

- Provision of new habitats in the trees after the fire for birds, reptiles, burrowing mammals and other wildlife.

Overall, fire is a catalyst for promoting biological diversity and healthy ecosystems. It fosters new plant growth and wildlife populations often expand as a result of an ecosystem being swept by smaller fires on a regular basis.
See also:

https://www.ahc.sa.gov.au/ahc-resident/Documents/sampson-flat-bushfire-recovery-biodiversity-fact3.pdf
(Good summary of bushfire recovery and biodiversity)

The disadvantages of wildfires are usually obvious when they impact urban areas and farming communities with the loss of buildings and other infrastructure, food crops, stock animals, wildlife and sometimes human life as well and the associated financial damage to local and national economies. The negative impact of a wildfire on an environment also includes:

- Loss of endangered species of both flora and fauna.

- Soil erosion if significant amounts of supporting trees and vegetation is lost, especially if heavy rainfall follows the period of fire. Heat from intense fires can also make soil particles **hydrophobic** i.e. rainwater then tends to run off the soil rather than to infiltrate through it.

- Additional large amounts of carbon dioxide and carbon monoxide gases as well as oxides of nitrogen being released into the atmosphere.

- Localised air pollution with the large-scale emission of particulates such as smoke and any natural or man-made toxins liberated by the fire. Smoke with a particulate diameter of less than one micrometre thick is able to travel vast distance and can penetrate many buildings, causing severe respiratory problems, especially amongst older people, children and asthmatics. Wildlife, stock animals and domestic pets also suffer similar problems with extreme smoke inhalation.

See also:

https://www.oregon.gov/osp/sfm/documents/airMonitoringreport.pdf
(Scientific study report on post-fire monitoring of harmful chemicals)

- Damage to the physical structure and purity of waterways due to debris such as partially-burnt vegetation, dead wildlife and stock, ash, eroded soil and toxins contained in the smoke. Released carbon compounds from the fire have been known to also react with chlorine in water-purification plants to produce harmful carcinogens thus destroying local drinking water of nearby towns and cities.

- Damage to surface property and health because of the fire-retardant chemicals dropped on the fire.

Wildfire prevention is better than having to combat a major fire when it starts.

See also:

https://www.unisdr.org/2000/campaign/PDF/Articulo_6_Australia_eng.pdf
(Wildfire prevention advice)

Too often, major fires become uncontrollable and fire authorities can only look to the evacuation of the local population and stock, build containment structures such as earthen walls and clear vegetation as firebreaks. Home and property owners are encouraged to:

- Report the build-up of undergrowth which may be a good source of fuel for a fire.

- Carry out good fire behaviour by not burning off sections of land, burning rubbish, leaving campfires unattended. Controlled burning off of undergrowth should be left to the local fire authority.

- Maintaining farm equipment and other tools to prevent sparks from exhausts and electrical appliances.

- Regularly monitor high voltage power cables to see if there could be any contact with trees and bushes.

- Build in areas where the vegetation has been cleared for a good distance around any construction such as the home or outlying sheds. For older establishments, clear fuel hazards around all buildings and access roads.

- Maintain good water storage in and around rural properties or do not cover or restrict access to fire hydrant points in urban areas.

- Do not attempt to fight any large-scale fire but report it and then evacuate to the nearest, open and safe areas. Most rural fire services have warning systems either by commercial media or mobile phone apps which are able to give fire warnings and instructions to evacuate to designated safety centres if needed.

- Have an emergency plan in case there may be a threat of fire and have at least two evacuation routes.

- Keep note of any fire warnings including the current fire danger status.

- Report any suspicious smoke, especially in natural forests and grasslands.

See also:

https://www.qfes.qld.gov.au/community-safety/downloadlibrary/Documents/GetReadyGuide-E.pdf
(A short but important guide on being prepared for fire)

https://www.lifehacker.com.au/2019/12/masks-smoke-bushfires-nsw-where-to-buy/
(Good article on selecting GOOD respirators for smoke)

http://www.bccdc.ca/resource-gallery/Documents/Guidelines%20and%20Forms/Guidelines%20and%20Manuals/Health-Environment/BCCDC_WildFire_FactSheet_FaceMasks.pdf
(Good information about using face respirators for smoke)

Many countries have a fire-warning colour coded system to alert communities about the current threat of wildfires. In Australia, the **Forest Fire Danger Index (FFDI)** was introduced by Alan McArthur and it has been in operation since 1967. Whilst there are some variations in different states both in Australia and the United States, the most common fire-warning codes are:

LEVEL COLOUR CODE	MEANING
LOW	Fuels do not ignite readily. Little chance of spot fires
MODERATE	Fires are not likely to become serious, and control is relatively easy.
HIGH	Fires may become serious and their control difficult.
VERY HIGH	Fires start easily and spread rapidly, difficult to stop.
EXTREME	Fires under extreme conditions start quickly, spread furiously, and burn intensely.

Table 6.5: Fire warning colour codes

See also:

https://www.maine.gov/dacf/mfs/wildfire_danger_report/fire_rating.html
(Further explanation of codes from the United States)

Following the Australian Black Saturday fires of 2009, a new category called CATASTROPHIC has been added to the roadside signs in several states to describe conditions more extreme than those of the older system. In Victoria, this alert is given a striped CODE RED designation and in other places such as Tasmania it is BLACK. This code means that any fires that start are likely to be so fierce that people and structures will be difficult to save and that evacuation is likely.

For general reference about fires, see also:

http://myfirewatch.landgate.wa.gov.au/
(Excellent interactive map which shows current fires in Australia)

https://www.google.com/url?sa=t&rct=j&q=&esrc=s&source=web&cd=13
&cad=rja&uact=8&ved=2ahUKEwjcw6CkwvTeAhVRFHIKHdadCDs4ChAWMA
J6BAgIEAI&url=https%3A%2F%2Fwww.aph.gov.au%2FDocumentStore.ashx%
3Fid%3D3d4e5dd5-9374-48e9-b3f4-
4e6e96da27f5&usg=AOvVaw1zXfE2OyjHe_qXWLRMUs6q
(Comprehensive report by CSIRO on fires in Australia)

https://www.climatecouncil.org.au/uploads/df9df4b05bc1673ace5142c3
445149a4.pdf
(Good overview of wildfires in Australia and overseas)

http://www.gtaq.com.au/Resources/Documents/Bushfires_final_txt_LR.
pdf
(Good general details about wildfires)

http://www.climateinstitute.org.au/verve/_resources/bushfire.pdf
(Global warming and bushfires in the future)

https://www.ncbi.nlm.nih.gov/pmc/articles/PMC4785963/
(Technical paper outlining the increase in bushfires in Australia)

http://www.climatecouncil.org.au/uploads/c597d19c0ab18366cfbf7b9f6
235ef7c.pdf
(Climate change and the future of bushfires in Australia)

https://www.preventionweb.net/files/3828_ForestfirehazardN3.jpg
(Good European fire hazard map)

https://www.wri.org/blog/2018/08/5-graphs-show-just-how-unusual-
years-wildfires-are
(Graphs showing increase in fire alerts in Europe)

Final Remarks

1. Some of the immediate problems associated with global warming are the increased hazards of weather extremes, floods and wildfires.

2. Extremes of weather are mostly to increase in some areas. These include more intense tropical cyclones (hurricanes) and tornadoes with greater frequency per season.

3. Floods can come at any time after heavy rain, especially in coastal areas which might also experience storm surges as an onshore wind pushes the sea onto the land.

4. Inland flooding can be caused by over flow from rivers, especially when levee banks are broken and the water rushes downhill onto the flood plain.

5. Lahars are floods of hot volcanic mud which can flow long distances down stream beds from an erupting snow-covered volcanic source.

6. In hilly regions, especially those heavily populated, flooding can be sudden and dramatic due to flash flooding after and during sudden storms.

7. Flooding can be controlled by dams and diversion channels and many regions of the world, especially those with arid landscapes such as in Egypt, rely on annual flooding for agriculture.

8. Wildfire (bushfires) may increase in number and severity with increased global warming and several countries such as Australia and the western United States and parts of Europe have had unprecedented wildfires.

More Applications

1. Do a safety audit on your locality by assessing the potential for severe storms, flooding and wildfires. One may have to research local meteorological projections about future hazard potential.

2. Read the advice given in this chapter (and websites) about emergency actions for storms, especially cyclones (hurricanes) and tornadoes. Devise a personal or family plan for such an event.

3. As well as having a designated shelter, water and food for family and pets, also have a portable radio with fresh batteries to hear progress warnings. Remember that communications, especially mobile (cell) phone systems may be destroyed.

4. If the hazards are going to be severe, evacuations may be advised by the authorities. Devise an escape plan for family and pets which involves several routes and the location of safe areas.

5. With small-scale non-violent flooding (say up to 1.5 metres above floor level), prepare for the flood by building a barricade in front of each door. This may consist of a panel of marine plywood glued to the door frame (doors closed) and floor using silicone sealer and light nails. Drape this from side to side and across the floor for about 50cm and then sandbag (sealed plastic bags of coarse sand).

6. To prevent water entering plumbing vents and coming up inside in baths and toilets, seal the vents with vertical 1.5 m lengths of PVC pipe.

Raised air vent

More Applications continued

6. Wildfires even at a long distance can cause major problems due to smoke inhalation. Purchase a family set of appropriate face masks rated for fine smoke, eye protection and have asthma medication and eyedrops handy. During severe days, stay indoors with windows and doors closed.

7. To stop flying embers from entering roof spaces, block up gutter downpipes with tennis balls or plastic bags of sand. Fill the gutters with water and if there is sufficient water pressure, place a long soaking hose along the roof. Close and seal all windows and doors. Have an escape route planned and EVACUATE if told to do so.

Further References

Blake, Conrad. 2016. Survival Tips and Tactics: Natural Disaster Survival Guide (Survival Prepping Guides) (Volume 1) 1st Edition. CreateSpace Independent Publishing Platform. **ISBN-10:** 1530592585.

Hubbard, Ben.2019. Natural Disaster Zone: Wildfires and Freak Weather. London; Hachette Children's Group. ISBN: 9781445165929 (aimed at the younger readers aged 9+)

Mantua, N.J., S. R. Hare, Y. Zhang, J. M. Wallace, and R. C. Francis, 1997: A Pacific Interdecadal Climate Oscillation with Impacts on Salmon Production. *Bull. Amer. Meteor. Soc.*, **78**, 1069-1079.

Nowka, James D. 2013. Prepper's Guide to Surviving Natural Disasters: How to Prepare for Real-World Emergencies. Living Ready. ISBN 144023566X (ISBN13: 9781440235665)

Queensland Fire and Emergency Services, 2019. Get Ready Guide. https://www.qfes.qld.gov.au/community-safety/downloadlibrary/Documents/GetReadyGuide-E.pdf
(A good general emergency advice program with specific reference to Queensland, Australia)

Reed, Dwan & Finnell, Marion. 2019. Flood Survival - The Complete Guide on What to do Before, During, and After a Storm. Los Gatos, CA; Smashwords. ISBN: 9780463349366.

Scott, P.T. 2018. Adventures in Earth and Environmental Science Volumes 1 & 2. Brisbane: Felix Publications. ISBN: 978-1-925662-19-1

Stein, Matthew R. 2011. When Disaster Strikes: A Comprehensive Guide for Emergency Planning and Crisis Survival. White River Junction, VT: Chelsea Green Publishing Co. ISBN: 9781603583220.

The Editors of *OutdoorLife*. 2016. The Natural Disaster Survival Handbook. Richmond, CA: Weldon Owen. ISBN13: 9781681881027.

Turner, Peter. 2019. Australia's Environmental Issues: Natural Disasters. Sydney, Redback Publishing. ISBN: 9781925860108

Chapter 7: Monitoring the Environment

7.1 Introduction

Since the mid-20th Century there had been a growing awareness that large scale mining, industrial and urban development needed closer monitoring because of their pollution, waste and ecological impact. Earlier resource exploitation and urbanisation tended to develop along a need's basis with little regard to the environment. This meant that natural resources such as fossil fuels, minerals, timber, food and resource animals were being harvested or mined as required with little monitoring or regard to the depletion of the resource. It was only when the situation became difficult for habitation or the resource began to disappear that some attention was given to the need for environmental management, resource conservation and better handling of waste and of the pollution of land, air and waterways.

Today, pollution and waste from mining, industrial and urban regions have become a world-wide problem. The consequences of living in closely-packed, industrialized and urbanized areas which in many places have become vast sprawling cities, still remains a problem many years after the initial squalor of 19th century industrial revolution. Environmental impact has occurred in all of the major spheres of the planet including:

- The atmosphere has become polluted with additional greenhouse gases such as carbon dioxide, nitrous oxide, methane and others as well as increased particulate matter and heat.

- The hydrosphere has suffered pollution at a local and global level in water quality, limitations on freshwater supplies, global ocean temperature rise and an increase in ocean flotsam of non-biodegradable wastes.

- The geosphere has been scoured and exposed in many mining and urban areas with increased top soil removal, desertification and salination of agricultural lands.

- The biosphere has suffered removal or extinction of species through climate change, urbanization, hunting or excessive harvesting and deforestation with corresponding habitat loss. Pollution in the other spheres also have had major impacts on the biosphere. Contamination of air by fumes from mining and the processing of ores have often led to destruction of vegetation through acid rain and toxic solutions leaching into the soil. Animal life has also been affected due to the consequences of resource extraction and processing. Waterways have become polluted with toxic solutions, especially those of heavy metals, arsenic and cyanide and from lowering of oxygen levels. The physical quality of water in many mining areas often is reduced by the amount of sediment and solid wastes which greatly increases the turbidity of the water. Moreover, pollution in only one aspect of an environment often causes enough stress on the biota that many species are affected and eventually the entire environment is lost.

There is now some limited realisation world-wide that all of the Earth's spheres need to be monitored and better managed, but this has come slowly and at a price. There is still some resistance to the changes that need to come about at both the national and international level. This is because of the economic momentum of resource companies, and the reluctance of governments and sometimes populations to maintain the past and current status of available resources, relatively cheap energy and in most modern, industrialised countries, a good standard of urban lifestyle. Currently, the world's population is over 7.4 billion people and growing rapidly, although the growth rate seems to have decreased in some industrialised countries. For a good view of the world's population see:

https://ourworldindata.org/wp-content/uploads/2013/05/updated-World-Population-Growth-1750-2100.png

With increasing numbers of people in the world and the general trend of urbanization as people move into the cities, these problems will be even more difficult to confront.

An interesting website gives an estimate of current populations and the world's numbers expressed as a digital readout:
http://www.worldometers.info/world-population/

Increased levels of pollution have forced most countries to pass various environmental protection laws. In Australia, the Environment Protection and Biodiversity Conservation Act 1999 is the main environmental legislation designed to protect and manage important flora, fauna, ecological communities and places of value. This includes the protection of world and national heritage properties, wetlands, migratory species and marine areas such as the Great Barrier Reef. Individual states also have separate legislation to protect the environment and control pollution. The Federal Environment Protection and Biodiversity Conservation and various Queensland state acts can be found at:

http://www.environment.gov.au/epbc and

https://environment.des.qld.gov.au/management/env-policy-legislation/

In the United States, the United States Congress has enacted a number of federal statutes for pollution control and remediation. These include the Clean Air Act for air pollution, the Clean Water Act for water pollution and the Comprehensive Environmental Response, Compensation, and Liability Act (CERCLA, or Superfund) for the cleanup of contaminated sites. There are also several federal laws governing natural resources such as the Endangered Species Act, National Forest Management Act, and Coastal Zone Management Act and the National Environmental Policy Act which is concerned with environmental impact of any actions undertaken by the U.S. federal government, involves all of these areas. A good site is from the US Environmental Protection Agency at:

https://www.epa.gov/laws-regulations

A comprehensive list of International protocols and environmental agreements and laws can be found at:

http://ec.europa.eu/environment/international_issues/agreements_en.htm

7.2 Monitoring the Atmosphere

In many industrialized countries, air pollution has become a regular way of urban life and there are indications that this spreads far beyond the sources of pollution.

Air pollution can be natural or man-made (anthropogenic). Natural, **primary air pollution** sources i.e. those which come directly from their source, include:

- Smoke particles, carbon dioxide, carbon monoxide and heat from natural fires sparked during dry seasons

- Dust from large areas of land such as deserts and land which have been depleted of vegetation to hold the topsoil.

Figure 7.1: Midday in Cairo, Egypt at the beginning of the dust season when air blows fine particles in from the Sahara Desert to the west of the city.

- Volcanic activity from major eruptions or daily emissions from vents and fumaroles including gases and very fine, razor-like particles which often produces a widespread haze of **'vog'** or volcanic **smog**, around volcanic areas such as Hawaii. Gases emitted can include sulfur dioxide (SO_2), carbon dioxide (CO_2), hydrogen sulfide (H_2S) and the hydrogen halides such as hydrogen chloride (HCl), hydrogen fluoride (HF) and hydrogen bromide (HBr). Carbon dioxide gas is particularly hazardous in the local vicinity as it is heavier than air and will flow downhill away from an active vent as an odourless, invisible and extremely deadly gas. Sulfur dioxide and the hydrogen halides readily dissolve in water to form strong acid solutions in waterways and as acid rain and when the halides fall on vegetation and crops. It causes mass destruction of the plants and any animal life which may eat them.

- Carbon dioxide from respiration of all living things and the reactions of natural acids on carbonate minerals and rocks.

- Oxides of nitrogen are produced in the atmosphere by lightning discharge which combines oxygen and nitrogen gases to form unstable nitric oxide:

$$\text{nitrogen gas} + \text{oxygen gas} \rightarrow \text{nitric oxide gas}$$
$$N_2 + O_2 \rightarrow NO$$

This then quickly reacts with more oxygen to form nitrogen dioxide gas:

nitric oxide	oxygen gas	nitrogen dioxide gas
NO	+ O$_2$	→ NO$_2$

Both of these nitrogen compounds are collectively known as nitrogen oxides or **NOx**.

- Methane gas (CH$_4$) is emitted as waste gas from animals following digestion and also by the anaerobic decomposition of vegetable matter. There is considerable methane gas trapped in the frozen soil or **permafrost** of arctic regions and this is emitted during periods of ice thaw.

- **Radon gas** (Rn) is radioactive and comes from the radioactive decay of natural deposits of radioactive substances such as the compounds of uranium and thorium. This gas is both odourless, colourless and as a noble gas is also chemically unreactive. It is much heavier than the gases in air, so it sinks into low-lying areas, especially buildings and is a major contributor to lung cancer in urban areas.

- **Volatile Organic Compounds (VOCs)** are those compounds which come from living things, especially from trees which can emit volatile chemical vapours on warm days. The characteristic blue hazes of many mountain areas such as Australia's Blue Mountains and the Great Smokey Mountains of the United States are due to the release of eucalyptus oils and other organic compounds. Poplar, willow and oak trees can also emit significant amounts of these vapours. They may also react with man-made gases to produce **secondary air pollution** products such as ground level ozone.

Figure 7.2: VOC haze in Australia's Blue Mountains in Summer.

Under normal conditions, with good air circulation, these pollutants do not significantly pose a world-wide problem although in local regions under certain climatic conditions they can pose a threat to human life including asthma, other respiratory diseases and cancers.

Much more important issues in air pollution come from the activities of humankind. Anthropogenic sources of air pollution come from the use of fossil fuels in the generation of energy, in transportation and in industry. In many cities of the industrial world, daily extremes of air pollution have become the norm and at times very dangerous for the citizens to walk the streets. Many of the cities in the Near and Middle East, especially in India and China suffer from extreme air pollution making respiratory diseases commonplace.

Figure 7.3: In the centre of Beijing, China on a bad day in summer.

The main anthropogenic sources of air pollution include:

- Particles, carbon dioxide, water vapour and heat from power stations which burn coal or oil to generate electrical power. These are considered to be the main air pollutants which pose a world-wide as well as a local threat. In some rural and developing countries, traditional biomass fuels such as animal dung, peat, timber and straw can produce localized air pollution. In mining centres, both on site where the ore is mined and processed and also in the surrounding countryside and centres of human habitation, dust containing fine particles of heavy metals, especially lead, have been found to have entered the bloodstream of many of the inhabitants. Children are especially prone to lead poisoning as their growing bodies absorb lead through contact and through their digestive systems because their immune systems are not yet fully developed. Lead is a cumulative poison and it is not excreted. Lead is a neurotoxin affecting the brain and so poisoning in childhood causes behavioural problems, learning difficulties and cognitive losses. It may also affect physical growth, blood cell development and the functioning of the kidneys. Lead ingested by pregnant women can also pass through the placenta to affect unborn babies. Whilst there is no safe limit, lead levels of 5 micrograms or more per decilitre (mcg/dL) of blood are known to be hazardous. Recent studies suggest that even lower levels may be harmful.

- Dust in mines, especially coal mines and asbestos mines, have led to many cases of **silicosis** and **asbestosis** causing lung cancers in miners and associated industry workers. In more recent years, it has been found that dust-related diseases are also common in industries where unprotected workers use cutting and sanding equipment which produce large amounts of dust from stone and manufactured materials.

- Similar pollution products also come from other furnaces such as factories and waste disposal incinerators. In many urban areas, backyard incinerators have been banned with an immediate reduction in the amount of particulate matter. Commercial incineration produces a considerable amount of solid ash which usually was removed by burial or as landfill but more recent applications uses this toxic ash as a raw material in the manufacture of more inert concrete, ceramics, asphalts and light-weight building blocks. The flue gases which are emitted would normally contain particulate matter, heavy metals, toxic organics such as **dioxins** and **furans**, sulfur dioxide, and hydrogen halides such as hydrogen chloride. The latter two types of exhaust gases form acid raid when they contact water. Flue gases are usually closely monitored, incinerator plants and factories use filters, scrubbers, and electrostatic particle precipitators to remove most of the toxic gases and particles.

- Transportation such as motor vehicles, ships and aircraft which use petroleum products in propulsion systems which use internal combustion engines as whole of part of their propulsion. As well as particulate matter as carbon soot, exhaust emissions include is nitrogen (N_2), water vapor (H_2O), carbon dioxide (CO_2) carbon monoxide (CO) from incomplete combustion, nitrogen oxides (NO_x) from excessive combustion temperatures and some hydrocarbons from unburnt fuel and lubrications. Petroleum once also contained lead additives to assist in the smooth operation of engines but concern over the amount of lead entering the blood from inhaling exhaust fumes led to its being banned and lead levels in the blood have consequently dropped in urban areas.

- Controlled burns of vegetation as in forest management for germination renewal and to remove potentially dangerous undergrowth prior to a wildfire/bushfire season. Some crops such as sugar cane were once burnt off prior to harvesting but this has been reduced or ceased in some areas. Now the unburned material is used as mulch in the cane fields, as well as being sold to gardeners.

- Waste fills in urban areas emit methane as part of the decomposition of the buried organic matter. There have been attempts to use this as a potential energy source.

- Some fertilizers in agricultural areas can break down and emit oxides of nitrogen (NO_x). These gases are also produced by high temperature combustion such as in coal-fired power stations and in internal combustion engines. With sulfur dioxide, they dissolve in rainwater to form **acid rain**.

$$\text{sulfur dioxide } SO_2 + \text{water } H_2O \longrightarrow \text{sulfurous acid } H_2SO_3$$

$$\text{nitrogen dioxide } NO_2 + \text{water } H_2O \longrightarrow \text{nitric acid } HNO_3$$

Acid rain in Europe has partially come from the emissions from coal-fired power stations, especially in Russia where the easterly winds blow the gases into Western Europe acidifying many of the northern European lakes as well as causing forest declines. In the Americas, gases from the power stations in the eastern United States have caused acidification in many of the Canadian lakes.

- Organic compounds with high volatility i.e. they are vapours or easily vaporized, from personal and industrial use can account for some of the air pollution in urban areas. Substances such as the fumes from paints, lubricants, industrial solvents, aerosols and ammonia from refrigeration are just some of the organic vapours that can pollute the air.

Particulate matter is often divided into two main groups according to its size:

- Coarse fractions contain larger particles with a size ranging from 2.5 to 10 micrometres (μm) and denoted as PM_{10} - $PM_{2.5}$.

- Fine fractions contain smaller particles with sizes at and below 2.5 μm ($PM_{2.5}$). Particles which are smaller than 0.1 μm are also called ultrafine particles. Most of the total mass of airborne particulate matter is usually made up of fine particles ranging from 0.1 to 2.5 μm.

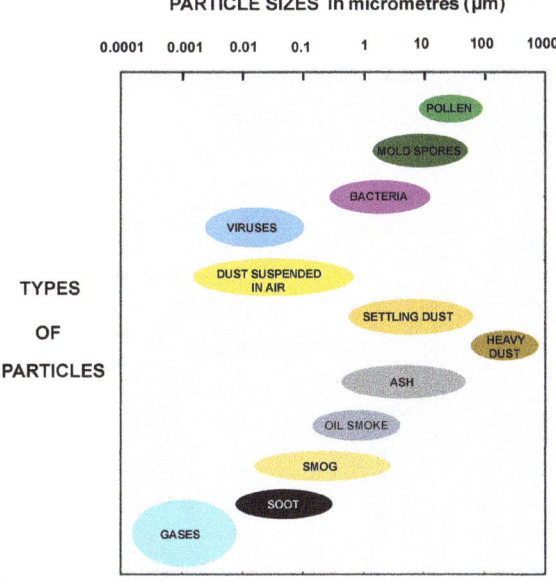

Figure 7.4: Particle size of some common airborne materials

In a modern, urbanized situation, such as a big city or regional centre, it is also important to consider indoor air pollution. There is a trend to have major high-rise buildings for commercial and domestic use which are totally sealed and rely solely on air conditioning. The quality of the air can be affected by accepted pollutants as: aerosols as insect sprays, perfumes, deodorants and air fresheners. Volatile varnishes and paints on walls and furnishings, synthetic materials, especially plastics as furnishings; synthetic carpets and their cleaning agents are highly polluting. Cigarette smoke is obvious, however mold and fungal spores which grow in damp, dark spaces; natural accumulations of radon gases from subsurface rocks, especially granites which contain traces of radioactive minerals are not so obvious. Commercial ozone air purifiers cause pollution, as does air conditioning which can accumulate bacteria in water condensation units causing diseases such as **Legionnaires Disease** to spread throughout the building.

Air pollution can be monitored using a variety of instruments for localized and world-wide conditions. Some of these instruments include:

- **Dust fall deposit gauge** is a bottle containing a known amount of water supported in a raised stand to prevent contamination from ground dust. A sample is taken after about a month and sent to a laboratory for chemical analysis. A dust fall rate is calculated by dividing the weight in milligrams of insoluble material collected by the cross-sectional area of the funnel in square metres and the number of days over which the sample was taken. The units of measurement are milligrams per square metre per day.

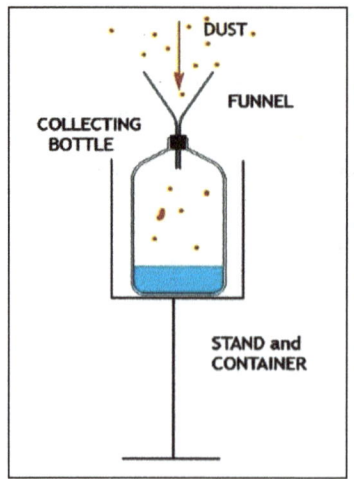

Figure 7.5 A diagram of a dust fall gauge.

- **Integrating nephelometers** measure visibility of the atmosphere which can be affected by concentrations of dust and **aerosols**, fine droplets suspended in air. They use the ability of these particles to scatter light. Three coloured filters of red (700 nm wavelength), green (550 nm), and blue (450 nm) are used to discriminate between various aerosols. Their light is captured by three photomultiplier tubes.

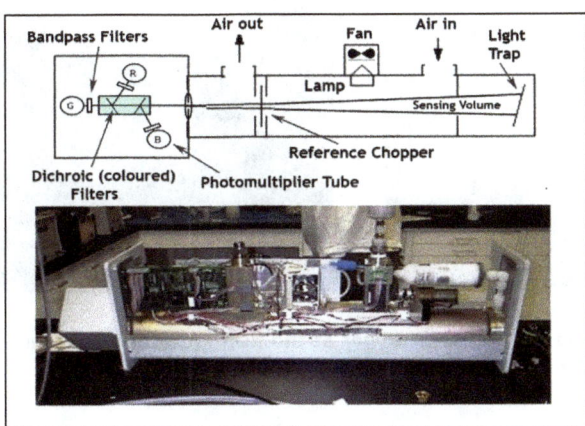

Figure 7.6 A diagram of an Integrating Nephelometer and an inside view (Photo: NOAA)

- **Carbon dioxide analyzers** measure the amount of carbon dioxide gas in the air as a **mole** fraction i.e. the number of carbon dioxide molecules in a given number of molecules of air after removal of water vapour, by measuring the amount of infra-red absorption by the gas. This is done by first collecting the gas at inlets placed high above the ground to limit the amount of extra CO_2 from ground organisms, soil reactions and machinery. Next, the air is slowly pumped through a small cylindrical cell with windows on each side. Infrared light is passed through the gas through one window and measured by an infrared detector. In the atmosphere carbon dioxide absorbs infrared radiation, contributing to warming of the Earth's surface.

Figure 7.7 & 7.8: The tall collecting tower at the NOAA facility atop Mauna Loa, Hawaii at left and the CO_2 gas analyser (Photo: NOAA)

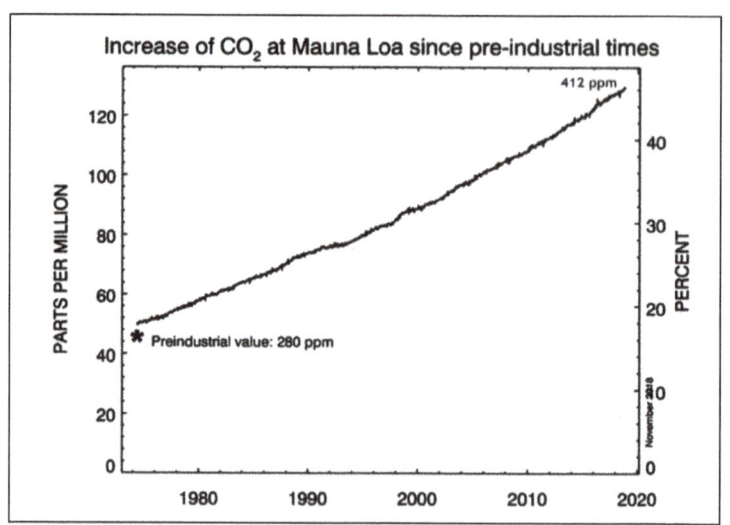

Figure 7.9: Graph showing the <u>increase</u> in atmospheric carbon dioxide levels measured at Mauna Loa. Note the projected increase for 2020 (Data: NOAA)

- **Methane detection and monitoring** has been an important part of the coal mining from its beginning. In more recent times, digital gas detectors using chemical analysis have been a mandatory piece of equipment in most mines and at natural gas wells. Methane comes from the decomposition of organic remains from coal, animal dung, sea-floor oozes and in marshlands. It is highly explosive and was known and feared by coal miners as firedamp, and is the cause of coal mine explosions. Atmospheric methane is relatively small, being around 1.8 parts per million but it is an important greenhouse gas because it is such a potent heat absorber. The concentration of methane in our atmosphere has risen by about 150% since 1750, apparently largely due to human activities. Methane accounts for about 20% of the heating effects by all of the greenhouse gases combined. Both natural and human sources supply methane to Earth's atmosphere.

 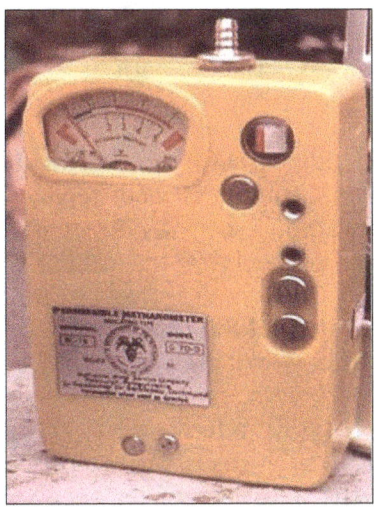

Figure 7.10 & 7.11: At left is the Davy Safety Lamp, invented in the 19th century to prevent mine explosions and its modern counterpart, the methane gas monitor (Photos: USGS)

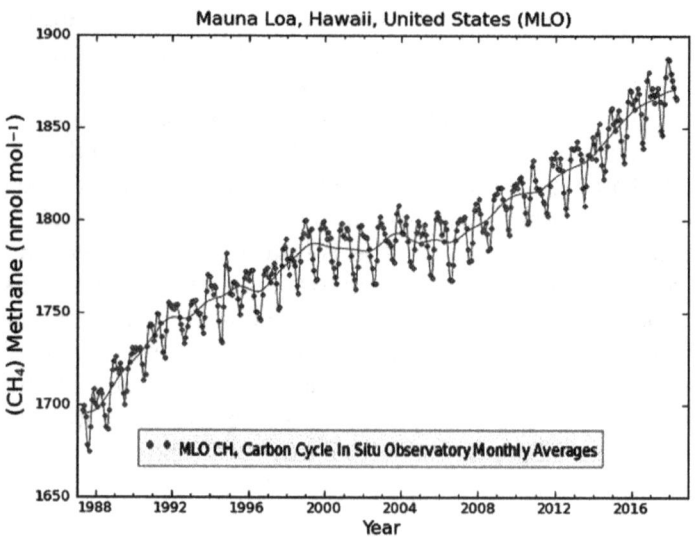

Figure 7.12: Graph showing atmospheric methane levels measured at Mauna Loa (Data: NOAA)

- **Radon radioactivity monitoring** is important in areas where this odourless, colourless and highly radioactive gas is emitted and concentrated, especially in confined spaces. Radon is one of the daughter products of the radioactive disintegration of uranium and **thoron**. This radioactive **isotope** of radon comes from the disintegration of the radioactive element thorium. In open cut pits and subsurface workings where radioactive minerals are mined or are associated with those other ores which are extracted, constant monitoring of the radioactivity is crucial to the health of all the mine workers. Radon gas is easily absorbed into the body through the lungs where the emissions of its decay, notably alpha particles can cause short-range damage to the lungs and other tissues of the respiratory system causing cancers. Radon is also emitted from many rocks, especially of granitic composition and forms part of the normal background radiation in many areas. Usually this is limited, but, because of radon's high density it often accumulates in low lying areas such as the basements and crawl-spaces of buildings where sufficient concentrations would pose a serious threat over time. Radon is also found in many thermal waters and springs it was once thought to be therapeutic.

As a radioactive substance, radon can be detected using passive radiation monitors which require no power source such as charcoal canister and **charcoal liquid scintillation devices** which absorb radon or its products onto charcoal which can then be analysed in the laboratory, where the radioactive particles emitted from the charcoal are counted directly by a radiation **scintillation detector**. **Alpha track detectors** are another type of passive detector in which a plastic film becomes etched by the alpha particles that strike it. The plastic is again analysed in a laboratory by being chemically treated to make the tracks visible, the tracks are then counted. Both of these types of detector are available commercially for home and business use. **Electret ion detectors** are another passive device but these are usually restricted to scientific use. They have a Teflon disc, which is statically charged but which produces an electrical current when struck by an ion generated from radon decay reducing the static charge on the disc. In the laboratory, this reduction of static charge is measured thus radon level is calculated.

Active devices requiring a power supply include **Continuous Radon Monitors (CRMs)** which measure the radon emissions over a time period, usually two days. There are several types, but most either collect air for analysis using a small pump or by allowing air to diffuse into a sensor chamber where the radiation level is measured by scintillation cells which detect the small pulses of light produced when they are struck by the radiation particles. Current or **pulse ionization chambers** are another type which produce small electrical charges when the radioactive particles strike gas molecules in the chamber, knocking off electrons. Electrical circuitry in the device then provides a summary report and often a recording over the time.

Radon gas emissions are measured in units of **picocuries**/litre (pCi/L) where one pCi/L is equal to 22.2 decays as measured by particle strikes per minute per litre. The United States Environmental Protection Agency considers that a level of 4 pCi/L or over would require some form of remedial or avoidance action.

Some useful sites on radon emission and detection can be found at:

https://www.arpansa.gov.au/sites/default/files/legacy/pubs/factsheets/RadonExposureandHealth.pdf
(Fact sheets on radon in Australia)

https://www.arpansa.gov.au/understanding-radiation/radiation-sources/more-radiation-sources/radon
(More general information from the same source)

https://www.admnucleartechnologies.com.au/files/sarad_application_note_radon_the_basics.pdf
(Some basic information about Radon)

https://www.admnucleartechnologies.com.au/files/sarad_application_guide_comaparison_of_radon_detection_principles.pdf
(Comparisons of detection methods)

- **Satellite monitoring** using optical photographs and spectroscopic imagery that can show the location, size and speed of large dust storms as well as an analysis of air composition. The Geostationary Operational Environmental Satellites-R (GOES-R) series of the National Oceanographic and Atmospheric Administration (NOAA) can monitor the particle pollution. The Joint Polar Satellite System (JPSS) also collects information about particles in our air as well as ground level ozone concentrations.

A general indicator of air pollution from a number of instruments monitoring a variety of air pollution parameters is usually given as an **Air Quality Index (AQI)**. Whilst this varies from country to country, it usually has various levels of air pollution effects and a warning colour code. In the United States and Australia, an AQI of 50 or less is considered healthy but the scale can go to over 300 which is considered extremely hazardous.

See also:

https://aqicn.org/map/australia/
(AQI for Australia)

https://www3.epa.gov/airnow/aqi_brochure_02_14.pdf
(General information from the US EPA)

There are many other useful websites on general atmospheric monitoring. Some of the better ones include:

https://www.qld.gov.au/environment/pollution/monitoring/air/air-monitoring
(Air monitoring in Queensland)

https://www.wmo.int/pages/prog/arep/gaw/measurements.html
(World Meteorological Organisation and locations of world air monitoring stations)

https://aqicn.org/city/brisbane/
(Real-time air measurements for Brisbane and the AQI)

https://www.environment.nsw.gov.au/aqms/aqi.htm
(A similar site for New South Wales)

https://aqicn.org/map/usa/
(Real-time AQI for the United States)

https://waqi.info/
(An interactive map to give the AQI for world centres)

7.3 Monitoring the Hydrosphere

Water quality and some of the ways in which it is sampled and tested to monitor natural water environments is a major part of any environmental study both in freshwater and marine habitats. Anthropological pollution of these waterways occurs due to such sources such as:

- Water is pumped out of mines, coal-seam gas and other industrial sites as well as runoff from tailings and overburden dumps. Good mining practice ensures that any water from mine or excavation sites does not run directly into natural waterways but is first pumped or channeled into holding ponds. There it can be either treated or filtered or allowed to evaporate so that the toxic solids can be removed for deep burial away from the water table. Overburden stock piles and stock piles of ores should be stored in areas which have been previously made waterproof using plastic or clay layers. Often, fine materials are also covered to prevent runoff after rain. It is one of the main occupations of mine environmentalists to see that any water from the site does not enter the local water table.

- Waste coolants from factories and power stations are often simply runoff into the sea or local rivers. Many modern power stations now have large ponds and cooling towers to remove the heat from their warm efflux water by evaporation cooling. Heat pollution in waterways has been known to increase the numbers of algae and bacteria in local waterways to the determent of other species.

- Untreated urban wastes and sewerage outlets contain mostly water with some solids and pathogens which can cause disease if they contaminate drinking water. These include viral infections including *Norovirus* and *Rotavirus* which can cause diarrhea and viral gastroenteritis. Bacterial infections including those from *Campylobacter*, *Escherichia coli* (*E. coli*), *Salmonella*, and *Shigella* can also cause gastro-intestinal diseases. In addition, there are also other waste products which are flushed down sewers such as plastic bags, which also enter natural waterways. In most developed countries, sewerage is treated at sewerage treatment plants with the solids and grease being removed by filters and traps with the liquid and suspended solids treated by natural organisms which break material down. In some extreme circumstances the effluent may be further treated chemically to remove any toxic minerals, especially soluble compounds of phosphorus and nitrogen, and harmful organisms. If treatment is efficient, the water which passes out of the plant is relatively clean and can go into the sea or river or used in irrigation of lawns and parklands. The solids can be deeply buried or composted and later used in agriculture.

- Natural parasitic infections are a common part of many rivers and other waterways. They are part of the natural environment but may be considered as pollution to the human body as they can enter through food or water and settle in the digestive tract causing diarrhea, loss of appetite and some major diseases. For example, *Cryptosporidium enteritis*, and *Giardia lamblia cause diarrhea*, certain single-celled amoeba species can cause amoebic dysentery and Schistosomiasis, also known as bilharzia, is caused by parasitic worms such as *Schistosoma mansoni*.

    See also:

    https://www.cdc.gov/parasites/az/index.html
    (Alphabetical Index of Parasitic Diseases)

- Excessive agricultural runoff includes excess chemical and organic fertilizers, animal wastes, pesticides and fine silt due to surface erosion of topsoil. Nitrate (NO_3^-) and phosphate (PO_4^{3-}) solutions are especially difficult to control as they are universally used, are soluble and can easily enter the environment. Excessive amounts of these compounds can act as nutrients to many water plant species which then can choke the waterways and remove oxygen from them.

See also:

https://www.epa.gov/sites/production/files/2015-09/documents/ag_runoff_fact_sheet.pdf
(Agriculture runoff fact sheet)

When investigating the sources of water pollution, environmentalists can look for a **point source** of origin such as a particular factory, power station, ships at sea etc. or they can consider non-point sources which are more general and often consist of multiple inputs such as runoff from agricultural lands, urban effluent into oceans and the like. There is also **transboundary pollution** which enters the environment in one place but travels a great distance to have an effect in a place far from the source of pollution. A good example of this is the incredible amount of man-made flotsam which it is carried across the Pacific and other oceans to pristine island communities where it is washed up as great heaps of plastic and other non-biodegradable material.

As every part of the world's water cycle is likely to be affected by contamination, both natural and man-made, particular environments may become the subject of study. These include:

- surface waters such as rivers, ponds, lakes and glacial meltwater
- marine waters both coastal and deep oceans
- sub-surface waters such as near-surface groundwater and deeper aquifers.

Where there is life in the hydrosphere, several important factors impact on the organisms within the water. These include:

- **Water temperatures** are important because aquatic species usually have a small range of optimum temperature. Plants can have a wider range but most aquatic animal life is **poikilothermic** meaning that their body temperatures change with that of the environment. **Homeothermic** animals such as humans, whales, platypuses, otters etc. that can maintain a constant body temperature through additional respiration processes, thus can withstand a wider range of water temperatures but even these animals have temperature limitations. Water temperatures change with the season and with the altitude of the body of water or stream. In the oceans, warm and cold currents influence the type of aquatic life that can exist in the various marine environments. In the Antarctic where water temperatures are just above freezing point, there is a rich abundance of both marine algae and fauna such as krill and fish which have adapted to the cold conditions. Water temperature can be measured simply with hand-held or remote digital thermometers.

- **Dissolved oxygen (DO)** content of a stream or ocean is vitally important to its aquatic life. It varies with water temperature, salinity, biological activity such as plant photosynthesis and organism respiration, and turbulence of the surface which also affects the rate of transfer from the atmosphere. The amount of dissolved oxygen in a water environment is a good indicator of the health of that environment.

Figure 7.15: Plastic containers and other flotsam washed up on an Hawaiian beach in the central Pacific Ocean (Photo: NOAA)

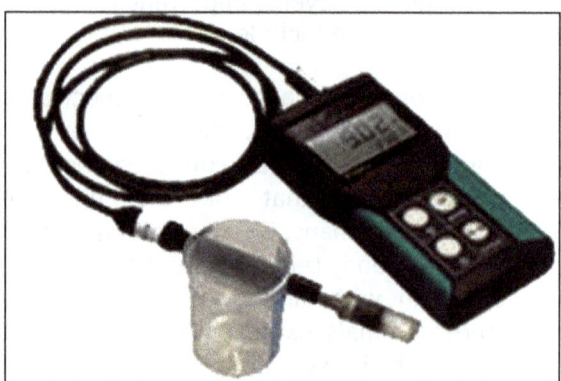

Figure 7.16: A field dissolved oxygen meter (Photo: USGS)

Colder waters usually have more dissolved oxygen than warmer waters simply because the solubility of most gases decreases as temperature increases, so concentrations are higher during the cooler months compared to the warmer summer months. Salinity also affects the dissolved oxygen content in water with it increasing with decreasing salinity. In some shallow seas, where salinity can increase as water evaporates, oxygen levels will decrease. In inland waterways, there can also a marked difference of oxygen content due to diurnal fluctuations. Day-time concentrations of dissolved oxygen are generally higher than those during the night because of the photosynthetic activity of aquatic plants.

Eutrophication (from Greek *eutrophos* for well-nourished) can occur when a body of water receives an excessive nutrient load, particularly dissolved compounds of phosphorus and nitrogen which results in an overgrowth of algae and micro bacteria. As the algae die and decompose, oxygen is depleted from the water. This lack of oxygen in the water causes the death of other aquatic animals such as fish. Dissolved oxygen content can be measured in the field using specialized digital meters or for a more detailed analysis, especially the impact of algae and mineral salts on the water quality by testing for **Biochemical Oxygen Demand (BOD)** of the water using more detailed chemical analysis in the laboratory.

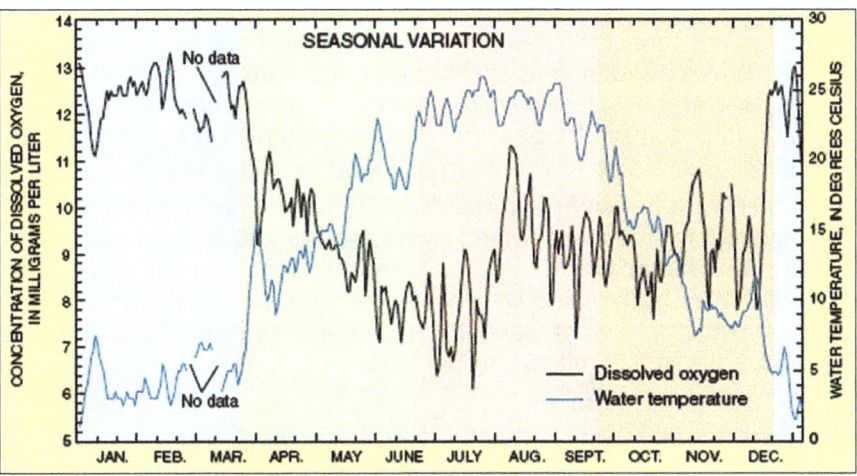

Figure 7.17: A graph showing how dissolved oxygen content is controlled by temperature with both a seasonal and a daily cycle. Note that this is for a river in New Jersey with summer being in June-July and winter in December-January (Photo: USGS)

See also:

https://water.usgs.gov/owq/FieldManual/Chapter7/NFMChap7_2_BOD.pdf
(National Field Manual for the Collection of Water-Quality Data)

- **Salinity** is the amount of salt (as chloride ions from sodium chloride - NaCl) in the water. Marine animals and plants have adapted to living in saltwater and so changes in salinity can have a dramatic effect on aquatic life. Whilst some marine fish, such as salmon, can adapt to both saltwater and fresh, most organisms are sensitive to any major change in the salinity of their environment. Changes in salinity in rivers can occur through land use, catchment geology and water flow. Use of excessive irrigation in agriculture can sometimes raise the water table which may bring salt solutions to the surface causing soil salination, when this runs off into local waterways, the salinity rapidly increases. The geology of the local catchment will determine what soluble weathering products will run into the local waterways and this is also dependent on rainfall and the flow rate of streams. In some places where inland waterways have not received their usual influx of fresh water, excessive evaporation will lower the water level and increase the concentrations of salts in the water.

  Electrical conductivity of water is commonly used to determine salinity and is measured in units such as **microsiemens** per centimetre (mS/cm or $mScm^{-1}$) at a standard reference temperature of 25^0C. Many instruments may also be calibrated for a simple measure of percentage of salt. According to the Australian and New Zealand Environment and Conservation Council (ANZECC 2000), the recommended default maximum values for lowland rivers are 125-2200 $mScm^{-1}$, upland rivers are 30 - 350 $mScm^{-1}$, and 20 - 30 $mScm^{-1}$ for lakes and reservoirs. Salt content of both inland waterways and marine zones can also be found by a total removal of water from a laboratory sample and the amounts of solids given in milligrams per litre.

Salinity status by total salt concentration		
Salinity status	Salinity (milligrams of salt per litre)	Description and use
Fresh	< 500	Drinking and all irrigation
Marginal	500 –1 000	Most irrigation, adverse effects on ecosystems become apparent
Brackish	1 000 – 2 000	Irrigation certain crops only; useful for most stock
Saline	2 000 – 10 000	Useful for most livestock
Highly saline	10 000–35 000	Very saline groundwater, limited use for certain livestock
Brine	>35 000	Seawater; some mining and industrial uses exist

Classifications from Mayer. XM. Ruprecht. JK & Bari. MA 2005. Stream salinity status and trends in south-west Western Australia. Department of Environment. Salinity and land use impacts series. Report No. SLUI 38

Figure 7.18: Salinity classifications for inland waterways (Data from the Western Australian Department of Water and Environmental Regulation)

- pH is a measure of the amount of the hydrogen ion in a solution and is based on the dissociation of water into hydrogen ions (H^+) and hydroxyl ions (OH^-).

$$\text{water} \rightleftarrows \text{hydrogen ions} + \text{hydroxyl ions}$$
$$H_2O \rightleftarrows H^+ + OH^-$$

Measurement values are given as the reciprocal of the logarithm (powers of 10) of the hydrogen ion concentration. For example, neutral water with a pH of 7 has 10^{-7} **moles** per litre of hydrogen ions, whereas, an acid with a pH of 2 has 10^{-2} moles per litre. An alkali such concentrated sodium hydroxide solution could have a pH of (say) 11 which means that its hydrogen ion concentration is 10^{-11} moles per litre which is very small. The pH scale ranges from 0 which is strongly acidic with a lot of hydrogen ions, to 14 which is alkaline and has fewer hydrogen ions and more hydroxyl ions (OH^-).

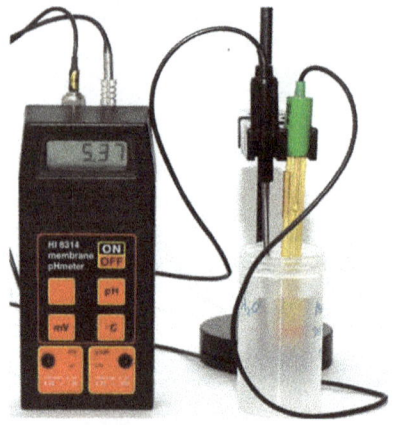

Figure 7.19: A portable pH meter showing a pH of 5.37 - a mild acid. (Photo: USGS)

Most living things, with the exception of some **extremophiles**, live within very small ranges of pH, usually around neutral 7.0. A slight change to these levels, such as with an increase in acid rain or the introduction of acidic runoff from a mine site will destroy any organism which cannot live within the new pH. pH is easily measured with simple Universal Indicator paper which is paper impregnated with a mixture of dyes which change colour within different pH ranges - green is neutral, red is acidic and blue is alkaline. For more accurate environmental readings, a pH meter may be used. These instruments have electrodes which measure the conductivity of the water; this being proportional to the number of hydrogen ions which will conduct an induced electrical charge.

pH	Level	Aquatic Life
14	ALKALINE	(Minimum pH) BACTERIA
13		
12		
11		
10		
9		
8		
7	Neutral	CLOWN FISH CORALS
6	ACID	SALMON
5		BASS CRAYFISH
4		PIKE TROUT ANGEL FISH PERCH FROG
3		CARDINAL TETRA
2		
1		
0		BACTERIA

Figure 7.20: A chart showing the minimum pH preferences for some aquatic life.

- **Water turbidity** concerns the clarity of the water due to the amounts of suspended solids present in it. These can be from a natural event such as sediment washed down the rivers or washed out to sea, turbidity currents falling off the continental shelf or upwelling of shallow sea-floor sediments. It can also be man-made such as effluent from factories, agriculture, soil erosion, dredging and the like. Turbidity concerns the amount of light that is scattered by material in the water when a light is shined through the water sample. The higher the intensity of scattered light, the higher the turbidity. Materials that cause water to be turbid include clay, silt, finely divided inorganic and organic matter, algae, soluble coloured organic compounds, and plankton and other microscopic organisms.

To measure turbidity, a water sample is collected in a bottle and the turbidity is measured by shining a light through the water very much like measuring suspended material in air. It is measured in **nephelometric turbidity units (NTU)**. In most normal environments, such as a stream or river, turbidity usually is seasonal. When there is a low flow rate, called the **base flow**, stream water is usually clear and the turbidity is also low, usually less than 10 NTU. After heavy rain or a flood cycle, sediments from the surrounding land are washed into the river making the water muddy and increasing the turbidity values. Also, during periods of high-water velocities water volumes are higher and material can be easily stirred up from the stream bed, causing higher turbidity. In streams, increased siltation can result in harm to habitat areas for fish and other aquatic life and an increase in algal growth can reduce the oxygen content of the water. Particles also provide attachment places for other pollutants, notably metals and bacteria. For this reason, turbidity readings can be used as an indicator of potential pollution in a water body. The World Health Organization (WHO) has suggested that the turbidity of drinking water should be less than 5 NTU and should ideally be below 1 NTU.

Figure 7.21: An hydrologist collecting samples for water turbidity analysis (Photo: USGS)

In still water, such as lakes and calm seas, turbidity can be measured with a **Secchi disk** named after Angelo Secchi (Italian: 1818 – 1878). This is a black and white quartered disk about 20 cm diameter which is dropped in the water attached to a cable. The marine version is simply a white disc of 30 cm diameter. Repeated trials are made with the averaged depth that the disk reaches before it disappears from sight is recorded and can be used as a measure of water clarity. If this depth in metres is divided into 1.7 a value called the **extinction coefficient** is derived which can also be used as a value for the water's turbidity.

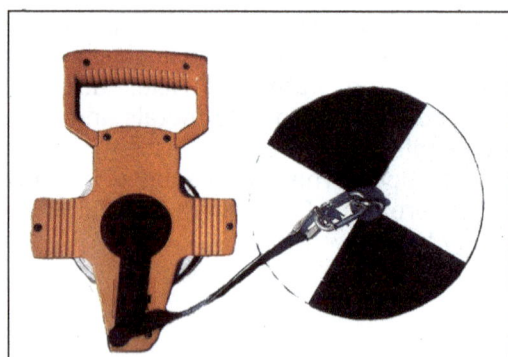

Figure 7.22: A Secchi Disc for measuring inland waterways (Photo: USGS)

- **Water hardness** refers to the mineral salts dissolved in water especially the amount of dissolved calcium and magnesium salts and was named as being hard to lather soap in washing. Temporary hard water is high in both calcium and magnesium dissolved minerals, especially those of the carbonates (CO_3^{2-}) or hydrogen carbonates (HCO_3^-) leached from minerals such as calcite and dolomite. Many rocks contain carbonates as secondary minerals or cement and these minerals easily dissolve in slightly acidic rainwater. Permanently hard water contains the sulfates (SO_4^{2-}) and chlorides (Cl^-) leached from such minerals as gypsum ($CaSO_4$) and salt (NaCl). Temporary hardness in water can be removed by boiling but permanent hardness requires the addition of a water conditioner or an ion exchange filter in larger operations. The presence of these ions can be detected using commercial water test kits and calibrated digital meters or the mineral content can be measured in milligrams per litre (mg/L) using quantitative reactions and chemical analysis. Simple indications of the presence of sulfates and chlorides in local water can be tested separately with laboratory solutions of barium nitrate and silver nitrate (CARE: stains skin etc.) respectively:

clear barium nitrate solution	sulfate ions		dense white precipitate of barium sulfate		nitrate ions
$Ba(NO_3)_2$	+ SO_4^{2-}	$\longrightarrow$	$BaSO_4$	+	$2NO_3^-$
clear silver nitrate solution	chloride ions		dense white precipitate of silver chloride		nitrate ions
$AgNO_3$	+ Cl^-	$\longrightarrow$	$AgCl$	+	NO_3^-

See also:

https://water.australianmap.net/physical_chemical/hardness/
(Comprehensive list of water details for Australia)

https://canvas.jmu.edu/files/73786992/download?download_frd=1&verifier=Ik5swkWdx0OGdtIuP4Y0iwJ3NznqUFESji7wHoui.
(Water hardness in the United States)

For some other good general references about water, including data see:

https://pubs.usgs.gov/sir/2011/5007/sir20115007.pdf
(Good water study report with data from USGS)

http://www.vic.waterwatch.org.au/resources/water_watch_module4_physical_and_chemical.pdf
(Good overall manual about water quality)

https://soe.environment.gov.au/science/soe/2011-report/4-inland-water/2-state-and-trends/2-2-water-quality
(Good maps showing water qualities in Australia)

https://www.epa.gov/sites/production/files/2015-09/documents/2000_06_30_305b_98report_98summary.pdf
(Water quality in the United States)

7.4 Monitoring the Geosphere

There are many aspects of the solid surface of the Earth or geosphere which are monitored on a daily basis. Much of it is done through remote sensing but field work is occasionally required for finer detail. As the time scale of changes in the geosphere is often in the tens, hundreds or even thousands of millions of years, most of the monitoring of the land surface and its rocks is involved with the physical changes which occur, notably movements along joints and fractures, seismic monitoring, volcanic activity and real-time changes to the surface features of the Earth.

Within the normal scale of time and of great importance to humankind are the changes which occur with soil and its agricultural use. Soil monitoring concerns the soil's chemical properties, its moisture content, its pH, its organic components and soil erosion. The chemical composition of soils is derived from the weathering and erosion of the parent rock which formed the soil. Minerals in the rock and soil are chemical compounds or ranges of chemical compounds which usually break down under chemical and physical attack by the weather. Chemical weathering breaks down the chemical bonds in the mineral's crystal structure with the formation of new secondary minerals and soluble ions which may later react with other ions to form other new secondary minerals. Most minerals chemically react at different rates and in different ways. **Ferromagnesian minerals** such as hornblende, augite and biotite found in many igneous rocks, are usually dark in colour because they contain high percentages of iron and magnesium ions which readily break down to form fertile red to black clay soils. Thus, good soils are derived from many of the basic lavas such as basalt and andesite. Given the right temperature and water, plants can be grown in fresh volcanic basalt within a year. On the other hand, soils derived from granites, sandstones and quartzite which contain mostly chemically-resistant quartz and some minor clays, are generally poor and require a considerable amount of effort and fertilizer to raise crops. In the case of the initial soil chemistry of a region, it is the nature of the soil mineral content which determines whether or not agriculture will be successful. After considerable use, soil chemistry may change. With excessive use, poor crop yield usually means that the soil has become infertile, provided that other factors such as temperature, rainfall and pests remain constant. If this occurs, then a full soil analysis would be in order and a planned program of new fertilizers or practices would be required.

Moisture content of soils is usually monitored on a regular basis by the farmer. Soils with poor drainage in a wet climate soon become waterlogged, especially those with a high clay content. Soils with a high sand content and on sloping ground will drain rapidly and quickly dry out. Soil moisture is vital for the regular growth of crops and its monitoring will decide how much irrigation may be needed. In a simple examination, a farmer can take a handful of soil, spit on it and roll it around to feel and observe how the soil is held together by its clay and moisture content. For more precise measures, a soil moisture meter is used to measure the moisture content at different depths within the soil to judge surface and root moisture availability. There are basically two types of soil moisture meters:

- **Water potential sensors** measure the difficulty in sucking up water from the soil or soil moisture tension. This gives a good indication of the plant's ability to also take in water through their roots by the transpiration pull throughout their xylem or water tubes. The most common example of this type of sensor is the tensiometer which consists of tube filled with water with a porous ceramic cup at its base. At the top of the tube is vacuum gauge which measures the pressure inside the tube. The tensiometer is pushed in the soil and a hand pump is used to pull a partial vacuum. As water is pulled out of the soil by plants and evaporation, the vacuum inside the tube increases but as water is added to the soil, the vacuum inside the tube pulls moisture from the soil and decreases. With the water in the tensiometer at equilibrium with the soil water, the gauge reading of the tensiometer represents the potential of the moisture content of the soil. Units are in kilopascals of pressure.

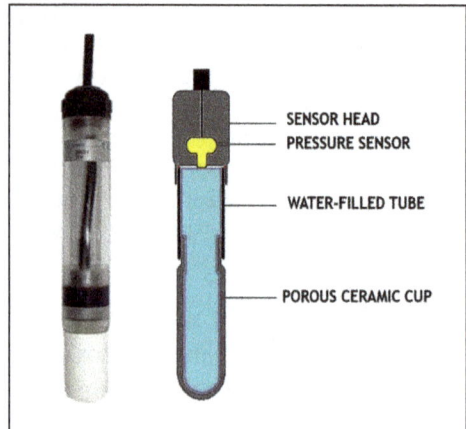

Figure 7.23: A tensiometer and a diagram showing its components (Public Domain)

- **Water content sensors** measure the soil moisture using electrical capacitance between electrodes in its sensor. They can measure soil moisture at a number of depths simultaneously therefore can give some idea of the movement of soil moisture in the section sampled. Their disadvantage is that they only measure the soil moisture in the immediate vicinity of the sensor so they are subject to error due to the electrical conductivity of mineral salts in the soil.

7.5 Monitoring the Biosphere

The close monitoring of the life on this planet is the major task for general environmental ecological management and protection. It also can be used for more specific functions such as environment impact surveys in preparation for large scale clearing, agriculture, urbanization, engineering construction and mining. It is good, responsible management for large companies and governments to undertake a full review of the total local environment prior to beginning any development. Usually it is part of any development application but sometimes the terms of the contract or mismanagement result in only a limited environmental survey or none at all being undertaken. This is especially true of incremental development where estates, shopping centres and small-scale engineering projects occur piece-by-piece, building-by-building over a long period of time. When this occurs, the natural environment is encroached upon slowly and mobile animal species migrate whilst others are lost.

In responsible mining projects, the company's environmentalists or contract environmentalist usually carry out specific functions after approval for the site to be developed but before the operation commences. This might include the following sequence:

1. Research the scope of the study area finding details of watersheds and drainage patterns of the site area, traditional and local owner boundaries and sites, potential risk to native flora and fauna, especially local endangered species.
2. Conduct a complete environment survey of the proposed mine site and some considerable distance surrounding it with regard to the water, soil and ecology of the site with specific surveys of important flora and fauna species and their individual habitats.
3. Begin collection of main static species and developing plans for escape corridors for mobile species. Plant species which are collected are cultivated in the company's nursery as are selected species of static animal species if possible, including eggs and cultures in breeding programs.
4. The mining company then removes the minimum amount of vegetation and topsoil needed and stockpiles the topsoil where it cannot be eroded or leached by water runoff.

After the mining operation has concluded, the post-mining environmental rehabilitation operation begins. This usually is an on-going operation with environmental rehabilitation occurring over small areas where mining has ceased whilst mining occurs in other parts of the lease. The procedure often includes the following sequence:

1. Backfilling any excavation after removal of any toxic tailings or dumps
2. Addition and contouring of the topsoil from the stockpile previously collected with special regard to drainage and special habit requirements
3. Re-planting of native plant species, especially grasses in open country and special habitat requirements such as logs, rocks etc.
4. Release of any maintained animal species and connection of animal corridors to undeveloped lands with encouragement of species return. This might also involve the capture and release of mobile native species.

5. Further long-term monitoring of the ecology and the environment (soil, air, water etc.) over time

In general, ecological surveys can take many forms depending upon the needs of the survey and the type of environment e.g. wetlands, rainforests, dry savannah, alpine, coral reef, deeper marine etc. The technology can also vary but the environmentalist will use the best available and vary the technique to match the parameters of the proposed survey site. These include the physical nature of the landscape, the climatic conditions, land-use and distance from base operations. Remote sensing using satellites, aircraft, drones and unmanned submersibles may be used in the preliminary planning stages. Eventually, however it will take direct field operations to obtain the fine scientific detail required for a comprehensive survey.

Ecological monitoring, being concerned with the living organisms within a particular environment adapted to a specific landscape and climate, can be initiated due to many environmental needs such as:

- **Ecological disaster, risk assessment and rehabilitation** monitoring following ecological disasters both man-made and natural. These might include surveys done after oil spills into marine habitats from ships or off-shore oil rigs, hurricane, fires and floods, or major disasters such as the Fukushima Daiichi nuclear disaster. These are usually quickly organized and comprehensive surveys with post-operative monitoring of all of the main environmental parameters as well as specialized monitoring of the ecology. Often these require an enormous amount of organization at the government level over a long time. The initial phase of such operation usually involves salvaging and care of wildlife species, removal and clean-up of all wastes and then the restoration and management of all aspects of the surveyed area.

Figure 7.24: Ecologists salvaging oil-covered wildlife on a beach following a major off-shore oil spill (Photo: NOAA)

- **Specialized longitudinal monitoring** may concern a focus on a particular species. Microbiological species may have special problems with health risks but these can be allowed for by using safe practice and sterilized conditions in the laboratory. A specific area can be sampled for microorganisms over a time period to monitor any changes in growth patterns within the environment and any spreading to others. Plant species are also easily monitored over time for a specific location or type of plant e.g. monitoring of introduced species such as the prickly pear, fire ant, carp fish and the cane toad in Australia and the kudzu vine, privet, African bee and the Burmese python in the United States. With larger, more mobile animals such as various species of bats, birds and small mammals, the monitoring may involve random capture within their known habitat, banding with a tag or leg-ring and then release.

Figure 7.25: Ecologist using a radio antenna to track tagged fish (Photo: US Fisheries & Wildlife Service)

Figure 7.26: Ecologists collecting aquatic specimens (Photo: USGS)

Continued trapping in the same and other localities and the monitoring of tagged animals often gives an idea of the animals' distribution, numbers and migratory behaviour. For localized studies the animal may be given an electronic, radio frequencies tag which can be monitored remotely by the signals sent out by the tag. More elaborate electronic monitors can also give some parameters of the animal's body signs.

- **Whole habitat monitoring** is usual when the ecology of a specific area such as a proposed mine site needs to be surveyed. This may occur over some time but it usually has a limited time frame before the development phase begins. The area to be survey may be approached randomly with piece-meal observation, sampling, and collection of major species or it may involve a planned operation in which the area is divided into sections by gridding or identification of specific habitat areas and then each section is studied in detail with appropriate observation, identification and if required, collection. In many cases, elusive animals may have to be counted using remote cameras or a detailed count of their remains such as nests, food scraps or waste droppings (scat).

Some additional notes and specific studies can also be found at:

https://portals.iucn.org/library/efiles/documents/2004-023.pdf
(Ecological monitoring of coral reefs – Australian Government)

http://www.envlaw.com.au/sqels5.pdf
(Environmental laws in Queensland)

https://openresearch-repository.anu.edu.au/bitstream/1885/130886/1/Making%20ecological%20monitoring%20successful_web.pdf
(Research and monitoring guidelines

https://smu.ca/partners/cbemn/documents/Env_Mon_Overview_and_Significance.pdf
(General notes on environmental monitoring)

https://dce2.au.dk/pub/SR97.pdf
(Case study of environmental monitoring in a gold mine)

https://www.greeningaustralia.org.au/wp-content/uploads/2017/11/Howard-Sands-Assessment-Framework-for-Rehabilitation.pdf
(Case study - Rehabilitation of Sand Mining in the Northern territory)

https://www.researchgate.net/publication/263506865_Mine_Rehabilitation_Mined_land_rehabilitation_-_is_there_a_gap_between_regulatory_guidance_and_successful_relinquishment_Rehabilitation_guidelines_-_mine_planning_and_regulatory_context
(Mined land rehabilitation with links to other similar sites)

https://environment.des.qld.gov.au/assets/documents/regulation/era-gl-receiving-environment-monitoring-program.pdf
(Good site for environmental monitoring)

Final Remarks

1. It became apparent over the last 100 years that the environment was complex and very sensitive to change. Knowing this, it became necessary to find ways of monitoring and measuring changes in the environment.

2. Because of the need to limit the overuse and pollution of the environment, many governments introduced laws and regulations for individuals and companies to follow.

3. The atmosphere has been monitored accurately using a variety of calibrated instruments to measure temperature, gas components, especially carbon dioxide, methane and particulate matter such as toxic solids and smog.

4. The hydrosphere too, is also sensitive to change and it has been used as a dumping place for wastes for generations. Temperatures of the oceans have increased as well as their acidity through massive dissolving of carbon dioxide. Freshwater environments are also very sensitive and these have to be monitored for temperature, oxygen content, nitrogen and ammonium compounds, other soluble toxic compounds and particulate matter.

5. The biosphere is monitored for the presence of foreign species of plant, animal. Bacteria and viruses are also monitored in air and water to prevent the spread of diseases, especially as climates change.

More Applications

1. Find out about the local monitoring station and how they monitor the air, water and biological environments.

2. With safety, monitor the local freshwater environments (lakes, streams etc.) for temperature (thermometer), sediment (make a home-made Secchi disk), Acidity as pH and salinity as chloride levels (use swimming pool test kits).

3. Reduce the amount of particulate matter in the atmosphere by ensuring that all engines using fuels are properly tuned. Report any factories, mills and power stations which give off large amounts of smoke. Keep open fires (e.g. BBQ pits) in proper surroundings and enable efficient burning by layering fuel to allow for air circulation.

4. Have facial masks handy for any days of extreme air pollution - ensure that they are appropriate i.e. labelled PM 2.5 or N95. Note that these are not protection for toxic gases within smoke or dust.

Further References

https://www.health.act.gov.au/about-our-health-system/population-health/environmental-monitoring
(many links about environmental monitoring)

http://www.neii.gov.au/
(A wide range of links about information and monitoring sites of the Australian environment)

Davies, P.E., 1994. National River Processes and Management Program Monitoring River Health Initiative: River Bio-assessment Manual. Commonwealth Environmental Protection Agency, Canberra

Grandy, J.; Asl-Hariri, S.; Paliszyn, J. (2015). Novel and Emerging Air-Sampling Devices". In Barcelo, D. (ed.). Monitoring of Air Pollutants: Sampling, Sample Preparation and Analytical Techniques. Comprehensive Analytical Chemistry. Elsevier. Chapter 7: pp. 208-237. ISBN 9780444635532.

Harter, T. (2008). Water Sampling and Monitoring. In Harter, T.; Rollins, L. (eds.). Watersheds, Groundwater and Drinking Water: A Practical Guide. UCANR Publications. Chapter 8: pp. 113-38. ISBN 9781879906815

National Oceanographic and Atmospheric Administration (NOAA), 2019. Monitoring Global and U.S. Temperatures at NOAA's National Centers for Environmental Information.
https://www.ncdc.noaa.gov/monitoring-references/faq/temperature-monitoring.php

Williams, R.; Kilaru, V.; Snyder, E.; et al. (June 2014). Air Sensor Guidebook" (PDF). U.S. Environmental Protection Agency.

Chapter 8: The Nuclear Alternative

8.1 Nuclear Energy

Nuclear energy is released by splitting the **nucleus** of atoms of uranium or plutonium by high speed **neutrons** in the process called **nuclear fission** within a nuclear reactor. This produces a **chain reaction**, as the neutrons from the nucleus of the atom are released by the fission of one atomic nucleus which then collides with other nuclei in other atoms which causes more fission, more neutrons, radioactivity and large amounts of heat. The heat produced by the reactor is then used to turn fresh water into steam which is used to spin turbines coupled to electricity generators. The electricity is then regulated and passed along high-tension lines to electrical sub-stations for conversion to the appropriate domestic supply.

The metal uranium, named after the planet Uranus, was discovered in 1789 by Martin Heinrich Klaproth (German: 1743-1813) in the mineral pitchblende, also called uraninite. This is a mixture of mostly oxides of uranium as UO_2 and UO_3 as well as oxides of lead, thorium, and rare earth elements. Uranium is the heaviest metal and last number of the 92 natural chemical elements. Its radioactive properties were discovered in 1896 by Henri Becquerel (French: 1852-1908), while working with phosphorescent material. The chemist Frederick Soddy (English: 1877-1956) first suggested in 1913, that the chemical elements each had different forms, called **isotopes**, which had slightly different weights whilst having the same **atomic number** and therefore the same chemical properties. This finding was based upon his studies of radioactive **decay chains** which showed that the unstable radioactive elements decayed through a series of isotopic changes, to the final, stable element, lead. Uranium has eight isotopes, some artificially produced within reactors, but naturally occurring uranium is found only as three major isotopes: uranium-238 of about 99% abundance; uranium-235 of 0.72% abundance; and uranium-234 of 0.0059% abundance. All three isotopes are radioactive, eventually decaying to lead with uranium-238 having a **half-life** of 4.4683×10^9 years. This half-life is the time taken for the radioactivity of the element to decay to half of its original level of radioactivity or the original mass of the isotope, with the emission of radioactive particles, energy and the formation of other elements known as **daughter products**. Uranium-235 can be used in nuclear fission and has a half-life of 703.8 million years.

Uranium is mined in about twenty countries, with half of the world's production coming from just ten mines in six countries: Kazakhstan with 36% of total production; Canada with 15%; Australia with 12%; Niger with 8%; Namibia with 8%; and Russia with 5%. The ores of uranium are usually mined by the open pit method because of the danger of radioactive radon gas which is often associated with the ores. After it is mined, the ore goes through a mill where it is first crushed. It is then ground in water to produce a mixture of fine ore particles suspended as a slurry. This is then reacted with sulphuric acid to dissolve the uranium oxides, leaving the remaining rock and other minerals undissolved. These wastes are then run off into a mine tailings dam. This leeching process can also be done within the ore body itself by drilling and pumping the acid into the fractured ore with the solution then being pumped to the surface. The uranium slurry is filtered and the uranium extracted by precipitation, dried and then sealed in drums as bright yellow **yellowcake** (U_3O_8) concentrate. This uranium oxide is only mildly radioactive.

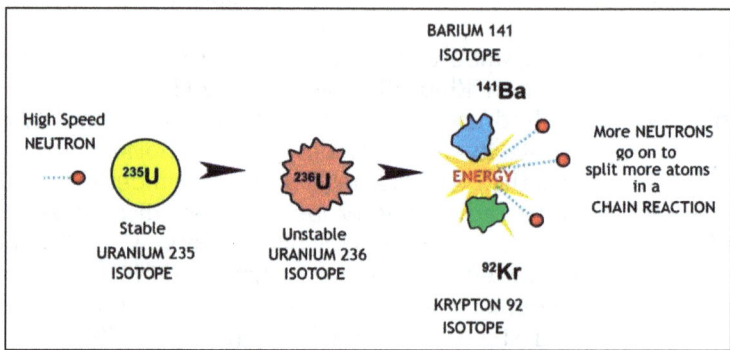

Figure 8.1: Diagram showing how uranium 235 is split to make energy. The numbers, such as 235, 236, 92 &141 are the atomic numbers of these isotopes

Figure 8.2: A specimen of uraninite ore (left) and the processed yellowcake (right) made from it.

The refined yellowcake must still be further processed because the powder contains a mixture of uranium-235 and uranium-238, with only the uranium 235 being useful for nuclear fission. The yellowcake must then be enriched before it can be used in most power reactors. In this process, the proportion of the uranium-235 isotope is raised from the natural level of 0.7% to about 3.5% - 5%. This is done by series of processes including the reaction of the yellowcake with nitric acid, ammonium hydroxide (NH_4OH) and hydrofluoric acid (HF) to form uranium tetrafluoride (UF_4) crystals. This is then treated with fluorine gas to form gaseous uranium hexafluoride (UF_6) which contains both UF_6 made from uranium-235 and UF_6 from the heavier uranium-238. The gas is put into a bank of centrifuge tubes which are then spun at incredibly high speeds, pulling the heavier uranium-238 UF_6 gas molecules into the centre of the tube for removal and potentially future conversion to plutonium for more fission. This leaves the lighter uranium-235 UF_6 gas molecules closer to the edges of the tube, where they can be extracted. The UF_6 gas extracted is then condensed to a relatively stable, white crystalline solid, which can later be vaporised in **autoclaves,** or high temperature pressure chambers, and converted to the uranium oxide (UO_2) and excess gas being converted to more stable UF_4 crystals by adding hydrogen.

The idea that nuclear fission of the heavier, unstable elements could be used as an energy source was suggested in 1938 by the Germans Otto Hahn (1879-1968) and Fritz Strassman (1902-1980). Work on the application of nuclear fission developed throughout the early 1940s in the USA and in Germany with the first successful experiment in which nuclear fission occurred on 2nd December 1942. This is when the first **atomic pile** (Chicago Pile-1) began operations at the University of Chicago under the supervision of Enrico Fermi (Italian: 1901-1954).

This first atomic pile literally consisted of a pile or stack of large graphite blocks which acted as **neutron moderators** to slow down the neutrons and thus control the chain reaction. The **fuel rods** of uranium 235 oxide are inserted into the top of the reactor pile. When sufficient amounts of uranium in the fuel rods are put together, past its stable limit or its **critical mass** of about 52 kilograms of U-235, then a chain reaction occurs producing a great deal of energy, atomic radiation and daughter products of other, smaller radioactive elements such as barium-141 and krypton-92, are produced.

This experiment was part of the Manhattan Project undertaken by the US Government to develop an atomic bomb during the Second World War. In the 1950s, wartime application of nuclear fission turned to more peaceful uses, especially electrical power generation. Today there are about 440 commercial nuclear power reactors operating in 31 countries providing over 11% of the world's electricity supplies.

As well as energy and daughter products, nuclear radiation is also given off. The type of radiation depends upon the nuclear reaction and there are a great range of radiation types and conversions. There are three main types:

- **Alpha particles** (α) consist of a helium atomic nucleus of two neutrons and two protons. These have total electrical charge of +2 and an **atomic mass** of 4 units compared to the mass of one proton = 1. They have high ionization ability, that is, they can easily remove electrons from other elements to form ions, but low penetration power of only a few centimetres in air. They will not penetrate skin or clothing but are dangerous if they are inhaled or ingested with contaminated dust, food or water and cause tissue damage such as cancers.

- **Beta particles** (β⁻) are high-speed, negatively-charged electrons. Positrons (β⁺) which have the same mass as an electron but with a positive charge also occur in some reactions. Beta particles have very low mass, less ionization ability but moderate penetration of a few metres in air, to the base of the skin layer, and a few millimetres of aluminium metal. They will cause direct burning of the skin externally but more tissue damage can occur within the body including specific organ cancers if ingested with radioactive dust.

- **Gamma radiation** (γ) consists of high energy electromagnetic radiation similar to X-rays but more penetrating and dangerous. They have no mass but great penetrating ability, and will pass through the human body causing severe burns and great tissue damage. They are stopped only by several metres of concrete and about 20 centimetres of lead.

There are several types of nuclear fission reactors which are used to generate electricity coupled with steam turbines and electrical generators. A typical nuclear power station may have its nuclear pile contained within a steel pressure vessel within a larger concrete containment structure. Within the central reactor core are the fuel rods which contain the uranium dioxide (UO_2) powder which has been compacted into cylindrical pellets which are firstly heated to a high temperatures and ground to achieve a uniform cylindrical shape. Such fuel pellets are then contained in metallic tubes made from either stainless steel or zirconium alloy. The zirconium rods are much preferred as they are corrosion-resistant and have low neutron absorption. The finished fuel rods are grouped into fuel assemblies which make up the core of a power reactor.

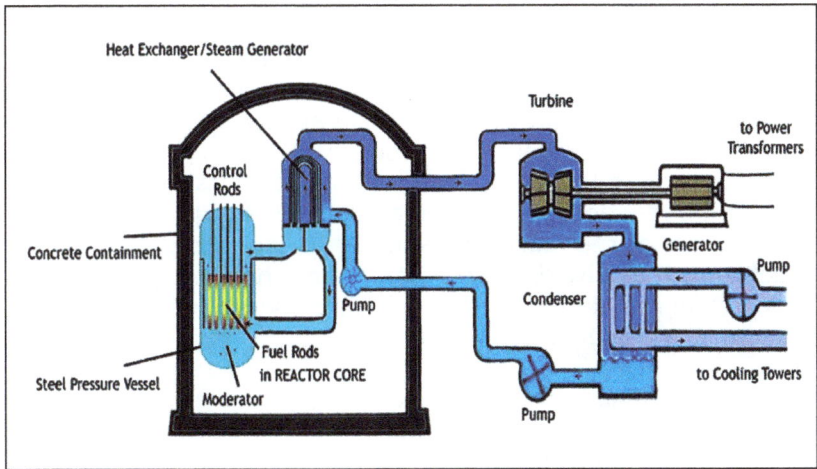

Figure 8.3: A diagram showing the main parts of a typical pressure vessel thermal reactor

In some pool-type reactors, **heavy water** is used as the **moderator** to slow down the neutrons of the chain reaction. This heavy water which consist of water molecules (H_2O) in which the hydrogen part is a natural isotope called **deuterium**, which has an extra neutron in its nucleus. These reactors use a large, deep, open pool filled with heavy water which acts as a radiation shield as well as the moderator. In the original atomic pile, graphite was used as the moderator but beryllium metal can also be used.

Figure 8.4: Looking into the core of a small, pool reactor used for research. The blue glow is due to Cerenkov radiation, which occurs due to electrons moving faster than the speed of light through the water moderator. (Photo: US Department of Energy)

Control rods can also be used to prevent the chain reaction happening too quickly. These can be made from an alloy of silver and cadmium or boron mixed with iron or carbon. These materials are good at absorbing neutrons and so reduce the number of neutrons available to continue the chain reaction. The control rods are moved up and down inside the reactor core to control the reaction. When lifted away from the fuel rods, they absorb fewer neutrons so the reactor gets hotter. If they are pushed down near the fuel rods, they absorb more neutrons and the reactor gets colder. By inserting them completely into the core, the control rods can used to shut the reactor down completely.

A small amount of nuclear fuel will make a large amount of **energy**. Energy is used over time, so usually the output of power stations is usually given in units of **power** which is the amount of energy used per time:

$$\text{Power (in watts)} = \frac{\text{Energy (in joules)}}{\text{Time (in seconds)}}$$

Nuclear power stations are not 100% efficient, so a lot of energy is lost to the environment and within the reaction itself. In general, a nuclear power station will produce about 1000 megawatts (MW) per day for only one kilogram of uranium fuel used. As a comparison, 1 gram of uranium will give about 1 MW per day, which is the energy equivalent of about 3 tonnes of coal or about 230 litres of fuel oil per day which produces approximately 0.25 metric tonnes or 250 kg of carbon dioxide.

On an historical note, the unit of energy, the **joule** (J), is named after James Prescott Joule (English: 1818-1889) who discovered the relationship between mechanical work and heat energy produced by it, and the **watt** was names after James Watt (Scottish: 1736-1819) who developed an improved version of the steam engine.

8.2 Advantages and Disadvantages of Nuclear Power

The advantages of nuclear energy include:

- Low pollution - nuclear power plants do not emit large amounts of greenhouse emissions such as carbon dioxide and methane associated with fossil fuels, although some plants use cooling towers to cool the core and so produce some heat pollution. There is usually little or no impact on local water supplies and no requirement for large areas of land storage.

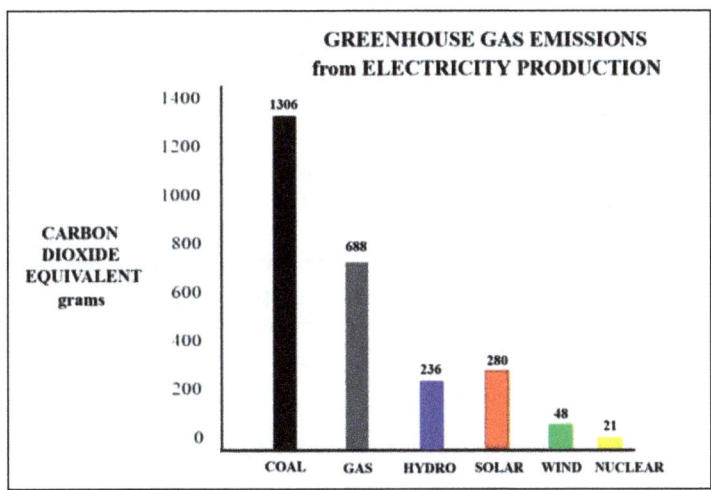

Figure 8.5: Graph showing the relative levels of CO_2 emission (maximum) from generating one kilowatt-hour of electricity from different sources (Data: International Atomic Energy Agency 2000)

- Low operating costs - nuclear power produces very inexpensive electricity. The cost of building these plants is high, but once completed, the cost of fuel and continued operation is relatively low. The normal life of nuclear reactor is anywhere from 40-60 years, depending on how often it is used so they are very cost effective.

- Reliability – there are over 400 nuclear power plants operating in over 33 countries producing electricity for their national grid. The first plant opened at Calder Hall on the north-western coast of Great Britain in 1956, and other countries soon followed. It has been estimated that at the current rate of consumption of uranium, the industry will have enough uranium for a reliable supply well into the future.

- More efficient than fossil fuels - the amount of fuel required by nuclear power plant is comparatively less than what is required by other power plants as energy released by nuclear fission is extremely large compared to that produced by the same amount of fossil fuel.

- Further potential - nuclear energy is not a renewable resource but, by using plutonium, a product of uranium fission in **breeder reactors**, more energy can be obtained from the amounts of waste nuclear fuel already available. Currently, breeder reactors are more expensive to build, require a higher critical mass, and often use liquid sodium metal, which is explosive in water and moist air, as a primary coolant. This makes such a reactor potentially more prone to accidents because of the dangerous coolant as well as radioactivity from core leakage. There is also a possibility of a safer source of nuclear energy in the future using **nuclear fusion**. This is the forced joining of the nuclei of the hydrogen isotopes deuterium and tritium to produce a helium nucleus and energy. Tritium is rare in nature but can be artificially produced in pool reactors, and has a nucleus consisting of one proton with two neutrons. At present research is attempting to find solutions for the containment of the very high temperature, electrically-charged **plasma** which would be produced in such a reactor.

The disadvantages of nuclear energy include:

- Environmental impact – this is perhaps the biggest issue in relation to nuclear energy and the mining, transportation, use and disposal of uranium and its daughter products. The process of mining and refining uranium has a number of issues, including the effects of radiation of the ore and associated radon gas at the mine site, as well as radiation dangers at all stages of its refinement. Transporting nuclear fuel to and from plants also represents a pollution hazard.

- Radioactive waste disposal - a nuclear power plant creates a considerable amount of nuclear waste per year. About 95% of the waste from electricity generation is high level waste (HLW), which gives off considerable amounts of dangerous radiation and high temperatures. Storage has always been considered a problem, and in the early days of nuclear energy, some was dumped at sea in sealed drums with potential for doing considerable damage to the ocean environment. Since the 1950's, most waste has stored in drums in deep mines such as stable salt mines, disused tunnels, in shallow pits covered with concrete, in drill holes in stable rock areas and other underground storage facilities. Some has also been dumped at sea.

- High cost – whist still considered relatively economical to operate, there is still an enormous cost in the construction of nuclear power plants. Moreover, there is also a high cost associated with the eventual clean-up of the amount of radioactive waste kept in many currently inadequate storage facilities. These administrative and environmental costs on top of the high expenses needed to build a plant, may also make it less desirable to use nuclear power in the future.

- Public opinion – there is considerable negative opinion in many countries which have nuclear power plants and in other countries which may consider the nuclear option. Environmental and health advocates are often active in preventing or limiting the governments to give the necessary permits for companies to set up new nuclear plants. The strong opposition to the nuclear industry has become a major deterrent to many governments in the trading, use and waste storage of nuclear fuel.

- Threat of nuclear weapons – there is a real or imagined threat that the use for breeder reactors can be used to produce weapons-grade nuclear material in some developing countries at a time when nuclear disarmament of the countries having nuclear weapons should be a priority. The **Treaty on the Non-Proliferation of Nuclear Weapons (NPT)** of 1970, is an international treaty whose objective is to prevent the spread of nuclear weapons and to promote cooperation in the peaceful uses of nuclear energy, and to further the goal of achieving nuclear disarmament. There are five countries having nuclear weapons which are signatories to the NPT: The United States, Russia, China, The United Kingdom and France. Israel, India, Pakistan and North Korea are also believed to possess nuclear weapons and are not signatories to the NPT. There is also some concern that unsecured nuclear material from any stage of the fuel processing and waste cycle may fall into the hands of terrorists and be used as a dirty bomb or general radioactive contaminant.

- Potential for nuclear accidents - whilst the nuclear energy industry has an overall safety record superior to that of fossil fuels, several accidents have demonstrated that the effects are far greater in terms of long-term suffering, destruction of the environment and greater human impact. The threat of nuclear accident is always part of having power stations close to areas of human habitation. Accidents can be caused by human error and natural disasters, especially if the plants are located in geologically unstable areas such as in earthquake zones. Many countries rely upon nuclear power, and in countries with large populations but confined borders, the risk of a nuclear accident is always a major concern. To date, there have been three major peacetime nuclear disasters:

  - **Three Mile Island** was a partial nuclear meltdown that occurred on March 28, 1979, in a reactor of Three Mile Island Nuclear Generating Station (TMI-2) in Dauphin County, Pennsylvania, United States. There was no loss in life as a result of this accident but it created alarm in the international public arena and cost a considerable amount in financial terms and confidence in the nuclear industry. The accident came about because of failure in valves operating the cooling system and the lack of sufficiently trained plant operators who were able to recognize the problem in time. The subsequent loss in coolant caused a partial meltdown of the reactor core as the heat suddenly increased. This also resulted in the release of radioactive gases, especially iodine into the immediate environment. The repair and rehabilitation of the site did not finish until December 1993, with a total cost of about $1 billion. Luckily, the radiation which was given off did not seem to cause any significant cancers in the local population.

  - **Chernobyl** was a catastrophic nuclear accident that occurred on 26 April 1986 at the Chernobyl Nuclear Power Plant in the city of Pripyat, which is now in Ukraine. The event occurred because of a sudden and unexpected power surge which led to the rupture in one of the reactor vessels. This was followed by a series of steam explosions exposing the graphite moderator of the reactor into the air, causing it to ignite. The fire sent a cloud of highly radioactive dust, steam and gas into the atmosphere over an extensive geographical area including parts of Russia, Ukraine and Belarus. For the next four years, over 350,400 people were evacuated from the contaminated areas, especially from the heavily contaminated country of Belarus. The heroic struggle to contain the contamination and prevent a greater catastrophe ultimately involved over 500,000 workers and cost the lives of 31 people immediately with another 28 dying of radiation effects later. Further long-term effects, such as cancers and birth-defects are still being investigated and human suffering will be felt for some time.

- **Fukushima Daiichi** nuclear disaster was an accident at the Fukushima Nuclear Power Plant in Fukushima, Japan, caused initially by a **tsunami** which followed the Tōhoku Earthquake on 11 March 2011. The reactors automatically shut down when the large ocean wave struck, but the tsunami destroyed the emergency generators which operated the cooling system, causing one of the reactors to overheat. This caused two other reactors to meltdown, the release of radioactive material and several hydrogen-air chemical explosions. It was later found that the causes of the accident could have been prevented but that the operating company had failed to meet basic safety requirements and have adequate risk assessment and damage prevention plans, and proper evacuation procedures. Though there were no fatalities caused directly by the accident, it is expected that there will be about 130-640 people who will die of radiation-related cancers in the decades ahead. As yet, the decommissioning of the plant is still on-going and the plant management estimate that it may take 30 or 40 years to complete this task.

As a result of these three nuclear accidents and several smaller malfunctions of nuclear cores around the world, several countries have started a phase-out of their nuclear power plants. Sweden (1980) was the first country to start a phase-out, but this was repealed in 2009. Italy followed in 1987, Belgium in 1999, and Germany began the closure of 8 of its 17 reactors in 2000. Later Taiwan and Netherlands postponed their phase-out intentions. Austria, Denmark and Spain have enacted laws to cease construction on new nuclear power stations and the debate continues in Europe about the need for any nuclear phase-out. However, since closing down some of their reactors, some European countries such as Germany have re-opened or started construction of new coal-fired power stations to make up their energy loss and to meet new demands.

Many countries including Australia, Austria, Denmark, Greece, Ireland, Italy, Latvia, Liechtenstein, Luxembourg, Malaysia, Malta, New Zealand,

Norway, Philippines, and Portugal have no nuclear power stations and remain opposed to nuclear power. Globally, more nuclear power reactors have closed than opened in recent years but overall capacity has increased.

8.3 Towards Safer Nuclear Power

Nuclear power has developed as a major source of power since the first commercial power station opened and currently, there are over four hundred nuclear power stations providing about 10% of the world's electricity. The International Energy Agency lists the use of nuclear energy as being the fourth of the world's energy sources:

Energy Source	World Use
Coal	38.3%
Gas	22.9%
Hydro	16.3%
Nuclear	**10.2%**
Solar, Wind, Geothermal, Tidal	6.6%
Oil	3.3%
Other	2.3%

Table 8.1: Ranking or world energy sources (Data: IEA)

Some countries such as the USA have a greater reliance on nuclear energy at 56% of its total energy supplies. Others such as Australia has no nuclear power but this country also has the largest reserves of uranium at 30% of the world's total.

One possible scenario would be to build new reactors at the site of the mining of uranium with the emphasis on being far from population centres. Processing of the fuel would also have to be done on this site as well. Limitations to this scenario would be the proximity to any national electrical grid for taking the power to consumers at some great distance and also the provision of freshwater supplies for the mine, processing plant and the power station. For example, the large uranium mine at Olympic Dam in South Australia has some of the biggest reserves in the world and is also at the end of an extension of the Australian national electricity grid. It is also near prohibited areas in which nuclear tests were once held and in which nuclear waste was (inefficiently) buried. Unfortunately, it is also in a remote area in extremely arid conditions and so freshwater supplies are a problem. It is not far from Lake Torrens, a large salt lake rarely filled with water which even then would be salty and unsuitable. This lake is about 30 metres above sea level and connected to the sea less than 100 km away by a series of usually dry depressions. Several plans have been put forward over the last 100 years to connect this lake to the sea and then progress on downhill to Lake Eyre, another salt lake further inland. Such projects designed to bring seawater to Australia's dry interior to change the arid climate were heavily criticised due to the vast amount of seawater which would be needed along a potentially low gradient channel and the excessive amount of evaporation in the area would exceed water inflow. Additionally, desalination plants on a large scale would be needed to provide freshwater for any mining, nuclear power plants and agriculture. Still, it is an engineering project of some interest.

See also:

http://www.cmar.csiro.au/e-print/internal/mcgregor_x2004a.pdf
(the idea of moderating climate by flooding central Australia)

The safe removal of nuclear waste would also have to be overcome. One area of potential for safer storage is **synroc** or synthetic rock, pioneered in 1978 by a team at the Australian National University in conjunction with the Australian Government's **Australian Nuclear Science and Technology Organisation** (ANSTO). This involves the combining of the nuclear waste with titanium minerals and water which is then heated and fused into a rock-like substance. This would then be buried down very deep drill holes within stable crystalline rock and then capped with concrete, effectively returning the radioactive material to the Earth.

See also:

https://www.ansto.gov.au/business/products-and-services/ansto-synroc-waste-treatment-technology#content-how-it-works
(Several videos about synroc and other nuclear technologies developed by ANSTO)

ANSTO has now finalised plans for production of synroc and the global use of their product. See:

https://innovation.ansto.gov.au/2019/04/09/new-global-first-of-a-kind-ansto-synroc-facility/

Elsewhere, nuclear power continues to be used, although still mainly near populated areas in highly-populated countries where space is a problem.

Considerable research is ongoing into the recycling of nuclear waste for the next generation of nuclear power stations which use and reuse the fuel in a closed cycle. Moreover, it is claimed that such waste would be unsuitable for nuclear weapons, removing another disadvantage of nuclear power. Unfortunately, many of the world's nuclear power stations are still using older technology and worldwide, the brilliance of nuclear energy potential given in the past has faded to some degree. Whilst new technologies such as the fast-breeder reactors show enormous potential, many countries are still not convinced that nuclear fission is the answer to the threat of global warming.

Another possibility which has received considerable research is the use of nuclear fusion. The resultant wastes would be helium atoms and excess neutrons and alpha particles. Normally this is impossible to achieve because of the strong electrostatic repulsion of the positively-charged hydrogen nuclei, but using ultra high temperatures this has been achieved in very limited amounts in research devices such as the Tokamak (Russian: *toroidalnya kamera ee magnetnaya katushka* - a donut-shaped magnetic chamber), designed in 1951 by Soviet physicists Andrei Sakharov and Igor Tamm. There are still many engineering problems associated applying this technology to energy production. Firstly, extremely high temperatures are required to start the fusion and as yet more energy is required as input than energy produced. Secondly, it is also difficult to contain the very hot, electrically-charged gas, or **plasma** produced. Whilst nuclear fusion offers the potential for cheaper, less harmful and cleaner energy, considerable works still needs to be done to achieve a reliable output of energy.

In general, however it must be remembered that the use of nuclear power should only be regarded as a stop-gap measure until less expensive and completely safe forms of environmentally-friendly power plants become mainstream and able to produce enough electrical power for future use. Currently, many nuclear power plants are beginning to show their age and building new nuclear power plants is very expensive. Moreover, they need to be built near plentiful supplies of freshwater and there is still the problem with transportation and storage of uranium fuel and the removal of waste. There will always be the threat that through human error, there may be more nuclear accidents.

See also:

https://www.world-nuclear.org/information-library/safety-and-security/safety-of-plants/safety-of-nuclear-power-reactors.aspx
(A detailed look at the safe use of nuclear power from the World Nuclear Association)

https://www.world-nuclear.org/information-library/nuclear-fuel-cycle/nuclear-wastes/radioactive-wastes-myths-and-realities.aspx
(More about the nuclear waste problem)

https://www.discovermagazine.com/technology/thorium-power-is-the-safer-future-of-nuclear-energy
(The use of safer Thorium isotopes as a nuclear fuel)

http://theconversation.com/nuclear-power-is-set-to-get-a-lot-safer-and-cheaper-heres-why-62207
(The potential for safer nuclear power)

https://www.world-nuclear.org/information-library/current-and-future-generation/nuclear-fusion-power.aspx
(Information on nuclear fusion as a source of energy)

Final Remarks

1. Nuclear power stations have been operating in many countries safely since the 1950's, using nuclear fission of uranium to produce electricity.

2. There have been several accidents in nuclear facilities around the world but most have been contained. Several however, such as those at Three Mile Island in the United States, Chernobyl in the Ukraine and at Fukushima in Japan have shown that such accidents have wide environmental consequences.

3. Nuclear fission involves the breakup of large, unstable and radioactive elements such as uranium into smaller daughter products, neutrons and harmful radiation such as alpha and beta particles and gamma electromagnetic radiation. A considerable amount of heat energy is also given off and this can be used to produce steam for use in electricity generation.

4. In a world where the emissions from the use of the fossil fuels coal and oil have caused global warming, the use of low-emission nuclear fission offers a good alternative if the threat of nuclear accidents, the use of nuclear weapons and the safe storage of nuclear waste can be achieved.

More Applications

1. Research the use of local nuclear power or use of nuclear material. Consider its potential as a power source.

2. If there is no local nuclear power industry, find out about State and National regulations governing or limiting the use of nuclear power.

3. Contact local environmental groups and local government representatives about future use of nuclear power in the state or nationally.

4. Contact local government agencies about the monitoring of natural and industrial nuclear radiation in the local area.

5. Research how radioactive sources are used for medical and industrial use as well as small-scale power production in transportation and research.

6. Find out about how radiation is used in the home apart from electrical power production.

Further References

Falk, J. (1982). Global Fission: The Battle over Nuclear Power. Melbourne: Oxford University Press. pp. 95-96. ISBN 978-0-19-554315-5. (Dated but a good look at the problems of the atomic age)

Findlay, T., 2010. The Future of Nuclear Energy to 2030 and its Implications for Safety, Security and Non-proliferation Part 1 – The Future of Nuclear Energy to 2030. CIGI,
https://www.cigionline.org/sites/default/files/part_1.pdf

Gore, Al (2009). Our Choice: A Plan to Solve the Climate Crisis. Emmaus, PA: Rodale. pp. 165-166. ISBN 978-1-59486-734-7.

Horvath, A and Rachlew, E., 2016. Nuclear Power in the 21st Century: Challenges and Possibilities. In Ambio - A Journal of the Human Environment. January 2016, Volume 45, Supplement 1, pp 38-49. Netherlands, Springer. ISSN 0044-7447. pdf available at:
https://link.Springer.com/article/10.1007/s13280-015-0732-y

International Atomic Energy Agency (IAEA), 2015.
Reference Data Series, No. 2: Nuclear Power Reactors in the World (PDF).
https://www-pub.iaea.org/MTCD/Publications/PDF/rds2-35web-85937611.pdf

IAEA Power Reactor Information System: Nuclear Share of Electricity Generation in 2018.
https://pris.iaea.org/PRIS/WorldStatistics/NuclearShareofElectricityGeneration.aspx

IEA, 2019. Nuclear Power in a Clean Energy System.
https://www.eenews.net/assets/2019/05/28/document_ew_01.pdf

World Nuclear Association, 2018. Plans for New Nuclear Reactors Worldwide.
https://www.world-nuclear.org/information-library/current-and-future-generation/plans-for-new-reactors-worldwide.aspx

Chapter 9: Alternative Energies

9.1 Energy Production – Where to Now?

There has been considerable argument for and against the continued use of fossil fuels and the increased use of nuclear energy; especially now that it is apparent that the burning of fossil fuels has contributed to carbon dioxide emissions and that this has caused an overall global warming with a subsequent sea level rise.

Currently, the world faces a dilemma, with the demand for more energy conflicting with the problems caused by the burning of fossil fuels and the subsequent problems of global warming. Energy will be in great demand in the future, but there is a need to prevent any further rise in atmospheric and oceanic temperatures.

The table below (Table 9.1) shows that the most industrialised countries use the most power. The data in the last column (Average Power/Capita), can be misleading as many industrialised countries such as China and India also have a large rural population which uses much less energy than countries with a high city population. Norway, for example, ranking at number 27 has a per capita power consumption of 2603 watts/capita and only a population of about 5 million, however, over 90% of its electrical power generation comes from hydroelectricity and not fossil fuels. Some of the alternatives to fossil fuels and nuclear energy include those which use water, wind, the Sun, biological sources and heat from the Earth.

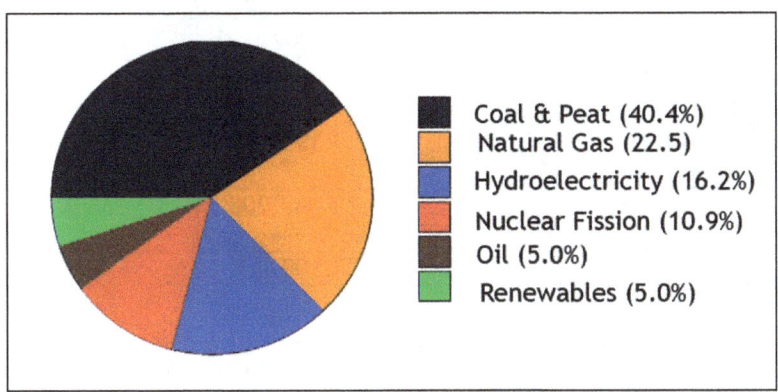

Figure 9.1: A pie graph showing the relative amounts of electrical energy produced by various fuel sources in 2012 (Data: International Agency http://www.iea.org)

Rank	Country	Electrical Consumption (MWattHour/year)	Population	Average Power Use/Person (Watts/capita)
1	China	5,650,000,000	1,360,720,000	474
2	United States	4,986,400,000	317,848,000	1843
3	Russia	1,016,500,000	146,019,512	801
4	India	983,823,000	1,242,660,000	152
5	Japan	910,700,000	127,120,000	941
6	Germany	782,500,000	80,716,000	1160
7	Canada	570,800,000	36,584,962	2185
8	Brazil	493,500,000	201,032,714	301
9	South Korea	482,400,000	50,219,669	1038
10	France	451,100,000	65,864,000	904
11	United Kingdom	353,300,000	63,705,000	822
12	Italy	327,200,000	60,021,955	781
13	Turkey	264,136,780	78,741,053	335
14	Spain	253,100,000	46,609,700	875
15	Taiwan	224,300,000	23,315,822	1080
16	South Africa	234,200,000	54,002,000	415
17	Mexico	232,000,000	121,409,830	101
18	Saudi Arabia	231,600,000	29,195,895	781
19	Australia	229,600,000	23,060,903	1610
20	Iran	216,290,000	77,800,000	317

Table 9.1: Showing the total power used by the top 20 countries. (Data from "COUNTRY COMPARISON: ELECTRICITY - CONSUMPTION". CIA. 2013)

9.2 Energy from Water

The energy provided by running water, both as a source of power and for transportation has been in use for a very long time. The main sources of energy from moving water are:

- **Hydroelectricity** – is not really a new, alternative energy as it has been used since pre-industrial revolution days to power mills to grind grain and perform other, simple tasks. Similarly, windmills have been used to perform relatively simple, mechanical tasks. Since the 1950's there has been a conscious effort to expand the use of water power to produce electricity which would not add to the pollution of the environment. Hydroelectric power generation is often the major source of energy in countries which have high mountains and large amounts of precipitation as rain or snow. It requires a large capital outlay in the construction of dams, pipelines and power stations, but once operating, costs are minimal and there is little damage to the environment. The water from these dams is piped downhill until it reaches sufficient speed to be passed into turbines within the power station. These are coupled to electrical generators which produce electricity.

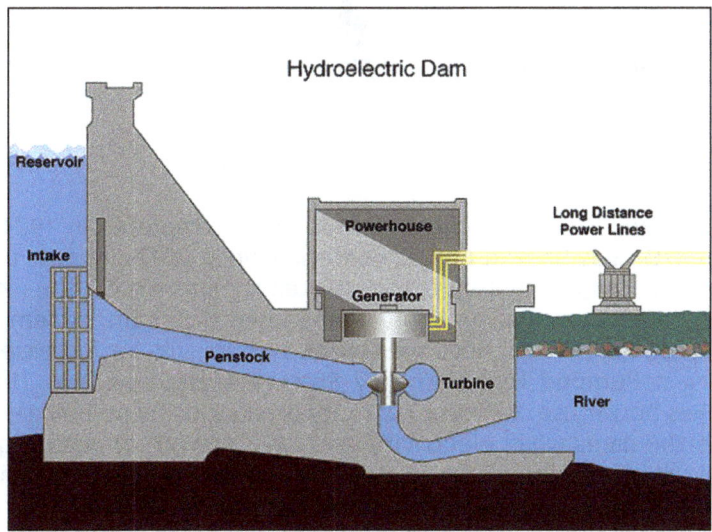

Figure 9.2: A diagram showing the operation of a hydropower station (Photo: Tennessee Valley Authority)

The water from the river which has been dammed usually continues on to be used further downstream. Only the lack of water and silting of the lake behind the dam are major problems to the generation of power but concerns have been raised in some countries about the restricted water flows downstream, the flooding of inhabited farming areas and the restriction of fish migration upstream. There is also a problem that some major lakes produced by dams may produce methane gas from organic sediment on the lake floor.

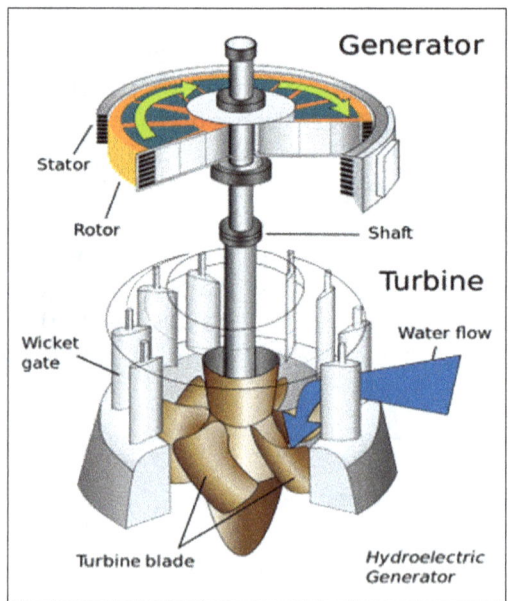

Figure 9.3: Diagram showing detail of the turbine/generator (US Army Corps of Engineers)

The Snowy Mountain Scheme in Australia, completed in 1974, and the Three Gorges Dam in China, operating since 2012 are just two of the world's major hydroelectric systems. In 2017, the Australian Government announced that the Snowy Mountain system is to be updated so that its 4000-megawatt system will be increased by 50% using a **Pumped Hydro Energy Storage (PHES)** system. This system makes better use of hydroelectricity production by pumping water back into the dams when electricity grid loads are not at peak requirement and then releasing the water for power production when more electricity is required. According to the Renewable Energy Policy Network, in 2015, hydropower generated 16.6% of the world's total electricity and 70% of all renewable electricity, with an expected increase of about 3.1% each year for the next 25 years.

- **Tidal energy** is a form of hydropower which uses the movements of the Earth's tides to produce electricity. This technology is still in its infancy, yet it has the potential for future electricity generation. Tides are more predictable and consistent than wind energy and solar power but as yet their use in electricity generation has suffered from relatively high cost and limited availability of sites. These sites must have a high tidal ranges or rapid flow velocities and this is a major limitation but there is some hope that improved technology will allow for a greater potential than previously expected.

Tidal mills have been used both in Europe and on the Atlantic coast of North America and the earliest occurrences date from the Middle Ages, or even from Roman times. In these mills, the incoming tide was contained in large storage ponds, and as the tide went out, it turned waterwheels that used the mechanical power it produced to mill grain. In modern times this idea has been used to generate electricity in a manner similar to that of hydroelectric dams, and the world's first large-scale tidal power plant was opened in 1966 on the estuary of the Rance River in France. It is a barrage system using a wall across the estuary and its 24 turbines reach a peak output of 240 megawatts (MW 10^6 watts).

Figure 9.4: The barrage of the Rance River, Brittany, France (Photo: Wiki Commons)

Currently, the Sihwa Lake Tidal Power Station in South Korea, completed in 2011, is the world's largest tidal power installation, with a total power output capacity of 254 MW. It is near Inchon, South Korea and consists of a large barrage across a coastal inlet. A variation on this is to build a barrage across larger ocean bays, creating an artificial lagoon as a tidal lagoon system. Currently one such system is under construction within Swansea Bay in Wales. The larger size of these lagoon systems means that they can be built as multiple barrages and thus make use of the different tidal heights at different times of the day, spreading out their operating times.

The **Tidal Stream Generators (TSG)** installed in these and similar projects, make use of the fast flow moving water to directly power turbines, in a similar way to wind turbines and using similar rotor types. These include the Propeller, Savonius and Darrieus blade types (see Figure 9.8) which are submerged just below water level. Tidal generators may be built into the structures of existing bridges or can be entirely submersed, thus avoiding concerns over the visual impact on the natural landscape. Land constrictions such as straits or inlets can create the high velocities needed to power these generators but usually only during times when the tides are running into or out of the inlet. Turbines can be orientated as either horizontal or vertical, and may consist of open or ducted blades and they are usually anchored near the bottom of the stretch of water where the tidal velocities are greatest.

The first generator of this type was the Seagen generator, installed in Strangford Narrows in Northern Ireland, in 2008. This generator produced 1.2 MW for between 18 and 20 hours a day while the tides are forced in and out of Strangford Lough through the Narrows. It was, however, only considered as a pilot program and its power generation capacity was considered insufficient so it is now being decommissioned. Many more tidal generators are planned for a variety of coastlines which have fast tidal races. These include the original site in Northern Ireland and near Darwin, Australia. Unfortunately, this technology suffers from limitations of the local marine habitat which is usually shallow, offshore and subject to collision with a variety of natural and man-made objects.

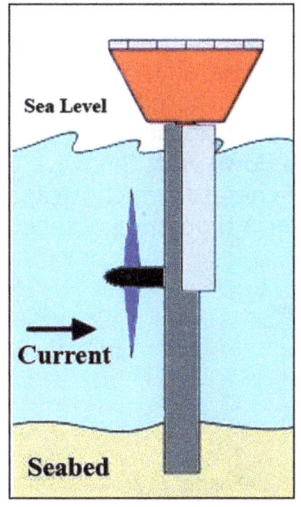

Figure 9.5: Diagram showing a Tidal Stream generator

- **Wave action generators** use the power of ocean waves created by winds and currents to generate electricity. A machine able to create wave power is known as a Wave Energy Converter (WEC). Wave-power generation is not currently widely employed commercially, although there have been attempts to use it since at least 1890. In 2008, the first experimental wave farm was opened in Portugal, at the Aguçadoura Wave Farm. It used three Pelamis-brand wave energy converters producing electricity of up to 2.25 MW. Unfortunately, due to some technical issues and problems with the parent company, the project was shut down only two months later.

Figure 9.6: One of the Wave Converters off the coast of Portugal (Photo: Wiki Commons)

Following further research into the use of wave power, the world's first grid-connected wave energy system began operations in 2016 offshore from Garden Island, near Perth in Western Australia. This system uses large, underwater buoys, or pods, which are anchored to the sea floor. The tube which connects these pods to their anchor also contains a vertical pump piston which moves up and down with the wave action. This piston is connected to an electrical generator which is capable of producing enough electricity to power over a thousand households.

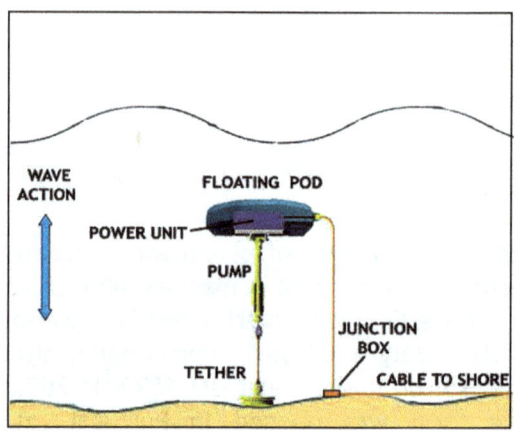

Figure 9.7: Diagram showing the operation of the wave action generator near Perth, Western Australia

The advantages of these wave action generators are that they:

- have little impact on the environment;
- can be beneficial by acting as artificial reefs and encouraging marine growth
- can be brought on-line as required with predictable wave action
- can be used in a greater number of off-shore locations near coastal cities

The disadvantages of these systems concern their maintenance, position near shipping routes and the risk of collision with ships and marine mammals.

9.3 Wind Power

Wind power like hydropower, has been used for centuries to drive small-scale industry such as grinding mills and to pump water. In more recent times it has been used on a larger scale to produce electricity, and most modern wind turbines can produce power with rated outputs of 1.5-3 MW. There are two main types of wind turbines: the classical **Horizontal Axis Wind Turbines (HAWT)** and the **Vertical Axis Wind Turbines (VAWT)**. The latter type includes the Savonius and Darrieus models, which have the advantage over the propeller HAWT wind turbine, in that they can operate in low winds coming from any direction. Propeller models need to be swung around into the wind to achieve maximum generation.

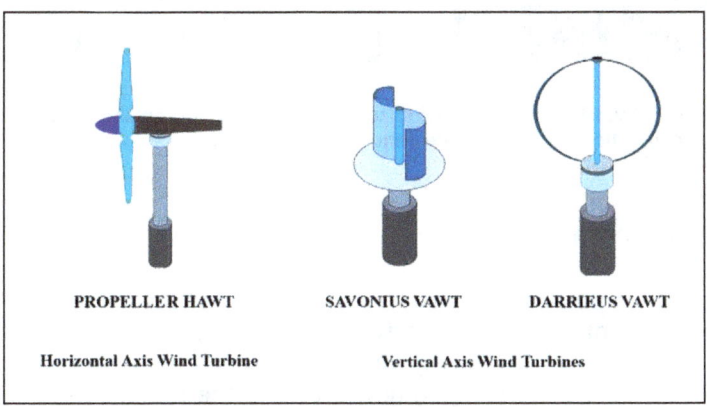

Figure 9.8: Diagram showing the most common wind turbines

There is still considerable potential for wind power however this would require a large-scale commitment to install towers over large areas as wind farms in high wind environments such as offshore or wind-swept plains. Even then, wind turbines only have typical operating efficiencies of from 16% to 57% annually. In 2014 the global wind generation was 706 terawatt-hours or 3% of the world's total electricity (BP Statistical Review – a terawatt is 10^{12} watts).

 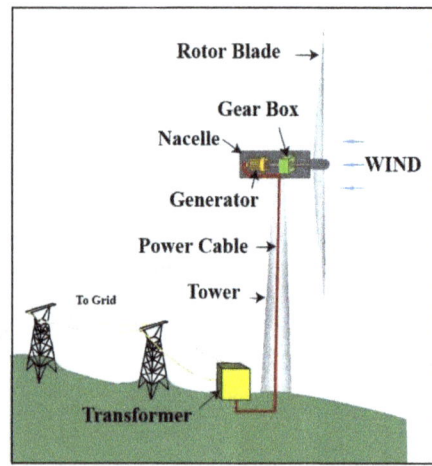

Figures 9.9 and 9.10: A propeller wind turbine in suburban England (left) and a diagram showing how such a turbine generates electricity

9.4 Energy from the Sun

Solar energy has perhaps the greatest potential for future energy needs, especially in countries which receive considerable amounts of sunshine. Solar technologies are broadly grouped as either passive solar or active **solar** technologies, depending on how they use the radiant heat and light from the Sun. Passive solar techniques attempt to provide the maximum heating or cooling effect for existing conditions, such as orienting a building in respect to the direction of the sun or building glasshouses in temperate climates to grow tropical plants. This could also mean selecting materials with favourable properties and designing the spaces within a building so that the heat of the sun will allow a better current of air throughout the structure. **Active solar technologies** are those which are constructed specifically to capture solar energy for direct heating or by way of conversion, to another form of energy such as electricity. Active solar technologies include:

- **Solar Thermal Energy (STE)** which uses solar collectors for heating water. Usually this is used for domestic or industrial hot water systems which heat water in metal or plastic tubes within a black, glass-enclosed box. This then stores the hot water in an insulated water tank which retains sufficient heat over night. Usually the water tank is placed above the collector so that the water can rise by convection when it is heated and the whole system is placed high up, usually on a roof, so that the hot water can run down by gravity. If mains water pressure is not available to replace water in the collector, then cold water must be pumped up daily. This is a simple operating system and home-made versions are easy to make, consisting of black plastic agriculture pipe coiled within a black-painted metal box covered with a sheet of glass or clear plastic and connected to a small water tank. Some domestic systems in colder climates which have limited daily sunlight also have an electric heating element in the water tank to boost the temperature at night.

Figure 9.11: A small domestic solar water heater

A similar system can be used in larger water heating applications such as the heating of swimming pools. In these systems, mats of heat-absorbent black plastic are used to cover a large area facing the sun. Heated water is then re-circulated by a pump to warm the pool sufficiently to allow swimming in colder seasons.

Figure 9.12: Solar pipes on a rooftop for heating a swimming pool.

- **Concentrated Solar Power (CSP)** is an industrial application of solar heating which generates a very large amount of solar radiation by using mirrors or lenses to concentrate a large area of sunlight onto a small area. Here, the extreme heat may be used for industrial purposes such as the melting and refining materials, high-temperature chemical research, or producing steam for turbines to power electrical generators. CSP has being widely developed in Spain, Morocco, the Middle East and in the United States. The global operational power stands at about 5000 megawatts (MW) annually.

Figure 9.13: A Concentrated Solar Power system used for research (Photo: US Department of Energy)

- **Solar electricity** is the production of domestic and industrial electricity using the rays of the Sun. These solar panels use light energy from the Sun to produce direct current (DC) which is a flow of electrons from the negative to positive electrical terminals. Developed from the discovery of the **photovoltaic effect (PV)** in 1839 by A.E. Becquerel (French: 1820-1891), the father of Henri Becquerel (French: 1852-1908), the discoverer of radioactivity. The photovoltaic effect produces a voltage, or potential difference, which is the electrical pressure which pushes electrons through conductors producing an electrical current.

In 1905, Albert Einstein (German: 1879-1955) described the photo-electric effect which is related to the photovoltaic effect, as being the removal of electrons off a metal in a vacuum when it is struck by ultra-violet light. Einstein suggested that the **photoelectric effect** demonstrated that light was composed of small bundles of energy, which he called **quanta** of light. These acted like particles, now called **photons**, rather than continuous waves, which would knock electrons out of metals if the photons had sufficient energy. This discovery led to the concept of **quantum mechanics** now accepted in physics earning Einstein the Nobel Prize in Physics in 1921.

The first solar cell using the photovoltaic effect, which was able to produce a very small amount of electricity was invented by the Bell Laboratories in the United States in 1954. In simple terms, solar cells are made from a thin wafer of **semiconductor**, a material such as the element silicon. This conducts electricity only under certain conditions. It has been specially treated with elements such as phosphorus and boron so that it becomes positively-charged on one side and negatively-charged on the other. When light energy strikes the solar cell, electrons are ejected from the atoms in the semiconductor material to form an electrical current when the two sides are connected by a conductor. The current produced is very small, only a few thousandths of an ampere (i.e. a few milliamps), so many cells are linked together on a flat surface called a photovoltaic module in series such that a negative side of one is attached to the positive side of the next and so on. This allows the current of all the cells to add together. The current produced by such a module is directly dependent on how much light strikes it and the length of daylight. Most domestic systems are only about 12% - 18% efficient.

Figure 9.14: A small 3-Kilowatt domestic photovoltaic module

For normal domestic electricity supply, the direct current must be turned into an alternating current, which provides energy by the to-and-fro motion of electrons, at the appropriate voltage using an electrical converter. This voltage varies from country to country with the USA using 110 V, Canada 120 V and many others using 220- 240 volts. The frequency of alternation also varies from country to country but it is usually about 50 to 60 hertz. The converters needed to change the direct, low current voltage from the solar panels to the main alternating domestic supply are a type of electrical step-up transformer. Whilst the converter increases the voltage, it loses some electrical energy in the conversion.

In many small applications, solar modules have been used to charge a wide range of items, ranging from pocket calculators, some domestic appliances and the International Space Station. However, the main problem with photovoltaic generation is that the amount of energy is relatively small compared to fossil fuel electrical generation, and that it only operates during days of good sunlight. Moreover, the manufacture of solar cells requires a considerable amount of energy and material and if they are used away from the grid of the domestic power system, they require banks of expensive batteries to store electricity for night use. The efficiency of these batteries is improving but they are expensive, contain toxic chemicals which are difficult to recycle and may only last 5-10 years.

Figure 9.15: A large photovoltaic farm in the desert (Photo: US Dept. of Land Management)

Despite the problems of high cost, relatively low output, difficulty in storage of power and their modest use in industry, the development of more efficient and cheaper solar photovoltaic cells is growing rapidly. Worldwide capacity of solar electricity reached 177 gigawatts (GW - 10^9 watts) by the end of 2014, which represents about 1% of worldwide electricity demand, with Germany producing over 7% of its national electricity demands (source: International Energy Agency). In addition, continued research into the storage of daylight electricity has shown some potential, especially in the use of cheaper, more efficient batteries and the production of hydrogen fuel by electrolysis. Electrolysis is the splitting up of the water molecule by electricity into oxygen and hydrogen gases which can then be used in fuel cells to produce electricity or in combustion engines as an environmentally-friendly fuel.

9.5 Storage of Renewable Energy

Some energy sources such as solar electricity will only operate during the day and other renewables, such as wind power and tidal power may have times when conditions are unfavourable for continual generation. Moreover, the demand for electricity in large urban and industrial regions also vary throughout time. Pumped hydro schemes are useful to some degree, but at the domestic level, lithium-ion batteries seem to have the best potential for smaller applications such as in household use and transportation.

In very simple terms, if solar voltaic cells are to be used domestically, they must be connected to an inverter which is a type of electrical transformer which converts the direct current charge produced by the solar panels into alternating current (AC) used in the home or work space. Appliances using DC such as a portable radio uses this direct flow as a source of energy. Alternating current consists of a flow of electrons which rapidly moves back and forth between the terminals. This energy is then used by the appliances such as those commonly found in the home. In addition, any energy from the solar panels must also be converted to match the electrical pressure or voltage used in the appliances. Any electrical power not used directly during the daylight hours can be re-converted to direct current and stored in batteries attached to the inverter-solar cell system. Batteries consist of many individual electric-chemical cells each having metallic positive (anode) and negative (cathode) terminals and some form of electrolyte - a chemical solution or paste which conducts electricity which is involved in the chemical reactions with the anode/cathode electrodes.

Electrochemical cells may be connected together in series, parallel or a combination of these. In a **series circuit**, the negative terminal of one cell is connected to the positive terminal of the next cell and so on and this arrangement gives an addition of the voltage of each cell with the amount of current flow equal to one cell e.g. if four cells each of 1.5 volts giving

2.0 amperes of current are connected in **series**, then the total output would be 4 x 1.5 or 6 volts (V) at 2 amperes (A). However, the current flow is only as good as the poorest cell in the series and lasts only a limited time.

In a **parallel circuit**, each cell is connected to the next using the same terminal so that anodes are connected to each other as are the cathodes. This arrangement gives a total voltage of only that of one cell but allows for more current to be produced. The system works even if one cell fails e.g. four 1.5 V cells connected together is rated at 1.5 volts but the current lasts longer. Many larger storage systems or banks will use cells connected in both series and parallel to suit the output and endurance required.

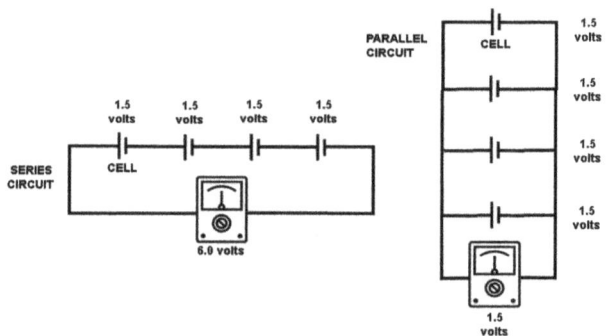

Figure 9.16: Series and parallel cells.

There have been many types of batteries produced since their invention by Alessandro Volta (Italian: 1745-1827) in 1800. He connected wire to either end of a stack of cells each consisting of a disc of copper separated from a disc of zinc by cloth soaked in saltwater. His pile or battery produced about 0.76 Volts (V) for each cell. Soon other electrochemical cells and their batteries were produced, each with an output of about 1 to 3.6 volts depending upon the nature of the metals used for electrodes and the chemistry or the electrolyte between them. The common lead-acid battery used in automobiles today was invented around 1859 and used lead and lead oxide as electrodes and sulphuric acid as an **electrolyte** to produce about 2 volts per cell. Usually six cells are used to produce 12 volts for the automobile's electrical system.

The most promising type of battery currently in use is the rechargeable **lithium-ion battery (LIB)**. These are now being used in small appliances such as cell phones and cameras, in some electrical vehicles and as a one to three-unit bank in the home in conjunction with solar panels or in larger banks for more commercial use. Lithium metal is not used as it is a very reactive metal and produces explosive hydrogen gas in water; rather compounds containing lithium ions are used in conjunction with other materials. Usually the **anode** or negative electrode, is made from a conductive material such as graphite, a form of pure carbon and the **cathode** or positive electrode can be one of three materials: lithium cobalt oxide, lithium iron phosphate or lithium manganese oxide. The electrolyte is typically a mixture of organic carbonates such as ethylene carbonate or diethyl carbonate containing mixtures of lithium ions such as lithium hexafluorophosphate. Such complexity makes them very sensitive and expensive, but used in a commercially produced bank they can give effective charge/recharge for anywhere from 5 to 10 years depending upon environmental conditions and how they are used.

See also:

https://batteryuniversity.com/learn/archive/is_lithium_ion_the_ideal_battery
(A good summary of lithium-ion batteries)

https://www.energymatters.com.au/residential-solar/battery-storage/
(All about the use of home solar cell- battery combinations)

Of considerable potential is the development of the **sodium battery** - sometimes called the saltwater battery. This is a wet-cell battery that uses a reaction with salt water electrolyte, air, a magnesium oxide cathode and a carbon anode to produce electricity. Just like any other battery, it requires chemical energy to produce electrical energy and that chemical energy needs to be replenished to keep the battery running. The unique characteristic of the salt water battery is that it uses the air as a cathode, so there is no need for separate half-cells for each electrode as in other wet-cell batteries. It may not be as efficient as a lithium-ion battery but it's cheaper, safer and its components are more easily recycled.

Another method of storing electrical charge, albeit for only a short time is the use of capacitors. These are common electrical components, found in most electrical devices, which store charge between two metal plates of surface separated by an insulator called a dielectric. Banks of capacitors can be combined to store large amounts of electrical charge which can be then used to produce a current to power devices. **Supercapacitors (SC)**, also called ultra-capacitors, are high-capacity capacitors which can store sufficient charge to be able to power electric motors such as in electrically-powered vehicles, cranes and elevators. They can be more quickly recharged than batteries but then can only be used for a shorter time.

9.6 Biofuels

These fuels have been derived from **biomass** or biological material such as agricultural products, wastes from sugar and timber mills, gases coming from landfill dumps and other waste gases. However, as fuels used in normal combustion engines, they still release carbon dioxide and water vapour. These fuels can be classified broadly into two major categories:

- **First-generation biofuels** which are derived from food sources such as sugarcane, corn starch and other plants. In these, the sugars present in the biomass undergo **fermentation.** They are then are distilled to produce bioethanol, an alcohol which can be used directly mixed with petroleum in automobiles or used as a whole fuel in combustion engines or used directly in a **Direct Ethanol Fuel Cell (DEFC)** to produce electricity. Many countries which have a large sugar cane or sugar beet industry such Brazil, the United States, the European Community, China and Australia, use some ethanol in petroleum. However, the efficiency of ethanol as a fuel is only about 66% that of normal petroleum and there is still the problem of carbon dioxide air pollution, the competition for the source as a food product; and the expense and use of toxic metals needed in fuel cell construction.

- **Second-generation biofuels** use non-food-based biomass sources such as agriculture and municipal wastes. These biofuels mostly consist of **lignocellulosic biomass**, which is mostly non-edible, low-value plant waste consisting of lignin and cellulose fibres. Landfill waste dumps generate methane gas as a waste product of vegetable decay. Some waste dumps have been sealed so that the methane can be extracted and used directly as a fuel locally. It burns more cleanly than coal or oil, producing less carbon dioxide for the same amount of energy. Despite being a favoured alternative to petroleum, production of second-generation biofuel has not yet achieved the economies needed to fully replace petroleum and in some countries in recent years, production has actually dropped because of the competition with food and the availability of cheaper petroleum.

9.7 Geothermal Energy

This is generated from the heat of the Earth in some volcanic regions such as the United States, the Philippines, Indonesia, New Zealand and Iceland, where subterranean heat sources have been used to heat water, warm greenhouses and in-home heating. More recently, such countries have used the high temperature steam from their geothermal springs to drive turbines for electrical power generation. Although the initial setup is expensive, and in some areas dangerous because of the proximity of volcanic steam near the surface, geothermal power is considered renewable and cost effective. Geothermal wells also release greenhouse gases trapped deep within the earth, such as carbon dioxide, sulfur dioxide, hydrogen sulfide, methane and ammonia. These can react with precipitation to form acid rain in the local district and hydrogen sulfide gas and sulfur dioxide are toxic and also have a very unpleasant smell.

Figure 9.17: Pipes carry the high-pressure steam to a conventional power station at the Wairakei Geothermal Power Station, north island, New Zealand.

On a smaller scale, there has been some limited success in non-volcanic regions which have deep, fractured crystalline rocks such as granite. These **Enhanced Geothermal Systems (EGS)** use high-pressure water which is injected and pumped deep within crystalline rocks. At a great depth of up to five kilometres, these rocks are very hot because of the natural geothermal gradient of about 25°C/kilometre of depth. In some places, natural deposits of uranium ores in the crystalline rock also provide additional heat. The water which is injected into these wells, percolates through the fractured rock, is rapidly heated and then pumped to the surface. Here it passes through a heat exchanger which transfers this heat into a secondary system of pipes to heat fresh water and turn it to steam which can then generate electricity using a turbine coupled with a generator. Such systems can be used for generating electricity for local use, but they can be expensive and the water coming up from such depth is highly charged with mineral salts and can sometimes also be radioactive.

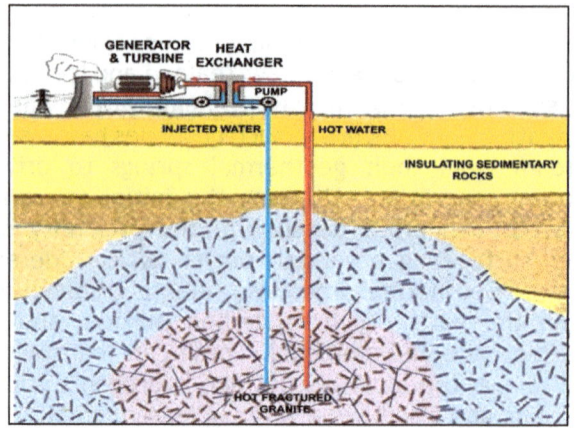

Figure 9.18: Diagram showing an Enhanced Geothermal System

Another geothermal system which uses low temperatures (say 50^0C to 350^oC) typical of non-volcanic heat sources makes use of generators using an **Organic Rankine Cycle (ORC)** often referred to as the Binary ORC as it uses two systems of heat exchange. In simple terms, a fluid, usually water, is passed down pipes drilled in the geothermal reservoir which heats the water. On the surface, this heated water is passed through a heat exchanger where it heats and vaporises an organic fluid called the working fluid - often isobutane or pentafluoropropane - within a second system of pipes. The expanded vaporised fluid is passed through a turbine attached to a generator which produces electricity. The fluid which passes out of the turbine is condensed as its temperature is dropped by another heat exchanger and cooling tower and then recycled back to the original heat source. Such a system can be used with any relatively low heat source such as may be produced by decomposing biomass waste dumps and solar collectors.

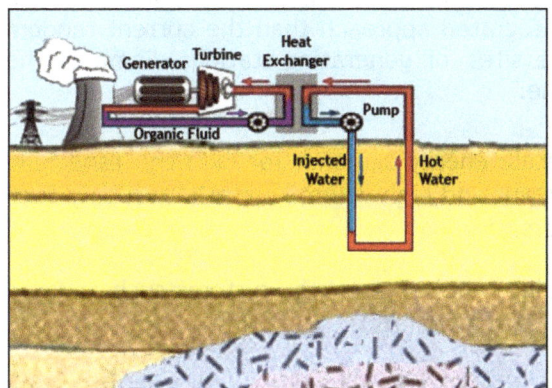

Figure 9.19 Simplified diagram of a Binary Organic Rankine Cycle power system.

9.8 Advantages and Disadvantages of Using Renewable Energies

There has been a great upsurge in the development and use of various renewable energies over the last 20 years, yet most of the world still relies on the traditional use of fossil fuel and nuclear energy as the main source of power. It is a matter of supply and demand. Renewable energies, except for hydroelectricity, have yet to be able to replace traditional power sources because renewables alone cannot meet the world's energy demands so far. The wholescale use of renewable energies is still in its infancy and several major problems are yet to be overcome, however, the use of integrated renewable energy systems offers great hope for the future.

In general, renewable energies are:

- able to provide energy without the fear of a diminishing supply

- do not create as much greenhouse gas emissions and are friendlier to the environment

- are easier to maintain requiring a smaller workforce

- provide independence from energy sources from other countries

- are cost-effective in the long term

However, there are some disadvantages of renewable energies which are yet to be overcome. Most of the problems could probably be overcome with a more united and integrated approach than the current random establishment of renewable sites or generating stations. Some of the disadvantages seem to be the:

- Inability to meet large-scale energy demands for industry, especially those which need a constant and high-volume electricity supply such as the aluminium smelters.

- High initial cost of building renewable energy systems such as large hydroelectric dams, solar and wind farms.

- Problems some have with the environment due to their construction sites and the use of construction materials which require large amounts of energy and use of non-renewable resources e.g. solar panels are based upon massive use of aluminium and silicon each of which must be extracted from the ground, refined and purified and then smelted using electricity and toxic chemicals which must be removed as waste. Solar panels may also contain a number of toxic materials such as cadmium telluride, copper indium selenide, and sulfur hexafluoride. Other environmental impacts include the destruction of wildlife and human habitats when hydroelectricity dams flood large areas, windfarms and solar farms take over large areas of useful land and tidal power stations affect marine habitats.

- Unreliability of some renewables as a constant power source as they are intermittent because they rely upon daily changes such as solar cells working only during strong sunlight, wind turbines operating only with strong winds. Other energies which are location specific such as hydroelectricity on river systems which have good flow rates and can be dammed without having massive evaporation of seepage loss, and geothermal power in regions where subterranean heat is economical to use and tidal power stations where there are great differences in tide levels.

- Lack of storage for those energies which only produce at certain times. To date, many storage systems rely on the use of batteries which rely on non- renewable resources for their manufacture, are very expensive, have limited life and provide some toxic waste products.

- Production of waste gases such as with the use of methane as a biomass fuel, which still produces waste gases such as the oxides of nitrogen in their production and carbon dioxide when burned to make steam for energy.

For example, the average energy consumption for an urban home in the United States was about 11,000 kilowatt hours per year (US Energy Information Administration data) and in an Australian home this would be about the same. A typical home using a large solar panel array generating 6.6 kW could generate about 10,000 kilowatt hours per year and with some additional power saving could meet their needs but there would have to be some change in social and domestic ways of life and there would need to be a lithium-ion battery storage system. A domestic battery system is expensive to set up with present costs at about $12,000 to install and the unit would last about 10 years. Lithium-ion batteries are also used in electric cars and they are beginning to be used in industry to maintain electrical power when grid power supplies are low. For example, the Hornsdale Power Reserve facility in South Australia which is adjacent to the Hornsdale wind farm, was promoted as the largest lithium-ion battery system in the world in 2017. It has a 100 MW output, 30 MW of which can operate for an extended period of three hours when the national grid load drops. However, lithium-ion batteries are expensive and use non-renewable materials, especially lithium, nickel, magnesium and cobalt metals which are also limited in their supply. Whilst solar electric systems have low greenhouse gas emissions in their manufacture, there may be some problems with their disposal unless some of the toxic metals used battery construction are recycled.

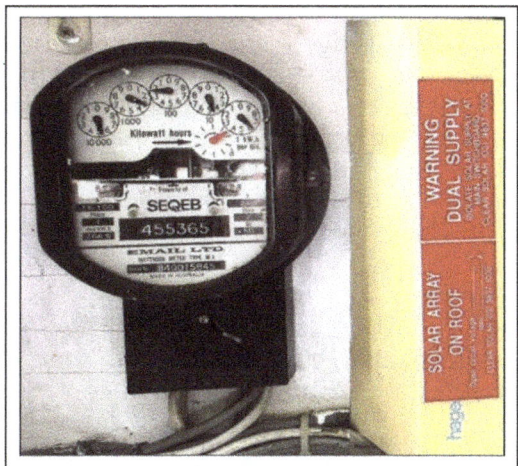

Figure 9.20: An energy meter in a home power box. The meter is read from the left and shows 53,246 kWh used to date.

The table below shows some of the power ratings of some home appliances. This rating is usually found on the back or under surface of the device on its compliance plate:

Name of Appliance	Power Rating (in watts W)
Air Conditioner (room)	1400
Air Conditioner (system)	2000 - 5000
Oven	1200
Hotplate (each)	1200
Toaster	2400
Microwave	600 - 1500
Water Heater (Electric)	4500
Clothes Dryer	3000
Space Heater, Radiator	1000 - 2000 (1000/bar)
Refrigerator (1 door)	1060
Refrigerator (2 doors)	1671
Freezer (upright)	1240
LED light bulb	5-15
Fluorescent Tube	40
Ceiling Fan	75
Desktop Computer/Monitor	190
Printer	10-40
Scanner	12
Television (LCD/Plasma)	270/310
Electric Jug/Radio/fan	1200 each
Coffee Machine	1000

Table 23.2: Power ratings for some domestic appliances

Note that for <u>power</u>, the rate at which energy is used per second, 1000 watts (W) equals 1 kilowatt (kW). Also, that 1 **kilowatt hour (kWh)** is the amount of energy used and is equal to 1 kilowatt being used for one hour. The amount of energy that is being used for a time, for a day or for a year can be calculated by noting the appliances being used and the time of their use and then adding the values. For example, on a hot day, a household cooking the midday meal may have the following appliances operating:

- A room air conditioner at 1400 W
- two hotplates on the stove at 1200 W each
- a refrigerator at 1060 W
- freezer at 1240.

This gives a total of 6100 W or 6.1 kW which, if used for one hour would be 6.1 kWh of energy. A standard 1.5 kW solar panel system a 1.5kW solar power system can make between 5.5kWh and 6.5kWh of electricity each day. This amount of electricity can supply 33% of an average home with normal use, however for a highly efficient home this system could provide 100% of the electricity required for daytime use. At night, battery power would have to be used. Alternative, one can find out how much power has been used during a certain time by comparing successive readings on the home's power meter (CARE! - Electricity).

9.9 Some Possible Home Applications

Some of these ideas may or may not be practical depending upon the domestic situation, but they may provide some stimulus for alleviating problems associated with energy. Many of them have been tried and tested using simple components:

- A simple hot water system can be made out of some black plastic hose purchased from agriculture supplies of even a long garden hose painted black. This is made into a tight coil and fitted into a shallow box lined with metal such as thick aluminium insulation foil and the interior painted a matt black. The box could be about 1 to 2 metres square and about 10 centimetres deep and made out of wood or metal sheet if one is handy with that medium. The box is then sealed with a glass or plastic sheet screwed or glued onto the top of the box. The ends of the tube should protrude through the sides of the box. The tubes coming out of this collector are attached to a large drum such as a heavy-duty plastic drum or a cleaned 100L metal drum as shown in the diagram below. This should also be painted black. Both the collector and storage drum should be securely placed up on a strong roof with the collector facing the sun with the drum higher than the collector.

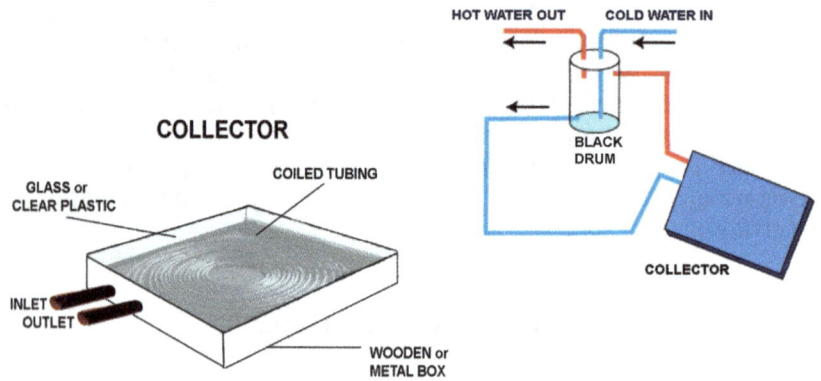

Figure 9.21 A home-made solar water heater

A unit based on this design was constructed back in the 1980's by a student of the author in an arid part of the country for a hot water shower in a remote area of a large cattle station. It worked too well and the water was often too hot, so an additional pipe with cold water had to be added to the shower outlet to mix with the hot water from the hot water outlet. The intake of cold water could come from mains pressure or hand-pumped from a nearby water tank, stream, pond or dam.

See also:

https://rimstar.org/renewnrg/diy_pex_solar_hot_water_thermal_dn.htm
(Homemade solar water heater)

- A similar system but without the storage drum and glass-topped box could be made on a larger scale and attached to the tiles or metal cladding of a roof to heat up a small pond or pool. The black piping can be laid out in parallel strips as in commercial pool heaters (see Fig. 9.12) using lengths of pipe and double right-angular bends also available at agriculture or hardware suppliers or a large coil could be made fixed to a supporting cross-shaped frame. A pump may have to be attached to channel the water up to the roof.

- Variations on this and the solar water heater may be useful at cooler times to warm glasshouses through secondary pipes or coils below the garden bed or even attached to a water bed for sleeping during cold nights.

- A solar oven for simple cooking, raising of dough in bread and the like could also be constructed. There are several variations, either using an insulated box internally lined with shiny foil and covered with glass sheeting or a parabolic reflector made from covering a curved shape with metal foil with the food container at the focus. For several designs see:

  http://solarcooking.org/plans/

- A small solar-powered battery-charging device for use on car batteries and charging small electrical appliance could be purchased but ensure that it is suitable for such charging. These are usually obtained from electrical stores or auto supplies. Also ensure that the solar panel and NOT the appliance being charged is in the sun.

  See also:

  https://www.instructables.com/id/Micro-solar-power-system/
  (Step-by-step micro solar electricity system)

- A biogas plant can be constructed using two plastic drums. This is probably a little beyond the average urban home handyman but useful for those living in rural districts and who have the skills associated with light engineering. The author has had no experience in this area other than being briefed as a young Army Officer about an Army Engineering civilian aid program in villages in Papua New Guinea in which clean 100 litre drums were used to make a gasometer such that methane gas produced from decomposition of waste pig manure and vegetable scraps would accumulate in an airtight, inverted container which would compress the gas by its own weight. The methane gas was then passed out through a one-way valve to prevent any flame from blowing back into the storage tank. Waste was fed into the lower part of the dual tank system below the level of the gas. The methane gas was then burned as a cooking fuel or run into a small electrical generator to make electricity for light and power. Residue waste material could be also flushed out and used as fertilizer. However, one should keep in mind that leaking methane gas is poisonous and explosive so any space between the tanks should be sealed, perhaps with a good layer of heavy oil.

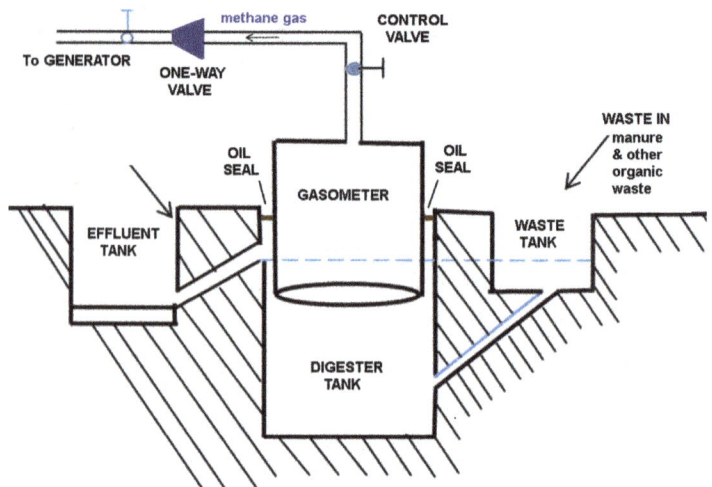

Figure 9.22 A home-made biogas generator

A home-made system is described in the following site but one must be warned about the explosive nature of methane gas and the need for good seals. See also:

https://www.instructables.com/id/Constructing-a-Medium-Sized-Biogas-Plant-Using-Kit/
(A homemade biogas system)

- A small-scale wind generator could be constructed from a simple fan or propeller and used car parts of an old alternator and regulator system connected up to car batteries to operate a 12volt system for small appliances. Larger scale wind turbines and commercial wind turbines are available but larger turbines may be noisy and many urban governments have regulations against such systems. Again, wind turbines may be more useful for rural homes and boats.

See also:

https://www.motherEarthnews.com/renewable-energy/wind-energy/diy-wind-turbine-zm0z17amz

and

https://openei.org/wiki/Small_Wind_Guidebook/What_Size_Wind_Turbine_Do_I_Need

- For land owners with a permanent water supply, a small-scale hydroelectric plant may be an additional option for powering some appliances. Remember that the water must have a continual flow and be owned by the landowner. There are several commercial systems available and many home-made sites which may or may not be suitable.

See also:

https://www.motherEarthnews.com/renewable-energy/homestead-hydropower-zmaz05fmzsel

and

https://www.smarterhomes.org.nz/smart-guides/power-lighting-and-energy-saving/hydro-power/

9.10 Conclusion

Currently the global community is in a transition phase position between the old world of excessive energy use at any cost and a new world of cleaner, safer energy and a future which is less uncertain. In the past the main goal of production was linked to the need for increased material wealth and profits. Today the world faces considerable uncertainty but with some hope. To date, the situation appears to be:

- Global warming has come about because of excessive greenhouse gases, mainly carbon dioxide with methane, in the atmosphere, largely helped by the activities of humankind, especially since the 18th century at the beginning of the industrial revolution.

- Global warming effects such as increased atmospheric and oceanic temperatures are causing many changes in the world such as sea level rise and the retreat of many glaciers, melting ice at the poles and record heat waves on land. All of these and other changes, coupled with humankind's other changes to the environment such as deforestation, industrial and mining use of valuable farming land, excessive fishing of the oceans, and poor waste management are causing social, economic and political upheavals.

- Industrial nations rely on large-scale mining, processing and manufacturing of goods for their economies. These activities require a large and continual exploitation of natural resources and the use of energy. Individuals who enjoy the benefits of these economies can practice less wasteful use of energy and recycling of resources and attempt to bring about social change. However, it is the large scale industrial and commercial applications at the national level which cause most of the global change.

- The world's industrial nations still depend upon the large-scale use of fossil fuels such as coal, oil, and gas and/or nuclear energy for their needs. Renewable energy sources are still at the lower end of total energy use.

- Most nations of the planet are now aware of the effects that humankind has had on the environment, especially the increase in global warming, and are aware of many of the issues of concern. It is hoped that, at the political level, these concerns will encourage reduction of pollution by fossil fuels and more research and development into the use of renewable materials and fuels.

In this transition period, more people need to be made aware of the actions of individuals and governments in causing these problems, and also being positive exponents of necessary change. Phasing out of all fossil fuels and industrial processes which put excessive greenhouse gases into the air will only come about in time. The pollution of the sea and land by mining, industry and agricultural chemicals will also require a great change in lifestyle and practice in the industrialised countries of the world.

Perhaps one scenario could be an intermediate period which may include the gradual reduction in emissions from the combustion of fossil fuels by the use of clean coal technology, natural gas and ethanol fuels, with further use of nuclear energy, possibly with the safe development of breeder reactors. This would gradually be phased out and replaced with large-scale use of renewable energies such as solar, wind and use of hydrogen as a fuel.

Whatever the future in the use of minerals, mined and processed materials and of energy resources, there will be a need for a corresponding shift in lifestyle, especially of those in an industrial world. Lifestyles may return to some of the simpler forms of transport, living with the environment rather than with artificial changes to it, and a better economy of resources.

Final Remarks

1. Energy supply has become one of the major demands in a world with increased population and industrialisation with most sources of energy still coming from fossil fuels and nuclear power plants.

2. There are other forms of energy which show promise for future large-scale use. These include some of the older energy sources such as water power and wind power as well as the new use of solar electrical cells and steam generators. Other alternatives to fossil fuels include the use of biofuels and geothermal energy in selective places.

3. Moving water has great potential in generating electricity by the use of turbines. Hydroelectricity has been used in many countries with abundant water supplies in high places. There is potential for further use of this source in new and innovative ways such as in Pumped Hydro Energy Storage (PHES) system. Other potentially-important uses of water power include the use of tidal and ocean wave generators.

4. Wind turbines are already a major contributor to the energy grids of many countries but some care needs to be used in siting such wind farms. They need to be in places of consistently high winds without interference to agriculture and shipping and yet be close enough to populated areas to be effective in their supply.

5. Solar electrical cells and solar-heated steam generators are also becoming a new feature in generating power. Again, as these are large-area systems, they need to be sited in areas where there is minimal disruption to agriculture. As they only generate power during the day, they need to be coupled with other ways of storing energy such as batteries or pumped hydroelectricity systems.

6. Biofuels derived from processed or waste organic matter also have some potential but they still rely upon their combustion in air and the production of carbon dioxide.

Final Remarks Continued

7. Geothermal sources have been used in some countries which have stable volcanic sources such as Iceland, New Zealand and America, but they have limitations in their use. Deep drilling into hot rock also has provided some local sources of energy but both sources of geothermal heat rely on the continuation of the heat source.

8. It will take time for the current traditional fossil fuels to be replaced as the renewable energies now being introduced will require considerably more development before they can match and exceed the amount of energy being produced by fossil fuels.

9. In the future, electrical energy will probably come from a number of integrated renewable energy sources rather than relying upon one main supply such as with the use of fossil fuels currently used in some countries.

More Applications

1. Use less electricity at home and in the workplace by turning off unwanted power supplies, lights, stand-by units and air-conditioning when not required. A time-switch on some appliances to be used only at specific times may help.

2. Home owners can look into solar power plans and if feasible obtain solar voltaic and solar heating systems. Look for reliable systems with good endurance and calculate whether batteries are necessary or not. Also check if there are any government rebates or subsidies for retired, handicapped, military or other groups.

3. Unit owners may look into a cooperative effort with others in the building to install solar units for group consumption.

4. Research the availability of relatively cheap electric vehicles including electric motor bikes or covered tricycles (auto ricksaw or tuk tuk) but check the local road licence rules. Alternatively use public transport if available and using electric vehicles.

5. Look into the expense of a stand-alone solar panel/food freezer (see caravan supplies) in case of power failures which are a possibility where local electrical supply is at threat.

6. Reduce energy use by installing skylights in the roof and making sure that there is a good ventilation system throughout the home. Use open windows draped with fine gauze steeped in water instead of air conditioning.

7. Install ceiling fans inside for cooler air flow and air vents in eaves and roof cavities.

6. In hot climates, research the use of solar air conditioning See: https://www.youtube.com/watch?v=cz-kquRmvqk&feature=youtu.be&t=58s

7. Where possible use the old-fashioned alternatives such as use of natural light, hand-operated food preparation and transport.

Further References

Australian Government, 2013. Your Home
https://www.yourhome.gov.au/energy/renewable-energy
(a general concept website on renewable energies with some good links at the end).

Blakers, A., Stocks, M & Lu, B., 2019. Australia: the Renewable Energy Superstar. Australian National University, Canberra, Australia.
http://re100.eng.anu.edu.au/publications/assets/100renewables.pdf
(Good website with some world comparisons)

Commonwealth of Australia, 2014. Australian Energy Resource Assessment – Second Edition. Canberra: ISBN 978-1-925124-22-4 (web).
https://arena.gov.au/assets/2018/08/australian-energy-resource-assessment.pdf
(An excellent review of energy in Australia)

Durant, Charlie, 2019. 10 biggest upcoming renewable energy projects in Australia. Engineeringpro.
https://www.fircroft.com/blogs/10-biggest-upcoming-renewable-energy-projects-in-australia-91233142448
(Good website showing some of the projects now underway in Australia with some good links)

Edenhofer, O. et al., (Edits), 2012. Renewable Energy Sources and Climate Change MitigationSpecial Report of the Intergovernmental Panel on Climate Change. New York: Cambridge University Press. ISBN 978-1-107-60710-1.
http://www.ipcc-wg3.de/report/IPCC_SRREN_Full_Report.pdf
(Excellent report from the UN about renewable energies)

Gorjian, S., 2017. An Introduction to the Renewable Energy Resources. Tarbiat Modares University (TMU)
Tehran, Iran.
https://www.researchgate.net/publication/317561674_An_Introduction_to_the_Renewable_Energy_Resources
(A presentation about renewable energies available as a pdf file)

IRENA (International Renewable Energy Agency), 2018. Global Energy Transformation: A Roadmap to 2050. ISBN 978-92-9260-059-4.
https://www.irena.org/-/media/Files/IRENA/Agency/Publication/2018/Apr/IRENA_Report_GET_2018.pdf
(Excellent coverage of the use and potential of renewable energies)

REN21, 2019. Renewables 2019: Global Status Report.
https://www.ren21.net/wp-content/uploads/2019/05/gsr_2019_full_report_en.pdf
(REN21 is a global renewable energy community of scientist, governments and industry. This is a very comprehensive document about the current situation of renewable energy in the world)

Wim Turkenburg (Lead Author), 2012. Global Energy Assessment -Toward a Sustainable Future. Cambridge, UK: Cambridge University Press. Chapter 11, Renewable Energy.
https://iiasa.ac.at/web/home/research/Flagship-Projects/Global-Energy-Assessment/GEA_-_Some_remarks_on_GEA_Wim_Turkenburg_Launch_GEA_Den_Haag.pdf

https://www.iiasa.ac.at/web/home/research/Flagship-Projects/Global-Energy-Assessment/GEA_Chapter11_renewables_hires.pdf
(Excellent pdf chapter from an extensive book).

Chapter 10: Renewable Resources

10.1 Introduction

Sufficient energy to use for comfort, transportation and for manufacturing needed materials is going to be more in demand in the future. However, even more basic to the future of humankind is the need for basic resources. The basic needs which have always been of major concern are water, food and shelter. With a rapidly growing world population, the availability and competition for these resources will become even more of a challenge for future generations. Currently, many of our valuable resources, especially in the construction and energy industries, are non-renewable and there will be a need to conserve or replace these in the future with renewable resources. Water and food supplies will also be at risk of being inadequate in the future if populations continue to soar. Better practices in the use of water and food will need to be developed.

10.2 What are Renewable Resources?

A renewable resource is any useful material, living organism or energy source which can be replaced or replenished in the same or shorter amount of time as it takes to be used. Most of these can come from the natural environment as **biotic** resources such as plants and animals or as **abiotic** resources such as renewable physical materials and energy. There are also renewable resources which can come from the recycling of man-made materials. Renewable resources can include:

- Fresh air is often taken for granted until it is polluted and steps are needed to obtain clean, fresh air.

- Water especially that used for drinking, agriculture and other domestic use.

- Animal life both from natural ecosystems and as domesticated animals used as sources of food and other material.

- Biomass such as plant life from natural ecosystems and from sustainable agriculture.

- Natural energies including water power, solar, wind, biofuels such as alcohol, and geothermal power.

- Recycled materials, especially metals, timber and petrochemical products.

- Human resources.

Figure 10.1: Mt Cook, South Island, New Zealand. A pristine example of air, water and land.

There are a number of important issues which come from the need for renewable resources. These include the:

- Realization that the world has come to depend upon non-renewable resources and that this level of dependence is on a vast scale which is going to be difficult to replace.

- Current use of renewable energy resources is very limited and whilst good progress is being made, it will take a long transition time before non-renewable energy resources are completely replaced. Renewable energies, for example, have not yet reached the capacity currently achieved by non-renewable energies powered by the burning of fossil fuels.

- Understanding that there will continue to be a need for a limited non-renewable resource industry to supply metals and other resources but that the depletion of these resources can be reduced with the help of recycling.

- Whilst there is always the threat of accidents in using atomic fission as a power source, these threats can be reduced by stringent safety procedures and the use of safe locations. Nuclear power may still be an option as a major energy supply until such time as safe nuclear fusion or renewable energies reach sufficient levels for future needs.

- There is a need to examine more closely the use and recycling of waste from urbanized societies. Waste should be seen as potentially a new resource. The recycling of metals, timber and petrochemical products such as some plastics, rubber products and others should become a major industry and a source of useful materials. Other wastes such a biological waste from agriculture and from sewerage also has the potential to be recycled or used in energy production.

- Use of closed-systems in future industries and power generation which will have zero gas emissions with no pollution of the atmosphere or hydrosphere.

- Realization that there will have to be massive changes in philosophy and practice in the way that modern societies think and live. Current urbanization due to mass migration from rural areas will need to be reduced in such a way that city life is more sustainable and that rural living can develop as a mutual way of life complementing that of the city by providing living space, employment and resources for an increasing population.

- The future of the development and use of renewable resources as well as the removal of non-sustainable pollution is a global problem and that industrial nations and developing nations each have responsibilities to come up to future global standards in maintaining a sustainable world.

- An acceptance that in urbanized societies there will be additional costs in the short term before renewable materials and energies become low-cost alternatives to current supplies.

It should be noted at this point that the terms renewable and sustainable are often confused. The former should refer to the nature of a resource, while the latter should refer to how it is used.

10.3 Ecosystems and Renewable Resources

The concept of turning to local and world-wide ecosystems as a source of renewable resources is hardly new. The use of such **ecosystem services** as a supply of food, water, materials and energy resources is more critical now in a world which been largely separated from these natural ecosystems by urbanisation and the artificial ways of thinking which comes from it. The availability of resources that have been developed through thousands of years of crop cultivation and animal husbandry has been taken for granted. In most urbanised societies, food, water and the other materials and energy needed for a comfortable living are obtained second hand from the local supermarket or urban utilities. Looking beyond these, one finds cultivated crops, farms of various sizes for livestock, mines and wells for fossil fuels and a few natural resources such as timber forests, stone and clay quarries and dams for water supplies. In pre-industrial revolution times when there were fewer people and a stronger connection between rural life, the use of the immediate local ecosystem was the normal practice. Moreover, as populations were much smaller, even in the cities, there was less stress placed on these ecosystems as a supply of natural resources. Today, people living in disconnected urban and artificial environments make up a large percentage of any nation's total population with United Nations statistics estimating that today, 55% of the world's population live in urban districts but that this will increase to 65% in 2050. With such a trend in urbanisation it would be appropriate to look further afield at the potential available in local ecologies. However, any such exploitation must be sustainable and within the ability of any ecosystem to cope with such a drain on its resources and still be renewed.

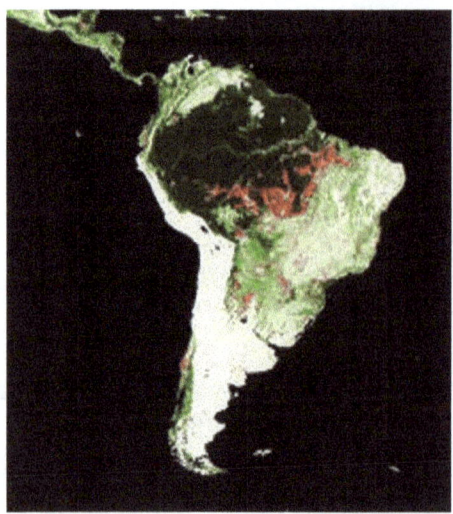

Figure 10.2: Satellite imagery showing deforestation in South America in red (Photo: NASA/USGS/UMD/SDSU)

Ecological systems are very fragile and many have been badly affected or destroyed in many parts of the world by the human activity with little regard to the husbandry and replacement of these systems. Whilst there have been changes in attitude over the last few generations about the need to reduce encroachment into natural ecosystems and to rehabilitate those which have been damaged, there is much to do before any real sustainable use of natural ecosystems becomes the best practice. Some of the current and potential resources from natural ecosystems are shown in the following table:

Ecology	Resources	Problems
Marine	Fish, salt, water (desalinated), algae, marine mammals e.g. whales & seals, minerals (seafloor e.g. manganese nodules), plankton, shellfish, prawns, shrimps & other crustaceans, offshore oil & gas, temperature moderation, carbon dioxide sequestration, tidal energy, medicines, recreational fishing, diving & swimming.	Depletion of fish stocks due to over fishing internationally. Destruction of coastal habitats due to port developments and polluted runoff of agricultural and industrial chemicals. Reduction in the ability of the oceans to absorb carbon dioxide. Competition and problems with off-shore oil and the effects of burning fossil fuels. Continued hunting of marine mammals.
Freshwater and wetlands (Rivers, lakes, ponds, marshes)	Drinking water, fish, crayfish & clams, recreational use, removal of wastes, birds, fertile sediment, carbon sequestration and transfer to seas.	Pollution of water ways by excessive agricultural and industrial wastes often with eutrophication of waterways by alga and other plants with the destruction of the habitat
Rainforest	Carbon dioxide sequestration, oxygen renewal, timbers, medicines, fruits such as bananas etc, nuts and beans e.g. coffee, cocoa, oils, gums, insecticides, spices. Recreational use.	Deforestation and clearing for crops and other uses such as urbanisation e.g. planting of nuts for palm-oil at the expense of the rainforest and many endangered species. Reduction also reduces oxygen regeneration, dries out the poor soil and climate.
Deserts	Sands, salt, other minerals such as lithium salts, borax, nitrates, phosphates, iodine, sodium carbonate, oil & gas, gold, metal ores of uranium, copper, lead, zinc, iron.	Excessive use of desert resources making many of them non-renewable. Salt and minerals from salt lakes, come from continual weathering and volcanic springs.

Grasslands and Open Woodlands	Grazing animals, cereal and other crops, honey, carbon sequestration, recreational use.	Unrestricted numbers of grazing animal species destroy the grasses and general ecology. Over hunting natural species to extinction.
Alpine	Water from ice & snow, pastures, grazing animals, recreational, birds, medicinal plants, mining of metals e.g. copper.	Global warming has reduced the number of alpine glaciers which are sources of drinking water and water for alpine meadows.
Other	Antarctica and the Arctic regions are cold deserts but rich in freshwater ice as well as many marine animals such as krill, fish and marine mammals, minerals, coal.	Treaties prevent the use of much of these lands as resources, especially in mining and hunting. Some Arctic places are still mined and over-fished and hunted.

Table 10.1: Some ecosystem resources both current and potential

It is important to note that the only the most common resources only are listed here and some such as whales and seals have been included only because they have been a resource and continue to be by some countries. Such activities as whaling and sealing and the use of other hunted species as resources should be prevented and restricted only to limited traditional hunting by indigenous peoples.

Figure 10.3: High pastures in Switzerland (Photo: Public Domain)

Figure 10.4: High desert near Uyuni, Bolivia with salt lakes rich in minerals such as lithium chloride (Photo: Matthew Scott)

Figure 10.5: Prairie grasslands in Wyoming, United States

Figure 10.6: Rainforest of the Daintree area, Queensland, Australia

Ecosystems also provide a range of services which help to maintain the health of the environment. Ecosystem services can be either:

- **Provisioning services** which are concerned with the production of food, such as game and seafood, fresh drinking water, soil, and some forms of energy such as tidal power and hydropower. Ecosystems also provide sources of timber, mineral salts for the building of shells of some organisms, natural fertilizers, medicines, herbs and spices, various animal products such as skins and also the provision of energy through hydroelectrical plants;

- **Regulating services** by controlling wastes, diseases, carbon capture or sequestration, promotion of reproduction by the transportation of pollen for flowering plants as well as spores and cysts of smaller organisms. Ecological regulating systems also assist in the purification of water by removing any sediment by the roots of aquatic plants which also remove bacteria and mineral salts. Other organisms such as some fungi can also remove toxic metal solutions from waterways. Some purification also occurs through filtration through by other sediments both on the surface during stream flow or through the ground by **infiltration** through porous soil and rock. Many micro-organisms also remove harmful bacteria and mineral salts by their ingestion or absorption. The oceans also play a major role in moderating the world's temperature and other climatic processes;

Figure 10.7: The moss *Funaria hygrometrica* can absorb an impressive amount of lead from waterways (Photo:USGS)

- **Supporting services** allow other ecosystem services to operate through nutrient recycling such as in the major environmental cycles such as the **Carbon Cycle** and the **Water Cycle** as well as the recycling of phosphorus and nitrogen. They are also vital in regenerating oxygen supplies through photosynthesis in terrestrial and aquatic green plants.

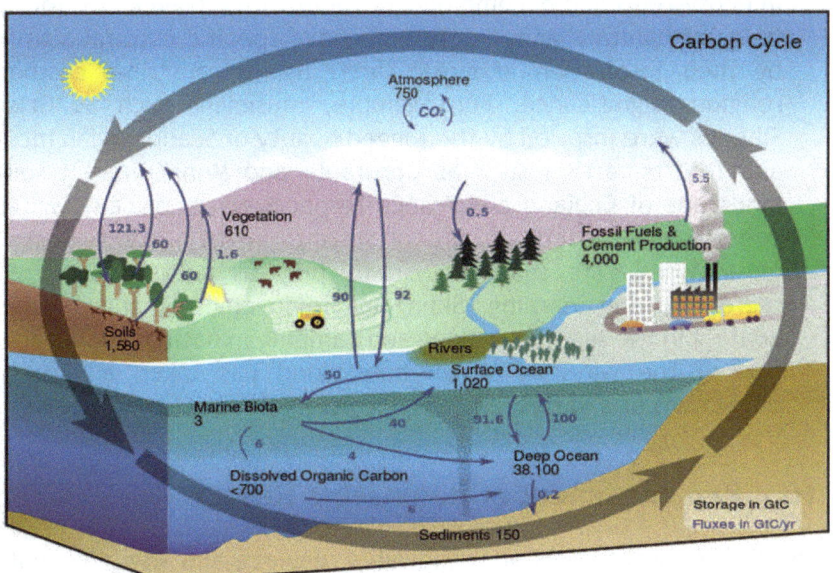

Figure 10.8: The Carbon Cycle showing how carbon is recycled

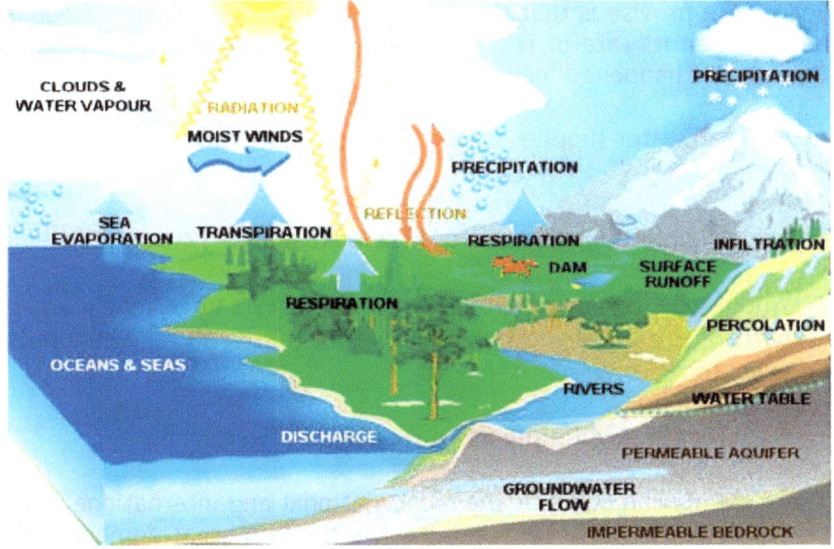

Figure 10.9: The Water Cycle showing how water is recycled in nature

- **Cultural services**, such as recreational, aesthetic, cognitive and spiritual benefits to humankind. Animals also find shelter and comfort in ecosystems and become part of them with a preference for specific types of habitat. Many ecosystems have specific cultural significance by their inspiration in music, arts, literature, tribal customs and religious significance. For example, musicians such as Grieg and Sibelius were inspired by the rugged beauty of Scandinavian mountains and forests, artists such as Constable and Monet were inspired by landscape of England and France respectively, and many writers and poets used a variety of outdoor settings in their works. Perhaps the most useful role that ecosystems play as cultural services is in recreation and learning. Skiing in alpine environments, surfing and boating in the sea and walking and camping in national parks are just a few of the ways by which humankind has taken advantage and enjoyment of a variety of ecosystems.

10.4 Replenishment of Renewable Resources

The major advantage of any renewable resource and energy is that it can be used and then replaced in a reasonable time. Unlike the majority of resources and energies which are currently used, they do not take millions of years to form but only short periods of time such as days, months or years. The proviso is that the rate of using the renewable resource must be less than its rate of replenishment. There are many examples where this has not happened including:

- Artesian water supplies drying up because too many wells have been sunk.

- Grain supplies depleted before the next season's crop.

- Wholescale logging of native hardwood forests in Tasmania and other parts of the world without concern that these will take hundreds of years to regrow.

- Fish stocks being totally exhausted due to overfishing.

These events can occur at the local, national and international level. For example, in the early 19th century, there was a great demand for whale products in Europe, America and Asia. These were used for such necessities as sperm whale oil for lamps, candles and cosmetics, whale oil lubricants and cooking oils, whale bone for fashion items and fertilizer and meat for human and animal consumption. Initially whales were taken locally from the North Sea, from the Atlantic coast off New England in America and off the coast of the Coral Sea in Australia. The invention of the small, steam-powered whaling hunter ship with its canon firing exploding harpoons made hunting for whales easier and by the 1860's the whale population had been so decimated by hunting that whaling nations had to go further afield. Whaling ships became bigger and able to stay at sea for months and ventured as far south as Antarctica and north into the Arctic Ocean. Soon whaling had become an international operation and there were few places where the whale population was safe from predation. To allow for the return of whale species to their optimum numbers, the International Whaling Commission (IWC) was established in 1946 to monitor and regulate the whaling industry. With numbers still low and some countries who were not signatories to the Commission's charter still operating, an international moratorium was set up in 1986. Whilst this has allowed for some regeneration in whale numbers over its many species, some limited whaling still continues today.

Other problems with the excessive use of natural renewable resources and difficulties in their replenishment occurs include:

- **Marine species** such as fish, squid, abalone, turtles and marine mammals such as seals, whales and dolphins are all at risk. As a case study, the Southern Bluefin Tuna has been the subject of wholesale international fishing with disastrous results. A popular fish for eating, it is considered to be an endangered species with estimated levels down to about 5% its optimum population. It has one of the longest migrations of any fish; ranging from the Indian Ocean to Australian waters. Whilst the Blue Fin fishing industry is managed by several nations, other countries are able to fish for the tuna as they come within reach of any coastline. The Commission for the Conservation of Southern Bluefin Tuna (CCSBT) allocates the amount of stock which can be taken in any season. In Australian waters most of the catch is taken as juveniles in the Great Australian Bight, fattened up in sea cages and then exported. Unfortunately, the emphasis on catching juveniles before they have had a chance to spawn means that re-building the depleted stock at sea of tuna becomes very difficult.

Figure 10.10: The Blue Fin Tuna, *Thunnus thynnus* (Photo:NOAA)

There has also been some controversy about some fishing companies using bottom trawling as a means to catch fish. In this method, large trawling vessels, sometimes working together, drag huge weighted nets along the bottom of the sea to catch any fish species available, especially tuna. This method is indiscriminate about the species of fish and bottom-dwelling marine life such as many mollusc species, which are taken. It also damages the sea-floor ecology. A more responsible way of catching fish like tuna is to use rods and multi-hooked lines which target the tuna only, this limits the number of fish caught and restricts the catch to only adult; juveniles being thrown back.

See also:

http://www.environment.gov.au/marine/publications/protected-marine-species-identification-guide
(More details about endangered marine species)

http://www.afma.gov.au/managing-our-fisheries/fisheries-a-to-z-index/southern-bluefin-tuna/at-a-glance/
(Southern Bluefin Tuna)

https://www.futurepolicy.org/biodiversity-and-soil/australias-great-barrier-reef-marine-park-environment-protection-and-biodiversity-conservation-acts/
(Great Barrier Reef protection acts)

- **Forests** and their timbers are an excellent renewable resource if handled properly. Unfortunately, much of the world's hardwoods and some of the other species also have been harvested with little concern for the future.

Figure 10.11: Foresters estimating a thinning program (Photo: US Dept. Agriculture)

There are natural causes of **deforestation**, climate change and the reduction of rainfall being the main natural cause, but it is indiscriminate logging of native forests which have done the main damage. Natural forests are logged so that the land can be used for agriculture or stock grazing, as part of urban expansion and for the valuable timbers. It has been estimated that natural forests, provide habitats for up to 80% of the terrestrial biodiversity as well as preventing soil erosion, desertification, and land degradation. Forests also assist in the maintenance of global climate stability by absorbing about 1.5 billion tonnes of carbon emissions per year. Unfortunately, about 5.2 million hectares of forest are lost through natural causes and land clearance every year. As an idea of the immensity of this problem, one hectare equals one thousand square metres and is about the size of an average sports oval. Excessive deforestation has a number of negative effects on the environment including: permanent removal of local plant and animal species by destruction of habitats; soil degradation and erosion with removal of top soils and destabilization of slopes; great reduction in the evapotranspiration rates which put water back into the atmospheric phase of the water cycle; local climate change by reducing atmospheric water content but increasing temperatures by bare-land absorption of heat and which heats up the air above; and increased surface runoff of water with less ground infiltration. Once removed, forests are difficult to replace.

Silviculture is the science of forest management and many countries have practices which involve the replanting of native forests and sustainable logging for timber resources. Such programs will be vital for the future as timber will play a major role as a renewable resource.

With growth rates which may be around 20 to 50 years, native hardwood forests are difficult to regrow. In many countries timber plantations have used faster growing softwood and hardwood species such as pine and eucalypt. Because of their fast rate of growth of around two to three metres per year, and their attractive appearance, Australian eucalypts have been replanted in New Zealand, California and Hawaii, parts of Europe, India and in many South American countries to stabalize soil, restore deforested areas and as a timber resource.

Apart from the use of designated tree plantations and replanting in areas of deforestation, logging in established new growth forests is still possible with sustainable logging practices. These include:

- Selective logging in which only one species within the forest is targeted over time with only a small percentage of this species being logged to allow for natural reseeding and growth.

Figure 10.12: Precise machine logging to reduce surrounding damage (Photo: US Dept. Agriculture)

- Minimal felling practice to prevent damage to surrounding trees and smaller vegetation. This can be done using advanced logging machines which hold the tree whilst it is being cut.

- Precise management plan for entering and exiting the forest to minimize damage whilst timber is removed.

- Support replanting in areas where past logging has reduced tree volume to below optimum growth values.

Bamboo is a type of fast-growing grass which is exceptionally strong, light in weight and hollow. It grows abundantly in most of the tropical and subtropical zones and it has been used in many Asian countries as a building material, as scaffolding on construction sites, as a source of fibre for cloth and twines, as piping for water and as a food source. It has also been developed as a valuable and superior alternate for wood composites, pulp and paper, stripboards, matboards, veneer, plywood, particleboard and fibreboard.

Figure 10.13: Giant bamboo grass has an incredible number of uses as an alternative to timber

See also:

https://www.google.com/search?q=forestry+and+depletion+of+trees&ie=utf-8&oe=utf-8&client=firefox-b#
(Good PowerPoint about deforestation)

https://www.rainforestinfo.org.au/good_wood/oz_pln.htm
(Native timbers of Australia)

https://www.business.qld.gov.au/industries/farms-fishing-forestry/forests-wood/native-forests/case-studies
(Good link Forest management in Queensland, timbers and case studies)

http://www.timberqueensland.com.au/Docs/Growing-Processing/Review-of-Qld-Hardwood-Plantation-Program-Summary-Report.pdf
(Hardwood plantations in Queensland)

https://www.rainforestinfo.org.au/good_wood/contents.htm
(Timber resources, forests and legislation in New South Wales)

http://www.agriculture.gov.au/forestry/policies/2020vision
(Australia's logging industry and legislation)

https://www.rainforestinfo.org.au/good_wood/os_ozsp.htm
(Australian eucalypts grown overseas)

https://www.sttas.com.au/forest-operations-management/understanding-our-forests/eucalypt-plantations
(Sustainable eucalypt management)

https://twosidesna.org/wp-content/uploads/sites/16/2018/06/Paper-production-supports-sustainable-forest-management2.pdf
(Use of paper in sustainable logging with good US and Canadian links)

https://www.fs.fed.us/forestmanagement/aboutus/lawsandregs.shtml
(Legislation in the United States concerning forestry)

http://www.fao.org/forestry/42658-0b8ddd1c5c20b4980467f2f4724f445a7.pdf
(Myths and reality of fast-timber plantations)

https://www.researchgate.net/profile/Pannipa_Chaowana/publication/271061918_Bamboo_An_Alternative_Raw_Material_for_Wood_and_Wood-Based_Composites/links/5636de6808ae88cf81bd14f8/Bamboo-An-Alternative-Raw-Material-for-Wood-and-Wood-Based-Composites.pdf?origin=publication_detail
(Use of bamboo as a timber alternative)

- **Geothermal resources** have been used for centuries in many countries where there are hot springs or the country allows for tapping into the heat in the rocks below. The Romans, Chinese, Native North and South Americans, and the New Zealand Maori have used hot mineral springs for bathing, cooking, and heating. Water from hot springs is still used world-wide in spas, for heating buildings, and for agricultural and industrial uses. Many people believe that hot mineral springs have natural healing powers. In countries such as New Zealand and Iceland which are volcanically active, geothermal resources are used in power stations to generate electricity. In other countries where hot granite and other rocks are heated by the radioactive minerals and they are within drilling depth, some limited geothermal wells have been used to generate small supplies of electricity by pumping water down to the heat source.

Figure 10.14: Hot spring used as a cooking pool by the Maori, the indigenous people of New Zealand at Rotorua, North Island.

Sustainability of geothermal resources depends upon the following factors:

- Temperature level of local subsurface environment, about the upper 10 km of the crust where temperature usually reaches values of some hundreds of degrees.

- Heat transfer processes in the upper crust which occurs by the processes of thermal conduction, convection in fluids, and advection of magma. Conduction is a slow but continuous process and convection occurs only if the geological setting involves fluids such as hot water and magma and is subject to change. This is perhaps the most important factor in the sustainability of current geothermal use where heat sources can be diminished by excessive use by additional injection of cold water in local heat sources. Advection is a relatively rapid process which occurs in volcanic areas and at other plate margins.

- Availability of reserves, and the time factor will determine how sustainable the geothermal resource may be and its potential use over time. Only a small fraction of the Earth's surface is available for geothermal application from deeper sources and lower depth extraction above 3 kilometres depth, may only have the potential for the short term e.g. about 30 years.

- Heat extraction technology depends upon the nature of the rock and the number of fractures within it. Most of the current geothermal resources are in naturally-fractured rock containing volatile fluids from heat sources close to the surface and use technology which is readily available. However, the technology exists which can be applied to man-made rock fracturing and injection of water to enhance the level of the geothermal development in particular areas with the proviso that usage is maintained at a sustainable level.

- Permeability of the rock will affect the carrier fluid, as steam, water, gas, or a mixture of these, and how the source is recharged. It would be important to maintain an appropriate balance between natural water recharge from nearby fluid intake areas on the surface and fluid produced by wells. If excessive fluids are extracted so that it exceeds the recharge rate, then like any subsurface fluid system, the geothermal source will become unsustainable.

- Injected and reinjected water in **Enhanced Geothermal Systems (EGS)** can be used to recharge the geothermal reservoir by acting as a carrier fluid. When this is used, an appropriate balance should also be maintained between inflow and outflow rates. As well, the thermal properties of the source rock should also be monitored so that the injected fluid does not lower and alter the original thermal characteristics of the reservoir.

- Nature of reservoir fluid also is a limitation of the sustainability of a geothermal resource since most geothermal fluids contain dissolved chemicals which can cause corrosion of the well casings and surface plants. In hot rock injection systems, some of the steam pumped out of the ground also contains some of the radioactivity of the granite heat source which increases the hazardous nature of the plant and reduces its sustainability over time.

The potential of the Earth's geothermal resources is enormous when compared to their current use and will be a well-needed addition to the future energy needs of humankind. There is considerable potential for use of geothermal resources worldwide in the near future as it has been envisaged that by 2050 geothermal electrical generation capacity may reach 70 gigawatts or about 5 times the present capacity.

Figure 10.15: Geothermal pipes and facility in Iceland (Photo: P. Honzatko)

See also:

https://orkustofnun.is/gogn/unu-gtp-sc/UNU-GTP-SC-22-11.pdf
(Sustainable management of geothermal resources – from Iceland)

http://citeseerx.ist.psu.edu/viewdoc/download?doi=10.1.1.455.2050&rep=rep1&type=pdf
(Good technical paper includes concepts of geothermal time scales and replenishment)

https://www.renewableenergyworld.com/geothermal-energy/tech.html
(Good general information about geothermal use from the United States)

10.5 Sustainability

The **sustainability** of any natural ecosystem can be defined as the dynamic equilibrium between natural inputs and outputs, modified by external events such as climatic change and natural disasters. When land or other natural resource is used by humans, one must consider the extent that the human activity would have in the interruption to the ecosystem's capacity to maintain its normal dynamic existence. If this definition of sustainability is applied to the use of land or other natural resources by humans it implies that there is no overall disturbance of the ecosystem over time. No agricultural system or resource extraction can be truly sustainable, with perhaps the exception of hunting and gathering at extremely low population densities.

The primary driving force for increasing land use is human population growth. In order to be sustainable, land use must display a dynamic response to the changing ecological and human population conditions. That is, in time and in the location of the human activity, the other features of the ecology and the human activity are maintained as a balance between the activity and local ecology. For example, sustainable harvesting refers to the use of agricultural land and the resources of rivers, lakes and seas in such a way as to ensure that over time no further net quantitative or qualitative loss of natural resources occurs.

10.6 Sustainable Harvesting

The progression of humankind from roaming small-group hunter-gatherers to sedentary, settled societies practicing agriculture and animal husbandry has been mentioned previously in this book. Having considered the nature of the environment which can both provide and support renewable resources it is worth looking at some of the aspects of harvesting of biota.

Harvesting is the process of gathering a ripened crop from prepared fields on a seasonal basis or it could also refer to harvesting of fish, prawns and other animal life, especially those kept in localised ponds or farms to ensure managed numbers and a better harvest. In many countries crop harvesting is done on a large scale using mechanised methods and usually for sale of the crop on the domestic market and for overseas export. In rural countries, much harvesting is done by hand often with the use of beasts of burden such as oxen and the crop is used for domestic use and some limited trading if there is an excess. Some peoples have been practicing sustainable harvesting for a very long time.

Many indigenous cultures such as the indigenous peoples of Australia have practiced limited harvesting of native crops, fish and other wildlife on a tribal basis for thousands of years. It is more than likely that Aboriginal and Torres Strait Islanders practiced some form of harvesting of cultivated native seed grasses, herbs and rotation of fish stocks in certain areas was also practiced. They also replanted yams and relocated other plant crops into areas more suitable for their growth and cultivation.

Figure 10.16: A fish farm cut out of the centre of a floating reed island belonging to the Los Uros, an indigenous people of Lake Titicaca, Peru.

It was certainly known from the early days of colonial settlement in Australia, that Aboriginal tribes practiced selective burning of the landscape in order that certain plants were regenerated and gave new fruit and also conservation of a selection of animals to be hunted by having taboo laws which restricted one tribe from hunting animals of their own tribal totem but allowed other tribes of different totem to hunt them within the same territory. Often these totem animals were painted as cave paintings to reinforce the identity of the tribe and to instruct the young people as to which animals they could not hunt.

Some useful articles about past and current indigenous foods and harvesting can be found at:

https://www.nintione.com.au/resource/NintiOneResearchReport_71_BushFoodGuidelines.pdf
(Aboriginal bush foods and ethical guidelines for their use)

https://www.firstpeopleslaw.com/database/files/library/A_Guide_to_Aboriginal_Harvesting_Rights.pdf
(A guide to Aboriginal Harvesting rights).

http://www.agriculture.gov.au/SiteCollectionDocuments/natural-resources/landcare/submissions/ilm-report.pdf
(Indigenous land management in Australia)

https://www.aph.gov.au/About_Parliament/Parliamentary_Departments/Parliamentary_Library/Publications_Archive/Background_Papers/bp9798/98Bp15
(Good links to indigenous land use and management in other countries including the US, New Zealand and Canada)

https://www.cdc.gov/diabetes/ndwp/pdf/Part_IV_Traditional_Foods_in_Native_America.pdf
(Traditional foods and practices of North American First Nations)

http://arnoldia.arboretum.harvard.edu/pdf/articles/1990-50-4-lost-crops-of-the-incas.pdf
(Lost crops of the Incas in South America)

https://www.nda.agric.za/docs/Brochures/Indigfoodcrps.pdf
(Indigenous food crops of South Africa)

https://www.iss.nl/sites/corporate/files/2-ICAS_CP_Demi.pdf
(A technical paper about indigenous food practices)

Figure 10.17: A modern harvesting of grain using a combine harvester and grain cart (Photo: US Dept. of Agriculture)

During the growing season for a particular crop, harvesting is usually very labour-intensive regardless of the methods used, and it occupies most of the farmers' time. Often there is a specific timeframe to be met such as harvesting before the crop becomes overripe, or the arrival of an expected climatic season such as the monsoon in Asia. Sometimes there are unexpected disruptions to the harvest: the onset of drought, a cycle of pest invasion, volcanic eruption, fire or flood which can destroy the crop. Good luck as well as good management is usually needed in farming regardless of the degree in sophistication of methods used. The most commonly produced crops are shown in the following table:

CROP/FOOD	TYPE	PRODUCTION (TONNES)	TOP COUNTRY of PRODUCTION
Sugarcane	Grass	1,800,377,642	Brazil
Maize	Cereal	885,289,935	United States
Rice	Cereal	740,961,445	China
Wheat	Cereal	701,395,334	China
Milk (cow)	Livestock	659,150,049	United States
Potatoes	Root Crop	373,158,351	China
Other Vegetables	Vegetables	268,833,780	China
Soybeans	Bean	262,037,569	United States
Cassava	Root crop	256,404,044	Nigeria
Tomatoes	Vegetable	159,347,031	China
Pig Meat	Livestock	118,168,709	China
Bananas	Fruit	107,142,187	India

Table 10.2 The major food crops of the world based on production (Data based on the Food and Agriculture Organisation FAO) of the United Nations.) See: https://web.archive.org/web/20110713020710/http://faostat.fao.org/site/339/default.aspx)

Diseases of crop plants have often been the cause of catastrophic events leading to mass famine, starvation and dislocations of whole communities. For example, the Great Famine of Ireland (Irish: *an Gorta Mór* or the Great Hunger) occurred between 1845 and 1849 when the **potato blight**, a virulent disease of that crop spread across Europe and into southern Ireland. It caused over one million deaths through starvation and disease in that country and led to mass emigration from Ireland to America and Australia. Crop failure in England also caused starvation and hardship in many villages and the government of the day encouraged emigration to the colonies such as Australia, Canada and New Zealand with assisted passages for whole families. Famines in rural communities often come about because of changed climatic conditions, natural disasters, pests and the reliance on a monoculture, that is, the cultivation of just one crop. When such events happen, there is not enough food to maintain the population at their normal level of subsistence and so starvation, death and migration are the result. Today, this is alleviated to some extent by better agricultural practices and the facilities of international aid agencies which are able to provide some temporary relief. However, in many parts of the world, natural and made-made disasters such as war, pollution and global warming has put considerable stress upon many agricultural communities with little hope in reclamation of the land.

In modern, industrialized countries there is still a tendency to grow and harvest crops as monocultures. This has led to some reduction in crop diversity and dependence on heavy use of artificial fertilizers and pesticides. This in turn has led to overexploited soils that are almost incapable of regeneration. By including a fallow year on fields every fourth-year growing legumes to re nitrogenize the ground has led to higher yields of sugar and banana monoculture areas in Queensland. In short, unsustainable farming of land degrades soil fertility and diminishes crop yield and with a steadily growing world population and local overpopulation, even slightly diminishing yields can be considered as a partial harvest failure. New fertilizers and farming practices such as minimal till agriculture and large-scale hydroponics are sometimes used to overcome the need for soil regeneration and international trade prevents local crop failures from developing into famines.

See also:

https://ourworldindata.org/famines
(Provides excellent data about famines and their causes).

https://www.ukessays.com/essays/history/causes-effects-and-solutions-of-famine-history-essay.php
(Another good site giving causes and some solutions)

https://www.nal.usda.gov/afsic/sustainable-agriculture-information-access-tools
(Good information and links about sustainable agriculture from the US Department of Agriculture)

10.7 Human Impact on the Environment - Revisited

Much as already been outlined in previous chapters of this book, but in conclusion it can be said that humans have had a long history of impacting the ecology of their places of habitation. It seems to be in the nature of humankind to modify ecology for its own needs. Sometimes this is for the general good and other times it is not. Humans rely on natural ecosystems to provide many ecosystem services such as provision of food, drinking water and shelter, pollination of crops, fertile soil, refreshing the air and cleaning water.

In the past, humankind tended to live with nature rather than against it, modifications to local ecologies were often beneficial by limited hunting, using biodegradable materials for shelter and clothing, in animal husbandry and replanting trees and crops. Such activities could be considered as closed loop systems because there was a considerable amount of internal recycling and interdependence within a particular ecology. Waste products produced by one species would be used by at least one other species and these waste products such as carbon dioxide, faeces and urine would eventually be converted into oxygen, food, and water. In such a system, there would have to be at least one autotrophic organism such as green algae or plant which would begin the food chain.

As civilizations developed however, humankind became more adept at using natural resources without much care for the effect that removal of these resources from the natural environment would have on the ecology nor the effects of pollutants which were put into the environment. After the Industrial Revolution, humankind became more concerned with gaining ever needed limited resources and modifying the environment for living space and resource acquisition. Waste products were easily dumped into a world which seemed to have unlimited capacity to absorb them. The sea and unused lands were seen as the ideal places for getting rid of unwanted wastes. Some of the negative ways that humankind has impacted on ecological systems include:

- Population increase has been as a matter of human survival but in many countries and the world in general, this has led to overpopulation where humankind is rapidly approaching the carrying capacity that our planet can sustain. This has come about because mortality rates have decreased, medicine has improved, industrial farming yields have increased and across the globe humans are living longer and increasing the total population. An increased population results in a greater demand for living space, energy and resources.

- Pollution of the atmosphere, hydrosphere and geosphere with the corresponding influence on the biosphere. This has led to climate change and global warming which in turn have other negative impacts on the world's ecology.

- Deforestation and the extinction of many species due to global warming, the spread of agriculture and urbanization and over harvesting of forest resources.

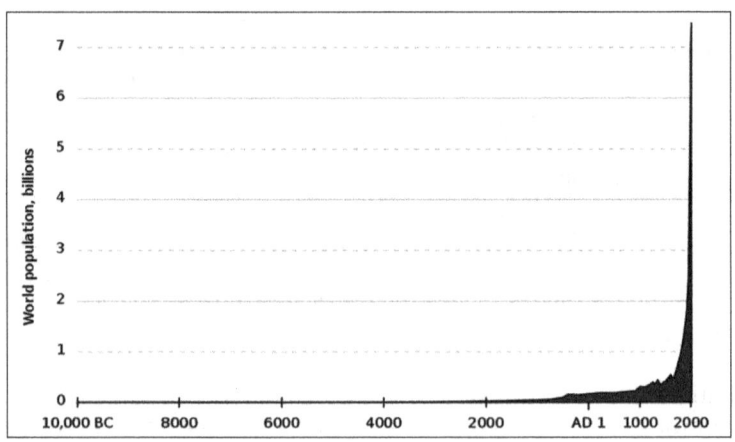

Figure 10.18: Graph showing global increase in population (Photo: Public Domain from data from the US Census Bureau)

- Depletion of fishing stocks and other wildlife populations through excessive harvesting often using large-scale and wasteful practices.

- Modifying the ecology by the introduction of invasive species. Sometimes this has been unintentional, such as the introduction of several varieties of rats, and insect pests through shipping, carp into the waterways of Australia and America and African bees into Brazil. Some species have also been imported intentionally as livestock, pets and as commercial crops of grains, fruits, vegetables and timber. Along with these useful species there have also been the introduction of species which have become pests such as rabbits, foxes, cane toads and the plants prickly pear and lantana to Australia. In the United States, introduced species include the European starling, ash borers and Burmese pythons.

- Mining activities which have become an international operation in most countries, especially in more recent times in so-called Third World nations which have in the most part been rural communities. Here, the environmental impact of many mining companies has been sudden and severe with both governments and their populations having little experience of the effects of modern technology and industrialization.

- Land degradation through over-cultivation and irrigation leading to salination and soil erosion. Excessive irrigation can cause the water table to rise which brings salt to the surface killing vegetation and the removal of grasses and trees greatly assists in the removal of topsoil and land in general through water runoff.

- An increase in local, national and global populations so that there is more demand for living space and resources with ever-increasing competition and tensions at all levels of civilization.

- Over-consumption of food, water and resources in many societies. Over consumption occurs when the demand for resources exceeds the ability of the local ecosystem to provide it. This also relates indirectly to manufactured resources as well with the subsequent drain on other natural resources and the production of waste.

Whilst the negative impact of humankind on the environment has been felt at the local, national and global level, there have been some important positive benefits of human activity since the advent of group societies. These include:

- Improved medical conditions and health care which provide a higher living standard and longer life in many countries.

- Reforestation and the establishment of national parks for flora and fauna. This assists in improving air quality, water purity and more recreational activities.

- Conserving water resources by the establishment of more rain-water harvesting, use of dams and artificial lakes for water storage and better irrigation methods such as drip irrigation.

- Soil conservation by planting more trees as windbreaks and to hold the topsoil, contour ploughing, terrace farming, crop rotation, windbreaks and minimal till agriculture.

- Use of renewable sources of energy such as hydroelectricity, solar energy, biofuels, wind energy and tidal energy which helps to conserve dwindling non-renewable energies.

- Eradication of invasive species and by restocking with native plants and animals from nurseries and animal reserves.

- Use of developed technology to solve many of the environmental problems and to limit further ecological disasters through the use of more efficient devices and systems.

It becomes now a race between the continued abuse of the environment by an ever-increasing world population, most of whom are not in control of global disruptions, and the development of new systems, ways of thinking and technologies to reduce the harmful consequences of humankind and get to a point where there is a balance between using resources and keeping the global environment in a healthy state.

10.8 Ecological Footprint

This is a measure of the health of an environment or of the global environment in general and concerns the amount of biologically productive area needed for all of the necessary resources that people use such as water, vegetables, fruits, fish and other animals, timber and natural fibres. It also involves factors from humankind's activities such as the amount of carbon dioxide from fossil fuels, space for buildings, roads and other infra-structures needed for living. Footprint and the capacity of the ecology to provide resources can be measured at the individual, regional, national or global scale. The ecological footprint of any specific location, society or individual will change every year according to the number of people, per person consumption, efficiency of production, and the ability of the ecosystem to provide resources. At a global scale, ecological footprint assessments show humankind's demand for what the planet can produce and its ability to renew used resources.

Ecological footprint can be measured for any activity or individual, local community, nation or globally by the measuring the number of hectares of biologically productive land and sea area that compete for biologically productive space. This includes producing renewable resources, accommodating urban infrastructure and roads, and breaking down or absorbing waste products, particularly carbon dioxide emissions from fossil fuel. The footprint then can be compared to how much land and sea area is available. Calculation methods can be standardized so results of various assessments can be compared.

In the last fifty years or so, attitudes have started to change and humankind is beginning to take a more positive approach to their impact on ecosystems. They have taken such actions as reducing industrialized wastes, beginning to limit the use of fossil fuels, creating national parks and more green spaces within cities, introducing environmental codes of practice and legislation and recycling materials and having a more concerned view of waste removal.

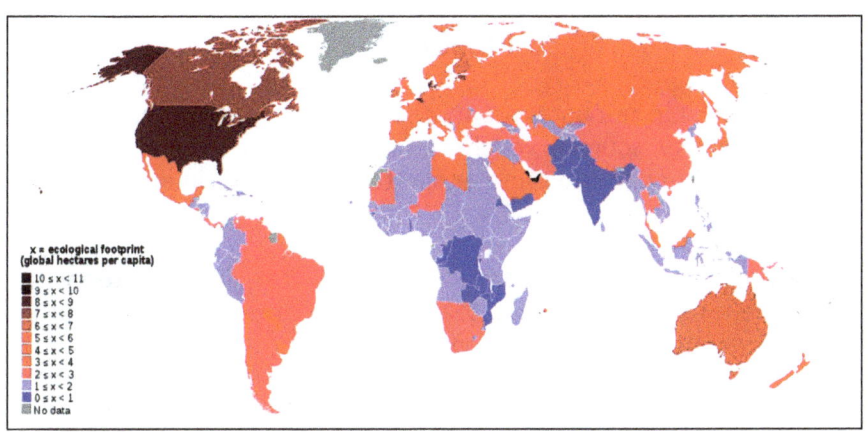

Figure 10.19: Map showing the ecological footprint of various countries in hectares per capita (public domain)

Some useful sites can be found at:

https://www.isa.org.usyd.edu.au/publications/documents/Ecological_Footprint_Issues_and_Trends.pdf
(Good summary of main points from the University of Sydney)

https://www3.epa.gov/airnow/workshop_teachers/calculating_carbon_footprint.pdf
(A DIY survey to calculate an individual's ecological footprint)

http://ecologicalfootprint.com/
(An interesting global Ecological Footprint calculator)

https://pdfs.semanticscholar.org/da4d/5edbab9332c60fe1b0d33f1f3caefcadeb01.pdf
(Another view about measuring ecological sustainability and some problems with the concept of Ecological Footprint)

https://atpsnet.org/wp-content/uploads/2017/05/rps22.pdf
(Indigenous water harvesting in dry areas)

https://cmsdata.iucn.org/downloads/indigenous_peoples_climate_change.pdf
(Indigenous cultures and climate change)

Final Remarks

1. Renewable resources are those useful materials, living organisms or energy sources which can be used and then replaced or replenished in the same or shorter amount of time that it takes to be used.

2. These include adequate amounts of water, food, materials for shelter, energy as well as general conservation of natural ecosystems and human populations.

3. In a world of rapid population growth, the need for ecosystem services as basic resources is becoming more critical and so it is necessary now to be able to develop renewable resources to fulfil population needs in the future.

4. Sustainability, or the use of renewable resources whilst maintaining constant levels, concerns all resources from land and sea. These include crops, animal resources and other useful natural materials.

5. Many of the past practices and uses of resources by humans were often developed in an age of plenty, but with far fewer people and simpler domestic lifestyle. Current industrial nations have long ago outstripped the availability of plentiful resources and so there is a need to look at the ecological footprint or impact of such actions.

More Applications

1. Whenever possible, conserve food, water, energy and other materials by reducing or using waste materials e.g. eat modestly and switch off energy appliances when not in use.

2. Use natural energies such as sunlight e.g. for drying clothing, and natural foods rather than processed foods.

3. Purchase food or other products which have been sourced naturally and which uses sustainable methods e.g. certain types of seafood.

4. Participate in conservation activities such as cleaning up waterways and seashores, protection of local wildlife, creating new and natural habitats and the planting of more trees.

5. Reduce the amount of household waste and recycle as much as possible, especially plastics, glass and paper products. Use organic wastes on gardens if possible.

6. Do a thorough examination of one's own ecological footprint - just how much do you intrude on the natural environment by your use of energy and resources. Try to find ways of reducing this footprint.

Further References

Bojang, F., (Edit). 2014. Sustainable Natural Resources Management in Africa's Urban Food and Nutrition Equation. In: Nature & Fauna, Vol. 28, Issue 2. Accra, Ghana: Food and Agriculture Organization of the United Nations. http://www.fao.org/3/a-i4141e.pdf

Chiras, Daniel D, 2001. Environmental Science: Creating a Sustainable Future - 6th Edition. Burlington, MA: Jones & Bartlett Learning. ISBN-10: 0763713163

Dahlquist, E. & Hellstrand, S. (Edits.).2017. Natural Resources Available Today and in the Future: How to Perform Change Management for Achieving a Sustainable World. New York: Springer International. ISBN 978-3-319-54263-8

Morgera, E. (Edit.), 2016. Research Handbook on International Law and Natural Resources. Cheltenham, United Kingdom: Edward Elgar Publishing. ISBN: 978 1 78347 832 3

Yang, Shang-Tian, 2011. Bioprocessing for Value-Added Products from Renewable Resources. Amsterdam: Elsevier Science. **ISBN:** 9780080466712

Yhaya, M.F., Tajarudin, H.A & Ahmad, M.I., 2018. Renewable and Sustainable Materials in Green Technology. New York: Springer International. ISBN: 978-3-319-75120-7

Chapter 11: More About Water

11.1 Introduction

Available freshwater for drinking, agriculture and industry only makes up about 2.5% of the world's water supplies. Most water exits as saltwater in the oceans and most of the remaining freshwater available is locked up in the ice caps in the Arctic and Antarctic and in the permafrost, the permanently frozen soil of high latitude countries such as northern Europe, Russia and Canada. Over 30% of the remaining freshwater is found below the surface.

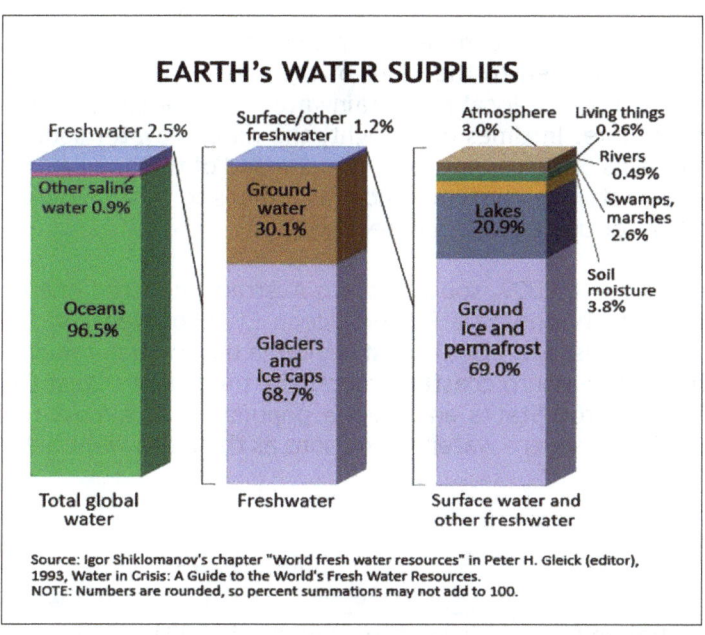

Figure 11.1: Diagram showing the main sources of the Earth's water (Photo: USGS)

Surface water which only comprises less than 0.4% of available freshwater is derived from rivers, lakes and from alpine glaciers. Many desert countries such as in the Middle East and in the deserts of Africa and Australia, have little or no available natural freshwater and must rely on subsurface water or desalination plants. Along the west coast of South America, parts of Peru and Chile rely on short streams which drain off the glaciers of the Andes for drinking water and agriculture. In these countries, water is a very valuable commodity and only limited populations can be sustained.

The availability of freshwater locally depends upon many factors such as rainfall, other climatic conditions, the use of dams and other water storage systems and population demands. In most urbanized areas, the population relies on water storage from the local dam or river weir to provide drinking water and water for irrigation. One major problem with an urbanized society is that it usually is far away from the primary water source such as a major dam and water-treatment plant. Because of this out-of-sight mentality, people may be wasteful in their use of water in their own consumption, watering of lawns and gardens and other domestic water chores.

In rural areas, people are more susceptible to changing climatic conditions and are more attuned to the state of their local water supplies because they must rely on the local river, rainwater tanks or sub-surface water as their only source. In times of drought, the natural water sources such as rivers and dams often cannot meet the demand of the population and then other water storage systems such as domestic water tanks or even imported water using tankers is needed.

Between 2001 and 2009, south eastern Australia suffered one of its most extreme droughts due to a Pacific Ocean El Niño event. In Queensland Australia, the Brisbane – Gold Coast region, a large urban region spreading along the north eastern coastline much like the Miami – West Palm Beach area of the United States and with a population of almost four million people, suffered severe water restrictions as the dams went below critical levels.

See also:

https://agupubs.onlinelibrary.wiley.com/doi/full/10.1002/wrcr.20123
(The Millennium Drought in southeast Australia 2001–2009)

This drought – flood scenario is common in many other parts of the world, especially in parts of Africa, India, Pakistan, China as well as in Australia and the United States. In the southern United States, the cyclone season often brings massive amounts of water to the coasts of Texas, Louisiana and Florida with subsequent flooding. In 2005, hurricane Katrina caused the flooding of much of southern United States and the city of New Orleans was inundated. In other parts of the world, notably Africa, India-Pakistan and south eastern Asia, water supplies involving several countries have often become the subject of international tension. In Egypt where the great Aswan Dam was built to regulate flooding along the Nile River further downstream, concern has been raised at the proposal of southern neighbours Sudan to build another dam on their part of the Nile to overcome the regular droughts which that country experiences. The Egyptians see this as a potential threat to their water supplies and electricity generation.

Figure 11.2: The flooded city of New Orleans following the passing of Hurricane Katrina in 2005 (Photo: US Coast Guard)

Figure 11.3: Photo taken from the Aswan Dam, Egypt looking south down the Nile River towards nearby Sudan.

See also:

https://soe.environment.gov.au/theme/inland-water/topic/australias-water-resources-and-use
(Good summary of Australia's available water)

https://pmm.nasa.gov/applications/freshwater-availability
(A good but short study of Australia's water from NASA with a useful map)

https://riskfrontiers.com/pdf/water-03-01149.pdf
(Causes, impact and implications of the Brisbane Flood of 2011)

https://forms2.rms.com/rs/729-DJX-565/images/tc_flood_risk_in_new_orleans.pdf
(Flood risk management in New Orleans)

11.2 A Tale of Two Rivers

In regard to the use of water as a renewable resource and the problems which are encountered through its excessive and competitive use, it would be interesting to compare and contrast two important river sources - the Murray-Darling river systems of eastern Australia and the Colorado River of the south-western United States.

The **Murray-Darling system** covers a large basin in the eastern part of Australia, covering much of the southern part of the state of Queensland, and most of the southern states of New South Wales and Victoria as well as the Australian Capital Territory and the eastern part of South Australia. The basin, drains around one-seventh of Australia and covers about 1,061,469 km² of flat, low-lying inland country which receives little rainfall. The two main rivers of this system consist of the Darling River in the north which flows south and the Murray River further south which flows west. The Darling River has many tributaries which flow west and south from highland regions of the Great Dividing Range of south eastern Queensland and eastern New South Wales. It has a total length of 2844 kilometres, and is the longest river in Australia. By world standards, the Darling River is shallow and narrow, drying up in places during drought and then widening dramatically when the rainy season in the north and east bring floods.

Figure 11.4: Map showing the Murray-Darling Basin system of south-eastern Australia.

Further south, the Darling River meets with the wider Murray River which flows for 2508 kilometres from the east, draining the high country of the Snowy Mountains. The Murray forms the state border between New South Wales and Victoria and there is much farming with extensive irrigation along its course. After joining with the Darling, the Murray River continues west and flows into the sea of the Great Australian Bight at Murray Mouth, South Australia.

Problems with the Murray-Darling system came about because of the overuse of its water for irrigation, especially for large-scale crops, such as cotton, wheat and further south, vegetables. It also provides water for stock such as cattle and sheep. The Murray River has also been subject to disputes between the New South Wales and Victorian governments about this extensive water use and the river has been dammed with several weirs. Near the headwaters of the Murray River, the mighty Snowy Mountains Scheme, consisting of several large dams and power stations, was developed for hydroelectricity in the 1950's. In the north, the area around along the Darling River is often plagued by drought and there has been considerable concern about the lack of water and environmental damage in some parts of the river.

The drought and the sudden blooming of massive amounts of blue-green algae have also dramatically reduced the amount of oxygen in the Darling River. This led to the death of very large numbers of fish in the many shallow waterways along its course. There has also been an associated drying of nearby lakes and their subsequent erosion by wind.

Despite large amounts of money being spent by both Federal and State governments, the Murray-Darling River system remains in a deplorable environmental condition. Following urgent calls from local governments and indigenous peoples, the federal Government is attempting to solve the problem.

See also:

https://www.agriculture.gov.au/water/mdb/basin-plan
(The latest Murray-Darling Basin Plan information)

https://www.pc.gov.au/inquiries/completed/basin-plan/basin-plan-government-response.pdf
(More information from the Australian Government)

Figure 11.5: Part of the drying Darling River in northern New South Wales, Australia (Photo: Public Domain)

Further technical information can be found at:

https://www.Humanrights.gov.au/sites/default/files/content/social_justice/nt_report/ntreport08/pdf/casestudy2.pdf
(The tragedy of the Murray-Darling System)

https://www.mdbrc.sa.gov.au/sites/g/files/net3846/f/mdbrc-exhibit-61-grafton-and-wheeler-economics-of-water-recovery-in-the-murray-darling-basin-5-march-2018.pdf?v=1530682071
(Water recovery in the Murray-Darling System)

https://theconversation.com/the-murray-darling-basin-plan-is-not-delivering-theres-no-more-time-to-waste-91076
(Problems with the Murray-Darling System

https://www.aph.gov.au/About_Parliament/Parliamentary_Departments/Parliamentary_Library/pubs/BriefingBook44p/MurryDarlingBasin
(Australian Government strategy and the Murray-Darling System)

https://www.mdba.gov.au/river-information/weekly-reports
(Weekly reports on the condition of the Murray-Darling system)

Across the Pacific in the United States, there is a similar environmental problem with the **Colorado River**, considered to be the most endangered river system in that country. The Colorado headwaters are in the Rocky Mountains at La Poudre Pass Lake in the state of Colorado. It then flows south-west into Utah and Arizona where it passes through the Grand Canyon. Turning south near Las Vegas, Nevada it then flows into in Lake Mead and then on to California before entering Mexico and the Morelos Dam. From this dam, the Colorado flows irregularly into the Mexican Delta and then into the Gulf of Mexico near Baja California.

Figure 11.6: The Hoover Dam on the Colorado River (Photo: US Department of Reclamation)

Most of its waters are diverted into the Imperial Valley of Southern California so that very little flows into Mexico. Along its course, water is used for hydroelectricity at such places as the great Hoover Dam, on the border of Nevada and Arizona where it forms Lake Mead. Unfortunately, this lake has never reached its full capacity due to high water usage, large evaporation rates and lack of inflow due to warmer climate changes. Further downstream, the Colorado River is used for large irrigation programs in California's Imperial Valley as well as providing drinking water for the large city of Los Angeles. The use of the Colorado River is rapidly outstripping its flow capacity and there is a bleak prognosis for its future and for the many towns, cities and the agricultural areas it services.

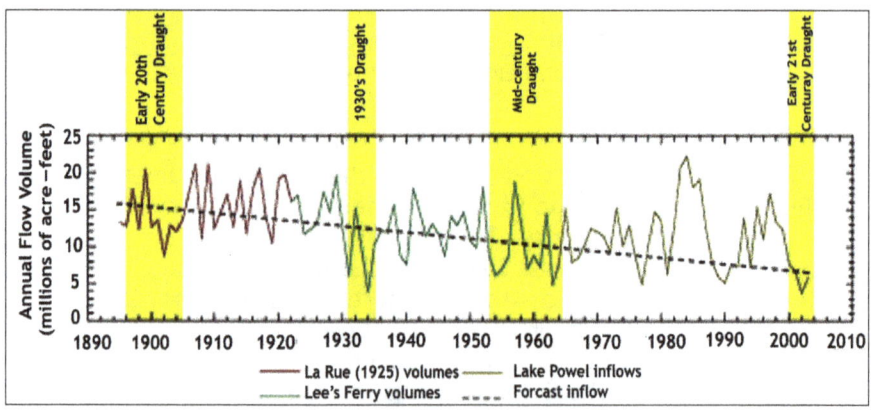

Figure 11.7: A graph showing the declining volume of the Colorado River (Photo: USGS)   (Note: 1 million acre-feet = approximately 1000 million cubic metres)

The US federal Bureau of Reclamation has started working with the seven states in the Colorado basin, namely Arizona, California, Colorado, New Mexico, Nevada, Utah and Wyoming to study the future of supply and demand on the Colorado, and to search for solutions that allow sustainable use of the river and a satisfactory **water budget** of water collected by the river compared to what is removed from it.

Further information can be found at:

https://www.cap-az.com/departments/planning/colorado-river-programs/colorado-river-issues
(Excellent site with many links about the problems of the Colorado River)

http://web.mit.edu/12.000/www/m2012/finalwebsite/problem/coloradoriver.shtml
(Good summary of the issues with many links)

https://www.coloradoriverresearchgroup.org/uploads/4/2/3/6/42362959/crrg_summary_report_1_updated.pdf
(Suggestions for the repair of the Colorado's water budget from the Colorado River Research Group)

https://www.usbr.gov/climate/secure/docs/2016secure/2016SECUREReport-chapter3.pdf
(Report from the Bureau of Reclamation)

https://www.epa.gov/sites/production/files/2015-03/documents/ca7-plan-colorado-river-basin.pdf
(Detailed report on water quality control by the US Environmental Protection Agency)

11.3 Subsurface Water

Subsurface water has become a vital resource for drinking, agriculture and stock in many countries including Australia and the United States. Many inland communities rely upon **sub-artesian** and **artesian** water as their only regular supplies of water, especially during times of frequent drought. Artesian water usually comes up from great depth, is rich in minerals salts and when tapped will come to the surface with its own pressure. It may also contain natural gas which is often ignited at the bore where it reaches the surface. Sub-artesian water comes from a shallower **aquifer** or water-bearing rock strata and must be pumped to the surface. These underground water supplies occur mainly in sedimentary rocks although some crystalline rocks such as basalt can yield water if they have been badly fractured.

Because they are made up of particles, clastic sedimentary rocks have various degrees of porosity due to the many open spaces, or pores, between the grains. They also have some permeability, meaning that they allow water to pass through them to some degree. Non-clastic sedimentary rocks and even some igneous rocks such as basalt and granite can also be porous and permeable, if they contain many cracks, joints and bedding planes which can hold water or allow it to pass. However, a rock can be porous but not permeable (or impermeable) if the pores spaces are not connected.

Porosity (Φ) is the ability of a rock to hold fluids such as water, mineral solutions, oil and natural gases, because of the many pores or interstitial spaces between its grains. The porosity of a rock depends upon:

- Rock type, as it is not the size of grains but the space between them which is the important factor. This is called **intergranular porosity**. For example, a mudstone will have a greater porosity than a conglomerate because there are many more but smaller, pore spaces within the finer-grained mudstone.

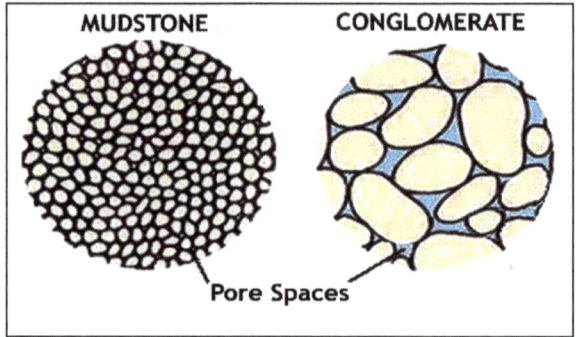

Figure 11.8: A diagram showing intergranular porosity

- Sorting of grains i.e. the variety and amount of grain sizes. Well-sorted rocks which have grains of about the same size, have greater porosity than poorly-sorted rocks because in these rocks, the smaller grains will fill up many of the larger pore spaces.

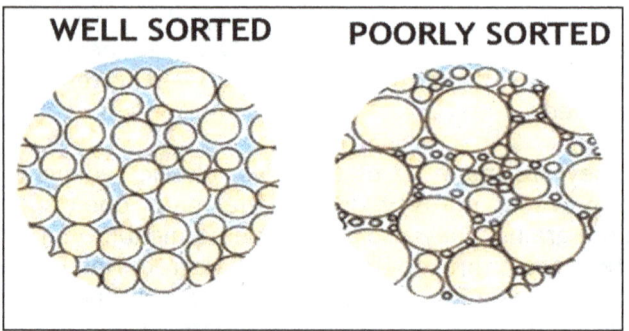

Figure 11.9: Diagram showing sorting of grains

- Cementation, or the minerals which hold the grains together will change the rock's porosity as good cementation which usually decreases the porosity of the rock by filling the voids between the grains with cementing material.

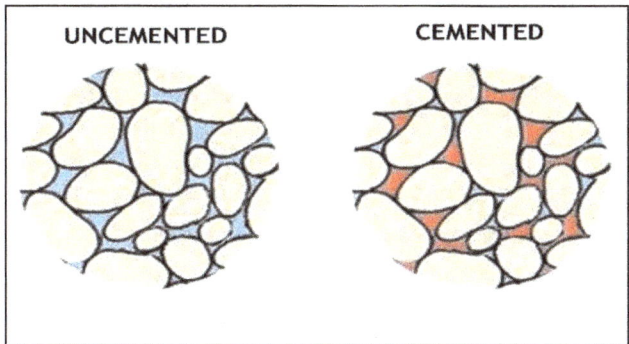

Figure 11.10: The effect of cementation reducing porosity

- Shape of the grains will also affect the porosity as grains which have good roundness and sphericity give a greater porosity than those which are more irregular in shape because the irregularly-shaped grains can interlock with each other reducing pore space.

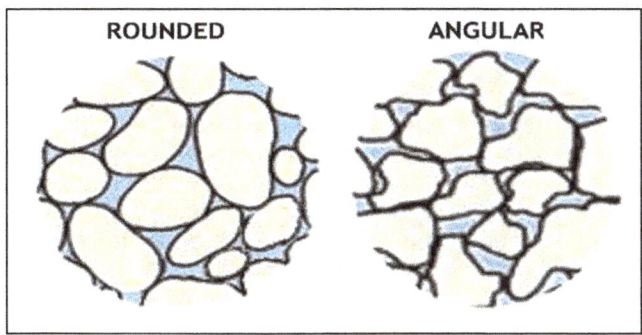

Figure 11.11: Shape of grains

- Compaction decreases the pore spaces as the grains are rearranged and pushed together more closely thus reducing the spaces in between them.

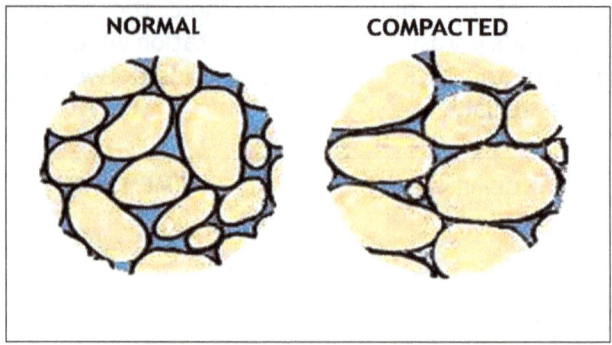

Figure 11.12: Effect of compaction

- Amount of fracturing increases the porosity by providing additional space caused by the cracks in the rock (fracture porosity). This also allows fluids to move through the rock more effectively.

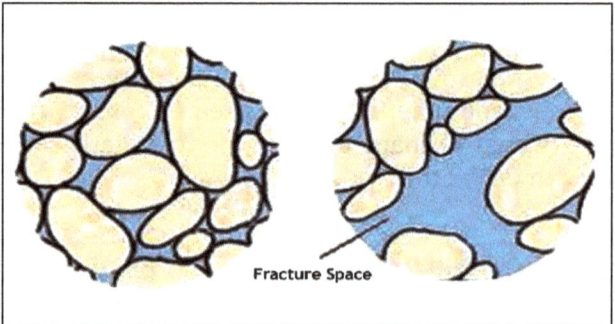

Figure 11.13: Fracture and pore space

In scientific terms, porosity is a measure of the total volume of the spaces between grains and is given as a percentage of that of the total volume of soil or rock:

$$\Phi = \frac{V_v}{V_T} \times \frac{100}{1}$$

Where V_v is the volume of the pore spaces and V_T is the total volume.

Some porosity values from several sources are given below as a rough guide to the ability of different rocks to hold fluids, but these will vary greatly with the amount of fracturing and weathering that the rock has undergone.

Material	Maximum Porosity (%)
Soil	>50
Silt	45-50
Sand	35-45
Gravel	30-40
Clay	50
Shale/Mudstone	1-10
Sandstone	5-30
Limestone	1-10
Igneous rocks	<1.5
Metamorphic rocks	Very low

Table 11.1: Some common porosity values

Permeability is the ability of a rock to allow the passage of fluids through it, either because the rock is fractured or has inter-connected pore spaces or both.

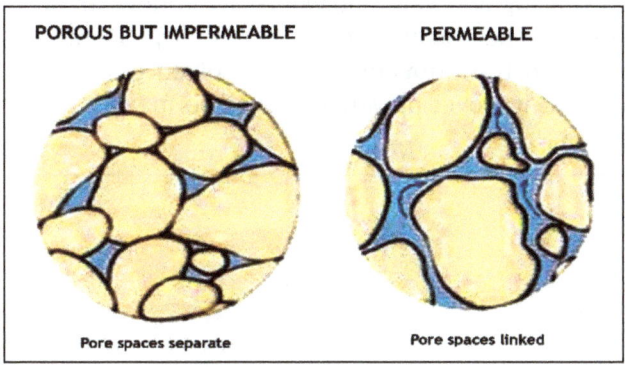

Figure 11.14: Permeability and Porosity

Rocks can be porous but not permeable simply because their pore spaces are not connected and the rock is intact and not fractured. Permeability measurements can be very complicated because the rate of flow of a fluid through a particular rock depends upon:

- Rock type, for example conglomerates and sandstones have high permeability because of their large, connected pore spaces. Basalts and some limestones also can have good permeability because of their high degree of jointing and fractures. Siltstones and mudstones usually have low permeability because their smaller grain size limits their connectivity. Non-fractured granites and other crystalline rocks simply do not have connected pore spaces or fractures to conduct fluids.

- Size of pore spaces will determine how much fluid will flow through the rock, provided the pores are linked. This contributes to the overall cross-sectional area available for the passage of the fluid through of the rock unit carrying it.

- Viscosity of the fluid, or its ability to flow, concerns the frictional effect which the rock has on the fluid or the inherent nature of the fluid as a sticky substance. Gases, such as methane, will flow better than water which will flow better than oil. **Viscosity** and therefore flow rate, also depends upon the temperature of the rock.

- Hydrostatic pressure or the force which will push the fluids through the rock. This is due to the **hydrostatic head**, or height difference between where water may enter the permeable rock, such as in a mountain range receiving high rainfall, and where it will flow out at some vertical distance below, such as at a spring in a valley. Sometimes, the head might be due to the **buoyancy** of a dissolved gas and its associated water and oil as it moves upwards through permeable rock layers in oil-bearing strata.

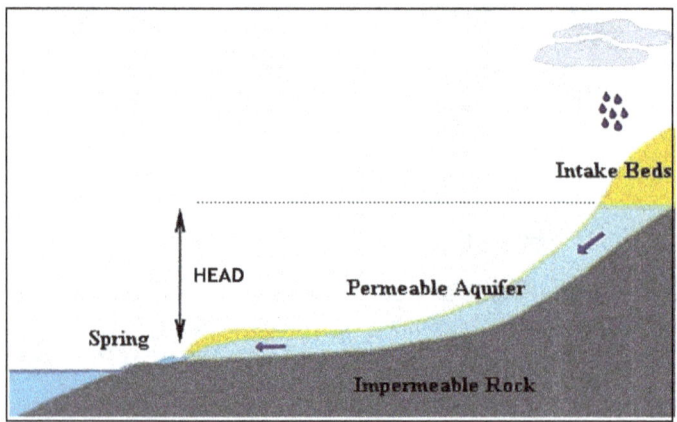

Figure 11.15: Diagram showing the concept of hydrostatic head

Groundwater systems of permeable and impermeable rock layers play a major role in the hydrological (or water) cycle. This is the natural recycling of water over the Earth. The study of movement of water in the ground through permeable rock units is called **hydrogeology** or groundwater hydrology.

In calculations involving the flow of fluids through permeable rocks, engineers and hydrologists use complex equations based on **Darcy's law**. This law states that the rate at which a fluid flows through a permeable medium, such as an aquifer, is directly proportional to the drop in the elevation between two places in the medium (the head) and inversely proportional to the distance between them. This does not include all of the factors and conditions which govern fluid flow within permeable rocks and soils.

In engineering geology and hydrology, the absolute permeability of a rock, or its **intrinsic permeability,** is calculated as the **coefficient of permeability**, also called its **hydraulic conductivity (k)**. This use can be complicated because the term is often used differently in different situations. Theoretically, hydraulic conductivity is a measure of how easily a fluid such as water can pass through soil or rock and has units of velocity, such as metres/second. It is independent on the type of fluid, the cross-sectional area of the aquifer and is a constant for specific rock or soil types. High values indicate permeable material through which water can pass easily and low values indicate that the material is less permeable.

Sometimes the equation for flow rate of fluids associated with Darcy's law is given as:

$$Q = - kA \, dh/dL$$

where:
- Q = the flow rate as a measure of permeability
- k = the hydraulic conductivity in units of velocity (e.g. metres/sec)
- A = the cross-sectional area of the aquifer
- dh/dL = the change (d) in the head of the water table. (i.e change in height of the heads dh divided by the change in the length dL between them. This means that dh/dl is a simple ratio of distances.)

The minus sign indicates that the hydraulic head decreases in the direction of flow (see Figure 11.15) i.e. a negative value will indicate flow from top of the head to the discharge point below.

The S.I. unit for the measurable value of permeability or **intrinsic permeability,** is the square metre (m^2) which relates to the area of open pore space at right angles to the direction of flow. This permeability can also be measured in a unit called the **darcy** (after Henri Darcy: French 1803 -1858). Not an S.I. unit, 1 darcy (D) permits a flow of 1 cm^3/sec of a fluid with viscosity 1 millipascal second (mPa·s) under a pressure gradient of 1 atmosphere per centimetre (atm/cm) acting across an area of 1 square centimetre (cm^2). One darcy is approximately equivalent to a permeability of 10^{-12} m^2. The millidarcy (mD) or 10^{-3} D is a more practical unit.

Some of the most common values for permeability are given in the following table:

ROCK TYPE	INTRINSIC PERMEABILITY (millidarcies)
Highly fractured rocks	10^5 to 10^8
Older sandstones - some fractures	10^3 to 10^4
Fresh sandstones	1 to 10
Fresh limestones & dolomites	10^{-1} to 10^{-2}
Granites - some fractures	10^{-3} to 10^{-4}

Table 11.2: Some values for intrinsic permeability

Groundwater systems of permeable and impermeable rock layers play a major role in the hydrological (or water) cycle of recycling water. The movement of water in the ground through permeable rock units is called hydrogeology or groundwater hydrology.

Permeable rock units, or **aquifers** which carry water and those which do not are called **aquicludes**. Conglomerates, sandstones, fractured rocks such as limestone and some basalts, as well as unconsolidated sands and gravels are good aquifers. Clays, claystones, siltstones and non-fractured crystalline rocks are good aquicludes. If a rock has low permeability, but will still carry water, such as very clay-rich sandstone, it is called an **aquitard**. Together, these rocks may form an artesian system and many countries rely on **artesian** basins as a source of useable water.

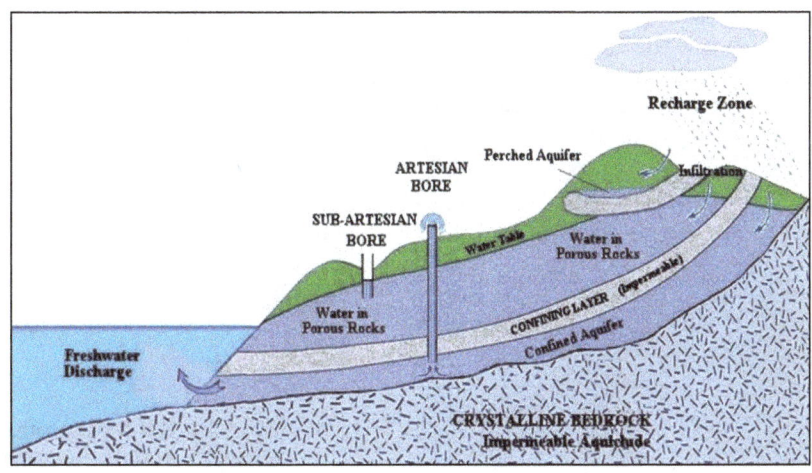

Figure 11.16: Diagram showing an underground artesian system

In an artesian system, water from precipitation enters or is infiltrated at the **recharge zone** and is carried down into the permeable rocks of the aquifers into the artesian basin. If this water has been trapped between two layers of impermeable rock as a confined aquifer, then drilling into it may result in an artesian bore. Here the water comes out to the surface from very deep underground, often from thousands of metres below, under its own pressure and it is usually hot, charged with minerals and sometimes is accompanied by natural gas. In some parts of the world, such as in the Great Artesian Basin of Australia, the most regular supply of water comes from these artesian bores, however it is only suitable for stock animals.

Artesian wells which have lost pressure due to use or sub-artesian wells which do not have sufficient natural pressure to come to bring water to the surface, usually are pumped by windmills or electrically-operated pumps. In the case of pumped sub-artesian water, it is important to remember that the pump should be operated at a good flow rate, usually determined by turning on the water flow valve or faucet to its maximum opened position. A slow rate of flow often allows the clay lining of the bore to collapse and block up the well requiring a new and expensive well to be sunk elsewhere.

Where the water is not confined, and simply percolates through permeable rock below, wells or sub-artesian bores may be drilled to obtain it. Recharging these aquifers depends upon regular precipitation in the recharge zone which may be thousands of kilometres from the well and in an entirely different climatic environment. Water from the recharge zone may travel at only a fraction of a metre per year so that much of the water coming from wells today may be many thousands or millions of years old. For example, in the Great Artesian Basin of Australia, recharge occurs in the wetter regions along the eastern coastal mountains of the Great Dividing Range which forms a rain shadow for the inland regions serviced by these wells. In the United States, a similar artesian system is the Edwards Aquifer which is one of the major sources of water for agriculture and industry in the state of Texas.

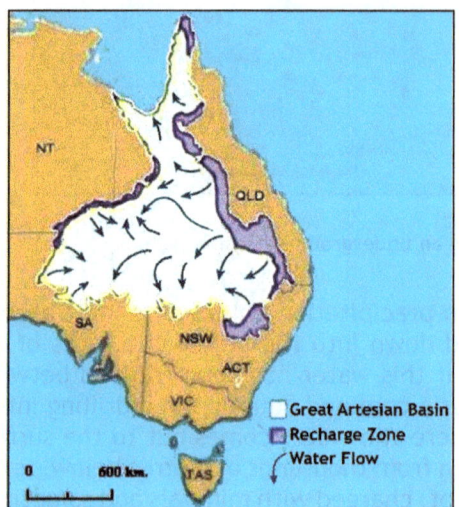

Figure 11.17: The Great Australian Basin of north eastern Australia

Figure 11.18: The Edwards Aquifer of Texas, USA

In places where extensive use is made of artesian and sub-artesian water for stock and agriculture, there is always the danger that use will exceed recharge rate and levels in bores drop or the bore dries up entirely. This can happen when there are long periods of drought in both the areas of use and the recharge zone. A drop in the water level in a well may happen for an isolated well or for several wells where the particular artesian aquifer is being used by many wells over a wide area. When this happens, users may note a change in the **drawdown** of the well. Drawdown is the drop in the level of water when water is being pumped out of the well or bore. Drawdown is usually measured in feet or metres and this can be monitored over time by doing regular pump tests of the well by applying a controlled pumping of the water and noting the response of the drawdown. By measuring the drawdown, farmers and graziers can detect problems with the well and with the aquifer below it. These problems may include encrustation of the well pipe, the pH level or its acidity, mineral content and other parameters of the water. Moreover, in some localities there is always the threat of contamination of groundwater through mining, coal-seam gas and use of pesticides in agriculture. Of more alarm in recent times is the control of water rights. Some big multinational pastoral companies may purchase land only because of its potential in supplying sub-surface water by the water rights which are available for it. This could lead to excessive extraction of water to the detriment of others using the same aquifer and even the complete depletion of the aquifer.

Figure 11.19: A common sight in many dry lands such as Australia and the south-western United Sates - a windmill for pumping up water.

Some useful sites about water resources can be found at:

http://www.hwe.org.ps/Education/Birzeit/GroundwaterEngineering/Chapter%203%20-%20Groundwater%20Flow%20to%20Wells.pdf
(Monitoring groundwater wells and pump testing - good resource)

https://www.dnrme.qld.gov.au/land-water
(Good general information about Queensland's water with many links)

https://www.ombudsman.qld.gov.au/ArticleDocuments/242/The%20Water%20License%20Report.pdf.aspx.
(Water licencing in Queensland)

https://www.csiro.au/en/Research/LWF/Areas/Water-resources/Assessing-water-resources/GABWRA/Overview
(Great Artesian Basin overview from CSIRO, Australia)

https://www.csiro.au/en/Research/LWF/Areas/Water-resources/Groundwater-resources
(Sustainability of Australian subsurface water CSIRO)

http://www.agriculture.gov.au/water/national/great-artesian-basin
(Specific details about the Great Artesian Basin with many links)

http://www.agriculture.gov.au/SiteCollectionDocuments/about/factsheets/water.pdf
(Australian water and its use – general overview)

https://water.usgs.gov/index.html
(Water resources in the United States - USGS)

https://water.usgs.gov/watercensus/AdHocComm/Background/GroundWaterAvailabilityintheUnitedStates.pdf
(Good data and information about available water in the United States)

https://pubs.usgs.gov/circ/circ1139/pdf/circ1139.pdf
(World's water resources – USGS report)

http://unesdoc.unesco.org/images/0013/001344/134433e.pdf
(Very detailed report on the state of the world's water from UNESCO)

11.4 Desalination

This is the removal of salts from saltwater or other water supplies which contain minerals or other compounds which make the water unsuitable for drinking. Many countries such as Saudi Arabia, Kuwait, Oman and the United Arab Emirates obtain water by desalination. In some other countries, the water derived from deep artesian bores is also unfit for human consumption and irrigation and could be made fit for drinking by desalination.

Whilst the use of desalination plants seems to be a satisfactory method of producing freshwater in arid regions, they can also cause problems for personal health and the environment. The seawater often used in desalination plants may have high amounts of boron and bromide, and the desalination process can also remove essential minerals like calcium necessary for well-being. In addition, the concentrated salt solution which remains after desalination is often dumped back into oceans where it increases its salinity and can affect the ocean's local environment. In addition, desalination is the most energy-costing method of obtaining water resources.

Traditionally, **distillation** has been one of the useful methods of desalination of seawater, especially in ships at sea. When water boils or is evaporated, the resulting vapour can be cooled and condensed into freshwater. This distilled water is free of the harmful salts but also contains no useful minerals found in normal drinking water. The extracted salt also remains in the distillation plant and has to be extracted physically. Any form of heating which will raise the temperature of water above its boiling point or allow water to evaporate quickly can be used in distillation, including direct solar energy or electricity provided from it using solar voltaic panels. Desalination can be achieved by a number of methods including:

- **Traditional simple distillation** in which contaminated water is heated by traditional methods such as fire or electricity and boiled to produce steam which is then passed through pipes and cooled back into freshwater which is then collected. Such techniques have been used for centuries on a small to modest scale and are still used today to make distilled water in work places and laboratories. The basic principles include boiling the water in a closed container, allowing the steam to pass out down a tube, often coiled, which is cooled either by air or a flow of cold water on its outside. The steam in the coil condenses to water which flows down into a collecting vessel.

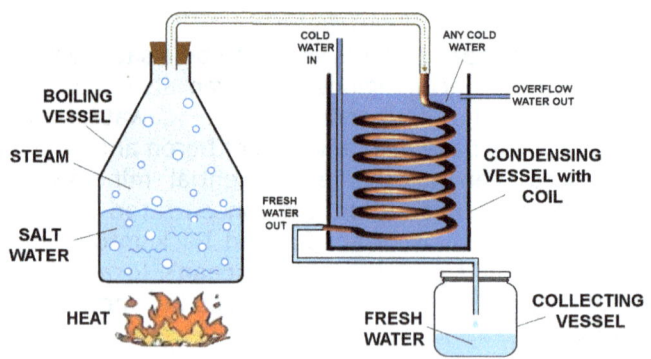

Figure 11.20: A diagram showing a simple distillation apparatus

Any suitable equipment can be used and at any size. For example, earthenware pots can be used for the vessels and black agriculture piping can be used inside a drum of water as a condenser. Many villages in some developing nations have made simple stills using basic equipment to turn contaminated water or seawater into fresh, drinking water. The system relies on a regular supply of heat either as an open fire, enclosed fireplace, stove or an electric hotplate. The cooling water could be any cold water drawn from the local source and the freshwater obtained can be collected in storage tanks. A good continual supply of distilled water can be made as long as the heat is applied and the steam condensed by a cooling device. This could be done during normal cooking times using the cooking fire. There can be many variations on this design, including some which are much simpler.

See:

Online Video 11.1: A simple home distillation unit.
Go to: https://www.youtube.com/watch?v=00kKPOs_FA4

and

Online Video 11.2: Making a simple distillation unit for a villiage. Go to: https://www.youtube.com/watch?v=zf4JrsqlIkU

- **Direct solar distillation** has also been around for many years for small-scale water purification, especially in survival kits both on land and at sea. Essentially, they consist of a covered pool of contaminated water in a black tray sealed in a glass or clear plastic container with a sloping roof. Exposed to the sun, the water from the container evaporates and rises to the sloping cooler roof where it condenses as droplets which run down the slope inside the container and are collected from a tray or shelf at its edge.

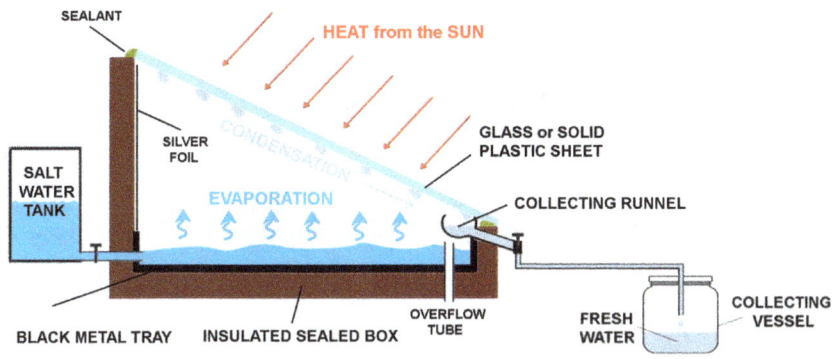

Figure 11.21: A diagram showing a simple solar distillation apparatus

Such devices unfortunately are often limited in their operation and produce only a small amount of water per day. The holding box is made air tight using a sealant such as silicone and is also insulated. During the day, the water from the saltwater tray evaporates and then condenses when it collects on the cooler the glass above. The amount of water depends upon the area and angle of inclination of the glass and surface area of the saltwater. It also depends upon the differences in temperatures between the hottest and coldest parts of the day.

There are many possible variations on such a model including larger pyramidal and spherical structures as well as arrays of complex and transparent tubes which perform the same operation of evaporation and condensation internally.

See also:

https://www.teachengineering.org/content/cub_/lessons/cub_water qtnew/cub_waterqtnew_lesson01_factsheet_solardistillation_v2_tedl. pdf
(A good technical summary of designing solar stills)

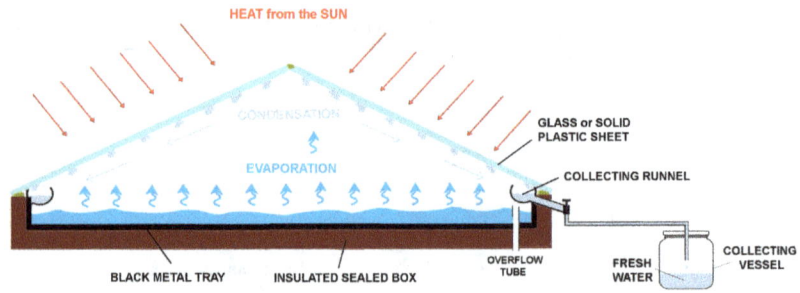

Figure 11.22: A diagram showing an improved solar distillation apparatus

- **Resin ion-exchange** filters are also used for small-scale water treatment or in association with larger processes of desalination. In the ion exchange process, water is passed through bead-like spherical resin materials and the unwanted ions, namely those of sodium (Na^+) and chloride (Cl^-), in the water are exchanged for other ions fixed in the beads. The resin beads are made of styrene and divinylbenzene containing sulfonic acid groups, will exchange a hydrogen ion (H^+) for any positive ion, or cation they encounter. These include sodium (Na^+), calcium (Ca^{2+}) and aluminium (Al^{3+}). Similarly, the resin beads made of styrene contain quaternary ammonium groups which will exchange a hydroxyl ion (OH^-) for any negative ion, or anions such as chloride ions (Cl^-). The hydrogen ion from the cation beads unites with the hydroxyl ion of the anion beads to form pure water. Whilst such filters are relatively inexpensive to set up, they must be renewed regularly and over the long-term can be expensive. Furthermore, they do not remove smaller particles nor lifeforms such as harmful bacteria.

- **Reverse Osmosis (RO)** forces water through a semipermeable membrane, i.e. a thin film which allows smaller particles to pass through, such as water molecules, but prevents the passage of larger particles such as mineral salts. **Osmosis** is a natural process whereby water and some mineral salts pass through a thin membrane due to differences in salt concentration called osmotic pressure. This is how water and nutrients pass through the cell membranes of living things.

In the industrial process of reverse osmosis, saltwater is forced by pumping it through semipermeable membranes which will allow freshwater to pass but not the mineral salts contained in the seawater. Reverse osmosis-plant membrane systems typically use less energy than thermal desalination processes but are still expensive to operate and only extract a small percentage of the water intake as freshwater. However, there is considerable potential for the use of solar voltaic panels to provide electricity for powering the pumps of reverse osmosis plants.

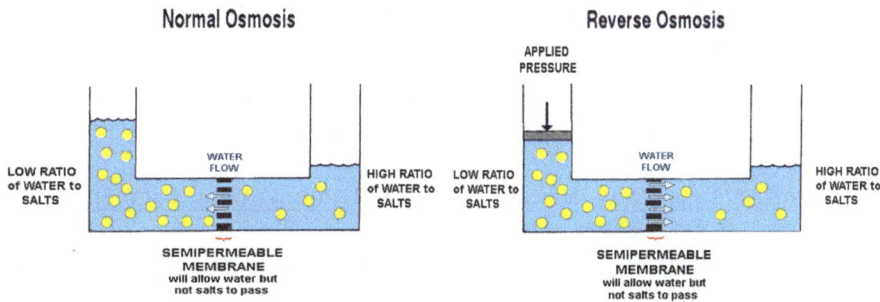

Figure 11.23: Diagram showing how normal osmosis is reversed in water desalination.

- **Electrodialysis (ED)** is another system for large-scale desalination which uses a multiple-layer of membranes with an applied electrical charge to extract mineral ions i.e. the charged atoms or groups which make up the contaminants in the water. Unlike the high pressures needed in reverse osmosis, electrodialysis uses only low pressure and electrical charge to separate the charged ions of the mineral salts contained within the water supply and is easier to maintain.

- **Concentrated solar stills** use a concentrated solar thermal collector such as curved reflectors to concentrate solar heat and deliver it to a multi-effect evaporation process for distillation. This system of multiple stages of heating and evaporation of water from the salt solution increases the natural rate of evaporation which would occur in a simple solar distillation unit. The concentrated solar unit is capable of large-scale water production in areas with plentiful solar energy such as in Australia, southwestern United States, North Africa and the Middle East.

See also:

http://www.clw.csiro.au/publications/waterforahealthycountry/2006/wfhc-DesalinationReport.pdf
(A technical article about the potential of desalination plants in dry Western Australia)

https://www.intechopen.com/books/desalination-and-water-treatment/solar-desalination
(A good site giving information about several types of solar desalination)

https://www.amtaorg.com/Water_Desalination_Processes.html
(Lists a variety of desalination technologies).

http://www.awa.asn.au/AWA_MBRR/Publications/Fact_Sheets/Desalination_Fact_Sheet.aspx.
(Gives a list of the desalination plants used in Australia and has several good links).

https://pacinst.org/wp-content/uploads/2006/06/desalination-grain-of-salt-a-cali-perspective-june-2006.pdf
(Excellent detailed document on desalination in California)

https://www.lenntech.com/processes/desalination/general/desalination-key-issue.htm
(Brief technical article with many good links)

https://www.usgs.gov/special-topic/water-science-school/science/desalination?qt-science_center_objects=0#qt-science_center_objects
(A quick summary about this topic)

11.5 Water Recycling

The recycling of water is vital to the dwindling supplies around the world and it already occurs in many major cities. It includes the treatment and recycling of waste waters such as from normal runoff, sewerage (black water) and water from domestic washrooms, laundries and industrial outlets (grey water). Many treatment plants can produce recycled water fit for human consumption which is often put back into local reservoirs. Other treated waste water can be used for agriculture, stock and for industrial use.

Treatment of waste water usually involves a number of steps to remove solids, organic matter, toxic mineral salts, bacteria and other pathogens. Steps might include:

- primary treatment including screening and settling, with removal of solids;

- secondary treatment which includes more settling, aeration and biological treatment using plants and microorganisms; and

- tertiary treatment which uses cleaning and disinfection processes such as the use of reverse osmosis and added chemicals such as chlorine or ozone and ultraviolet radiation to remove harmful pathogens.

With thorough and multiple water treatments, sewerage and grey water can be made suitable for drinking and some of the wastes which are removed may also have commercial application. For example, much of the solid sludge initially removed can be used as fertilizer. It is also possible to remove much of the nitrogen from sewage and produce ammonium nitrate, another valuable fertilizer.

Figure 11.24: Sewerage and water treatment plant in Tennessee, United States (Photo: USGS)

See also:

https://www3.epa.gov/npdes/pubs/primer.pdf
(Primer for municipal waste water treatment from the US Environmental Protection Agency)

https://www.who.int/water_sanitation_health/hygiene/om/linkingchap6.pdf
(Good article with a table showing some of the common household treatments to treat water and their effectiveness)

https://www.unescap.org/sites/default/files/Policy%20Guidance%20Manual%20on%20Wastewater%20Management.pdf
(Extensive details about waste water management from the United Nations)

http://www.iwa-network.org/wp-content/uploads/2018/02/OFID-Wastewater-report-2018.pdf
(Good selection of case studies of cities using waste water to solve their water shortages)

11.6 Hydroponics

Hydroponics refers to the many systems of growing plants without soil. It basically means growing plants in an aerated, nutrient solution and it is a method by which crops such as vegetables can be intensively grown using minimal water and space under controlled climatic conditions.

In the early 1980's, the author was Head of Science and Agriculture in a school in the dry northwestern part of the state of New Wales, Australia. It had not rained for five years and most of the small-scale crops and local grasses had died. Graziers were lopping branches off trees to feed what stock remained alive and most of the local dams were dry. To save the seedlings and plants of the school farm, a 2.5 metre square glasshouse was constructed. This was connected to the school's sub-artesian bore and an automatic watering system. A hydroponic system was constructed within the glasshouse using benches which were slightly sloped and which supported long tubes of flexible black plastic sheeting containing perlite, an inert and sterile crushed volcanic rock. Water percolating from a tank in the higher end of the bench contained a hydroponic solution of nutrient minerals made up from the school's chemistry laboratory. The water flowed slowly through the tubes which had small openings every 10-20 centimetres. In these openings were placed small pots which contained the agriculture seedlings and plants. At the end of the sloping benches, the water was collected in a drum to be recycled back to the tank at the top of the slope. Periodically, the amount of nutrients was checked using an electrical meter which had been calibrated from the original solution by measuring its electrical conductivity. This system operated fairly successfully but was labour intensive. In hind sight, a small water pump could have been used to pump water back into the reservoir tank.

This basic idea had come from another system used by the author when in a mountain city of further south at an elevation which gave snow in winter. As an elevated town in the interior of the country, cold winters required the importation of fresh vegetables at considerable cost and dubious quality. Accordingly, a simple hydroponic system was set up inside where the large window faced the sun. The room was also continuously heated by an oil heater. In this system, salad and other vegetables were planted in sterilized sand in large pots arranged on a stepped grandstand arrangement close to the glass of the window. Every two days the pots were watered with a fresh hydroponic solution and the plants grew rapidly and well allowing for several crops to be harvested over the winter.

See also:

http://www.greenhouse.cornell.edu/crops/factsheets/hydroponic-recipes.pdf
(Some useful mixtures for hydroponic solutions)

Hydroponics has been used on a much larger scale for commercial use for some time. Whilst expensive in the initial stages, the benefits of such a system using long, sloping above-ground trays and a sterile planting medium all housed within large plastic-covered greenhouses include:

- greater yield of seedlings;

- faster growth with often several crops per season;

- guaranteed amounts of nutrients;

- minimal use of recycled water;

- minimal evaporation and so conservation of water; and usually

- no unwanted soil diseases and pests.

In an urban setting, plastic PVC drainage piping can be used to grow vegetables at home. Lengths of 90 millimetre (mm) diameter PVC, used as downpipes from gutters, were cut into three metre lengths. Holes 60 mm in diameter were cut into the pipe every 10 centimetres using a keyhole saw. Into these holes were placed disposable plastic party tumblers filled with coarse sand which had been sterilized by spreading it out onto a plastic sheet in the sun for a few hours. Holes had also been made in the bottoms and lower sides of these cheap plastic tumblers.

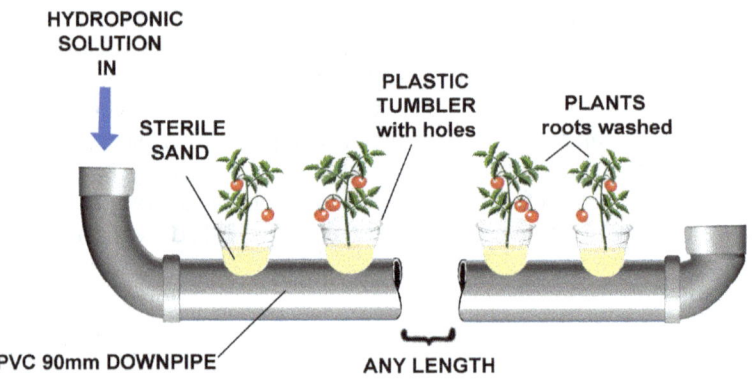

Figure 11.25: Diagram showing a simple hydroponics system using PVC pipe.

This basic system can be modified to suit individual circumstances. For apartment dwellers with a balcony facing the sun, several pipes can be added to a vertical frame holding pipes parallel to each other up the frame. For a continual flow of water, the pipes are connected together as a vertical, parallel arrangement with the pipes being filled with sand or any light-weight, absorbent material. The tumblers can be dispensed with and the plants grown directly into the holes. The solution can be dripped into the upper-most inlet from a tank and collected in another at the base. The collected waste solution can then be returned to the tank by hand or by pump. If this arrangement seems over complicated then hydroponic growing can be done simply by using garden pots standing in a trough of hydroponic solution which is regularly topped up. Evaporation can be minimized by having a lid for the trough with holes cut to allow the addition of the pots. This is a very simple arrangement which can be improved with drip systems, overflow tubes and recycling pumps but often, the simplest systems are the most effective.

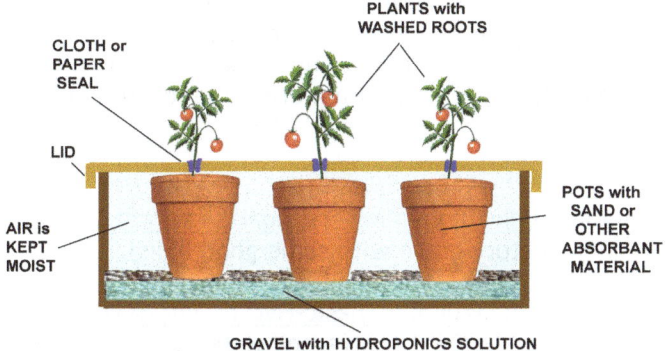

Figure 11.26: Diagram showing a simple hydroponics system using pots standing in a trough.

A simple greenhouse made from light-weight timber such as bamboo or even PVC piping. A cover of plastic sheeting or shade-cloth will also assist in reducing evaporation and the effects of pests and wind. Shade-cloth covers will also assist in reducing intense heat from the sun. In such controlled conditions and on a large scale, hydroponic trays can be arranged in stacks giving a vertical garden of several layers which would also provide a productive crop over much smaller area.

Figure 11.27: A large-scale hydroponic farm within a greenhouse

For the individual grower, adequate hydroponic mixtures in crystalline form can be purchased at the local gardening shop but any suitable plant food can be used. For a more organic variation, hydroponic setups using water flow in trays with minimal loss of water through surface evaporation can also be stocked with normal high-quality garden soil or potting mixes but water flow must be maintained and the whole garden regularly fertilized to replace the natural mineral salts leached out by the water.

See also:

https://www.niu.edu/communiversitygardens/_pdf/Hydroponics.pdf
(A simple introductory guide with some good links)

https://gthydro.co.za/upload/Basic-Hydroponic-Systems.pdf
(Good, simple explanation of different systems)

https://www.dpi.nsw.gov.au/__data/assets/pdf_file/0007/385576/Leafy-Asian-veg-final-Low-Res.pdf
(Excellent notes on growing leafy vegetables by hydroponics)

http://1.droppdf.com/files/uZzEM/hydroponics-for-the-home-grower-2015-.pdf
(Very detailed set of notes on growing by hydroponics)

http://www.e-gro.org/pdf/E305.pdf
(Calculating amount of fertilizers to use)

http://www.greenhouse.cornell.edu/crops/factsheets/hydroponic-recipes.pdf
(A very detailed set of notes about hydroponic mixes for appropriate plants)

11.7 Future Consequences and Water

If predictions about global warming are correct, then water will be a problem in the future; both as a scare commodity in some places and an excess in others. It is assumed that the semi-arid margins of dry land will become hotter and drier and those regions on the borders of cyclone and storm regions will themselves be subject to these storms. Moreover, storms are likely to become more intense as the nearby oceans heat up. Marginal regions which could become hotter and dryer include much of inland Australia, already the world's driest continent, the southwestern part of the United States, the central parts of the western Andean slopes, northern, central and southwestern Africa and far inland Asia. Desertification has been a consequence of natural climate change for millennia, with large parts of northern Africa for example going from extensive grassland to desert. With anthropogenic climate change, this process will be accelerated and arid conditions will encroach on lands which are currently inhabited and useful for crops and grazing. Moreover, as semi-arid lands begin to have more frequent and longer dry seasons, overgrazing by stock will further add to the problem as topsoil is disrupted and blown away. With additional care, poor farming practices such as over grazing can be reduced or reversed by planting desert crops and building barriers to retain topsoil. Water practices can be also improved and there may be some situations where engineering projects may be used to store or even introduce new water supplies.

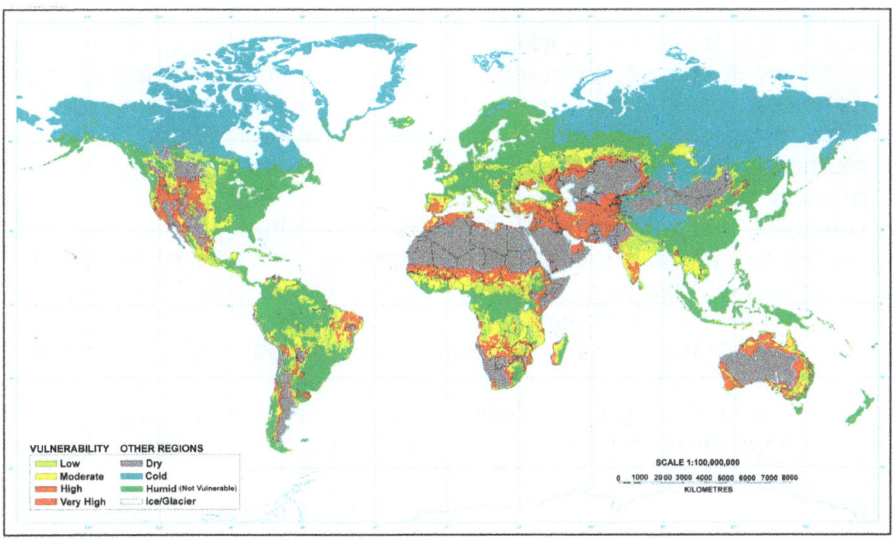

Figure 11.28: A map showing world water vulnerability (USDA)

When it comes to water and its storage, there are certain conditions which need to be considered:

- source of the water;
- collection of the water;
- evaporation from its surface; and
- accessibility.

With these factors in mind, there will need to be further development of major engineering projects worldwide in order to maintain agriculture and lifestyles in water-deprived regions.

See also:

https://www.dni.gov/files/documents/Special%20Report_ICA%20Global%20Water%20Security.pdf
(General overview of water security in the world)

https://www.frontiersin.org/articles/10.3389/fenvs.2018.00150/full
(An article on water transfer megaprojects planned worldwide)

11.8 Water Future in Australia - Bradfield Revised

Australia is the driest continent on Earth and has always been a land of extremes of climate. Currently (2020), the country has just experienced its hottest year and many regions continue to be in drought with the added disadvantage of also having unprecedented wildfires, especially in the southeast. Meanwhile, the monsoon rains have arrived in the north and parts of the country there are being flooded. It is predicted that the climate inland will become hotter and drier whilst parts of the coastal regions in the tropical north will become inundated with extreme rainfall. If this occurs, then innovative ways of reducing the effects of drought and flooding need to supplement current methods of combating weather extremes. This may include projects which will take some time to complete and which may have many difficulties to overcome in their construction. Such projects have been considered in the past, but were thought to be unrealistic with the engineering practices at the time, too expensive or requiring long-term planning and therefore impractical for governments being short of funds and having short-term vision.

One such project, the **Bradfield Project**, proposed the collection and transfer of excessive water from the north coast of Queensland into the dry interior of the Murray-Darling River system which has already reached crisis point and had stopped flowing in some sections. In 1938, Dr John Bradfield (Australian: 1867-1943) the engineer who built the Sydney Harbour Bridge, proposed the diversion of some of the far northern Queensland coastal rivers over the Great Dividing Range into the inland rivers that would then flow south into Lake Eyre in the dry centre of the country. It involved diverting water from the upper reaches of the Tully, Herbert and Burdekin and possibly the Flinders Rivers, over to the western side of the Great Dividing Range and into the Thomson River on which would eventually flow south west to Lake Eyre. This would be done using a series of dams, tunnels, channels and pumps which would also provide hydroelectricity. It would also allow irrigation of large areas of land to the west of the Great Divide along the Thompson and Flinders Rivers right down to the New South Wales/South Australian borders. Bradfield also suggested that such a scheme would produce a large inland waterway in the usually dry, salt expense of Lake Eyre which would then moderate the dry climate of the inland as well as reducing the risk of floods in the northern coastal rivers. Bradfield's scheme was heavily criticized at the time because it was considered not to be practical as it would require a very large amount of initial capital and would have large ongoing running costs which would make the project uneconomical. Furthermore, the massive inland evaporation rates would greatly reduce the amount of surface water which could be diverted inland. It was also suggested that Bradfield's modelling was based on European trends and that his data of the land relief and the amount of water which could be diverted were both greatly inaccurate. With such criticism and lack of interest, the Bradfield Scheme was abandoned in 1947.

See also:

https://www.docdroid.net/Dlp2CkP/41993.pdf#page=4?abcnewsembedheight=600
(A 1941 reprint of a report by Dr. Bradfield about his scheme)

A revised proposal for an inland diversion in several stages based on the Bradfield Scheme was proposed in 1981 by a group of local landowners and politicians and studied by the Queensland Water Commission. They believed that some of the rivers of the north coast of Queensland which received high rainfall and regularly flooded, could be diverted inland and piped south into the Warrego River which ran into the Murray-Darling System and used for irrigation in southwestern Queensland and western New South Wales. In more recent times, the Murray-Darling System has suffered greatly from over use and is currently an ecological disaster. Unlike the old Bradfield Scheme, this plan suggested that with the building or upgrading of various dams and the additional building of pipelines would allow most of the water to flow by gravity rather than requiring massive pumping. In addition, some of this water could be used for pumped hydroelectricity to supplement the main hydroelectricity potential of the new dams and pipelines. As yet, whilst various State Governments and politicians have revisited the proposal, there has little action because of the costs involved. There is also some concern that this new proposal favours the interests of large pastoral companies with little concern for small farmers along the northern coastal rivers, indigenous peoples on both sides of the Great Dividing Range and the ecology of the northern coastal rivers.

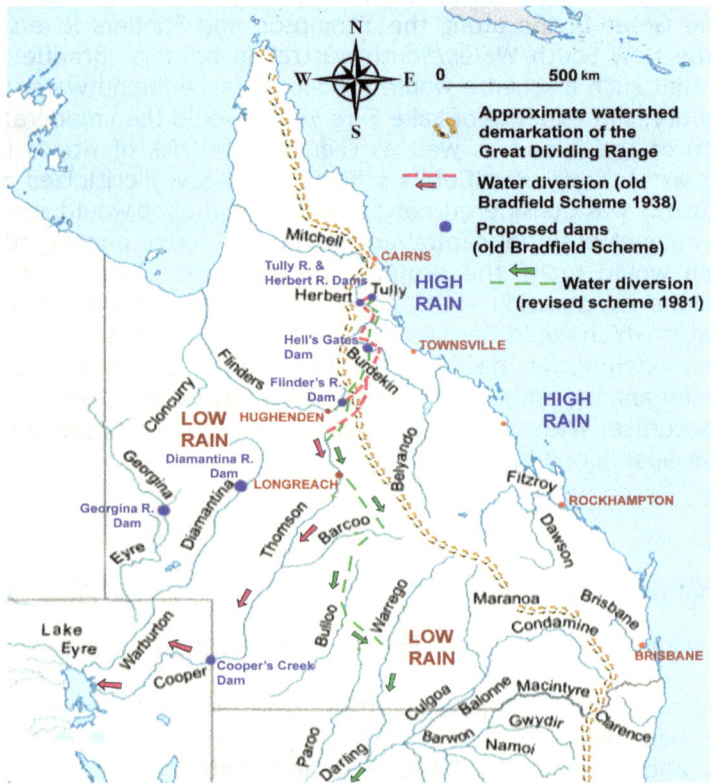

Figure 11.29: Simplified map of the old and new Bradfield Schemes

See also:

https://www.parliament.qld.gov.au/documents/TableOffice/TabledPapers/2007/5207T995.pdf#page=1&zoom=auto,-78,326
(Report on the revised Bradfield Scheme 1981)

https://www.mda.asn.au/Source/ckfinder/files/Briefing%20Paper%20Bradfield%20Scheme.pdf
(Short report with links on the Bradfield Scheme 2019)

https://www.etheridge.qld.gov.au/downloads/file/427/qld-bulk-water-opportunities-statementpdf
(Bulk water opportunities in Queensland, Australia 2017)

https://researchonline.jcu.edu.au/33823/7/Defining%20the%20North%20Final.pdf
(Water transfer schemes for northern Australia)

11.9 Diminishing Water in America

It is also predicted that the western and southwestern United States will face a growing water problem in the future, as the region becomes hotter and dryer due to global warming. The west has long been a major source of grain and cattle, but overgrazing and excessive use of surface and subsurface water has put the region at risk.

Several states in the western region can be considered as arid, semi-arid or dry subhumid based on their climate and soil types and as such are vulnerable to continued desertification. The most severely affected areas are in New Mexico, Texas and Arizona where overgrazing has often led to soil erosion. In addition, excessive withdrawals of groundwater for irrigation and town water supplies often exceeds the ability of the aquifers to replenish, resulting in a decline in height of the water table. For example, over-pumping of water from the **Ogallala aquifer** in the southwestern United States, has assisted in the desertification of parts of Nebraska, Kansas, Oklahoma and Texas. Excessive surface irrigation has also led to salination of some surface crops which further depleted in agricultural production and contamination of groundwater.

The Ogallala aquifer, also known as the High Plains aquifer, is the largest groundwater source in America, underlying the western states of Colorado, Kansas, Nebraska, New Mexico, Oklahoma, South Dakota, Texas, and Wyoming. The aquifer accounts for about 30% of the irrigation in the United States but in recent times more water is being drawn out at a greater rate than it can be recharged. One of the major problems is that there has been little control of water extraction in the past resulting in excessive irrigation of crops in these states. Water usage in the United States is the concern of local authorities with some limited control by the state and federal governments.

See also:

https://pubs.usgs.gov/sir/2017/5040/sir20175040.pdf
(US Geological Survey report on the Ogallala Aquifer)

Figure 11.30: The location of the Ogallala Aquifer and major rivers of the western United States

In more recent times there has been more control applied in the use of pumped wells, especially in Nebraska and Kansas where the state authorities have limited the use of wells by landowners. In addition, with the knowledge that this aquifer is slow to recharge and that it could not recover from massive depletion, some farmers have applied better practices in agriculture. They have found that their wells can be maintained at moderate levels using such practices. These include using more drought-resistant crops, applying minimal till cultivation and pasture cropping by intermixing crops with grasslands. Despite some local successes, the level of the aquifer still continues to drop and the problem has now become one of great interest to the Federal authorities. The Ogallala Water Coordinated Agriculture Project, is a multi-disciplinary collaborative effort funded by a joint venture of the US Department of Agriculture with the National Institute of Food and Agriculture. This collaboration is focused on developing and sharing practical, science-supported information relevant to best management practices for optimizing water use across the Ogallala region.

Further to the southwest, there is also a dwindling water problem with the **Colorado River**. A series of severe droughts in recent times has led to above-average water extraction from Lake Mead, the huge reservoir behind the Hoover Dam on the Colorado River. Between 2000 and 2012, major reservoirs in the basin drained by the Colorado have dropped to dangerous levels with Lake Powell, further upstream from Lake Mead, falling to just one-third of its capacity in early 2005. The local watershed is becoming warmer with an earlier snowmelt and a general overall reduction in precipitation. Average reservoir storages have declined by about one third putting further pressure on local water supplies and hydroelectricity generation with possible dire consequences for the future if water usage at present levels is maintained.

Figure 11.31: Lake Mead behind the Hoover Dam (Photo: Public Domain)

Some predictions for the western and southwestern United States have always been a concern and there have been many water-sharing projects between drainage basins completed in the past. Some of these have had disastrous effects on the local ecosystems from which the water has been drawn. The Los Angeles Aqueducts for example, which were completed in 1913 to carry water from the Owens River to Los Angeles and surrounds, virtually wiped out the farming communities along the Owens River. The Colorado River Aqueduct carries water due east from the Parker Dam on Lake Havasu which is downstream from Lake Mead. This aqueduct with its pumping stations runs due west through the Mojave and Colorado Deserts and is the main source of drinking water for lower California. The San Juan-Chama Project is located in the states of New Mexico and Colorado and consists of a series of tunnels, pipelines and channels which transfer water from the drainage basin of the San Juan River to the watershed of the Rio Grande River further east. This project supplies water for irrigation and municipal use to cities along the Rio Grande including Albuquerque and Santa Fe. There have even been suggestions of transferring water from Canada into the northern river basins of the United States.

https://pdfs.semanticscholar.org/1c0b/8f628d1ea4abd0c307e979d6262352f2d4ce.pdf
(Massive water-transfer schemes proposed for North America)

Whilst there have been some projects involving water transfer between river basins, the main problems in the United States concern water control and its use. In the future there will need to be more control over allocations of sub-surface and surface water with an increase in state and federal laws and supervision. Moreover, irrigation and water use in general seems to show little concern for water conservation and best practice in farming and grazing. These will become more of a problem as the west and south west of the United States become hotter and dryer.

See also:

https://www.c2es.org/site/assets/uploads/2018/10/resilience-strategies-for-drought.pdf
(Generalised suggestion for water conservation in the United States)

https://waterforfood.nebraska.edu/-/media/projects/dwfi/documents/resources/2019-agricultural-water-transfers-report.pdf?la=en
(Good article on agricultural water transfers in the western USA)

https://water.usgs.gov/ogw/aquifer/101514-wall-map.pdf
(Major aquifers in the United States from the USGS)

https://pdfs.semanticscholar.org/1c0b/8f628d1ea4abd0c307e979d6262352f2d4ce.pdf
(Major water transfers in the United States)

https://science2017.globalchange.gov/
(Special report on the future of the United States with global warming)

11.10 Water Storage and Evaporation

After the problem of lack of rainfall or snow-melt, evaporation from the surfaces of water storage dams is the next major consideration. In some arid and semi-arid regions considerable amounts of water are lost through evaporation from individual farm dams, major reservoirs, channels and natural waterways.

Evaporation occurs when the water molecules from the water's surface are removed and become free water vapour molecules. Evaporation rate is a very complex process because the rate of evaporation is controlled by many interrelated factors including:

- solar radiation
- water temperature
- surface area;
- air temperature;
- air pressure;
- wind speed; and
- humidity.

Calculations of evaporation rate are complex and often involve parameters which vary greatly. Estimations of the local evaporation of a small farm dam or a local lake is best made by empirical methods i.e. doing a trial by immersing a graduated measuring stick into the dam so that the water level is at a set height. After a good time, interval, such as a week, the decrease in water level can be measured. The evaporation rate could then be estimated by scaling this value up to the rate in millimetres (mm) per month. This value will naturally vary from month to month depending upon the climatic factors such as rainfall, humidity and temperature and a detailed record over time can be kept

Another way of estimating evaporation rate is to consult evaporation maps for the local district by referring to such maps from national authorities. These will give the evaporation rate for each month based on past data collected over a wider area and the rate for the local district can be found. Some of these maps include:

http://www.bom.gov.au/watl/evaporation/
(Evaporation map for Australia. Click on the box to get an appropriate month or season)

https://www.circleofblue.org/wp-content/uploads/2013/02/USGS_estimation-of-evapotranspiration-across-the-conterminous-unites-states.pdf
(Scroll right down to evapotranspiration map in centimetres/year)

http://www.waterandclimatechange.eu/evaporation/average-monthly-1985-1999
(A world interactive map for evaporation rate – select the appropriate month)

Controlling evaporation from dams and reservoirs, especially if they are large, has been a challenge for engineers for a long time. Most often, the body of water exists in semi-arid regions and often presents a large surface area exposed to local winds and solar radiation. There have been many ideas for reducing evaporation from medium and large dams including:

- Siting new dams in more appropriate places so that the dam lakes are deep and have a reduced surface area. This would be possible when dams are built in gorges and steep-side valleys. Naturally the nature of the rock and strata in such places is also important to prevent seepage from the sides and floor of the reservoir. There may be an option for storing water in old quarries or open-cut mines provided that there is no residual toxicity from the mining operations and that there is minimal loss through the rock strata.

- Chemical coverage by monolayers such as certain fatty acids have been suggested by some experts but these have not been fully tested and if applied to a surface would need to be done regularly and at some expense.

  See also:

  http://www.ecosmagazine.com/?act=view_file&file_id=EC135p33.pdf
  (Brief notes on monolayers)

  http://www.urbanwateralliance.org.au/publications/UWSRA-tr6.pdf
  (More detail and research on monolayers)

- Use of windbreaks around the edges of the reservoir to reduce the effects of surface wind in evaporation. This seems to be a good, practical suggestion provided that the windbreak is not more expensive compared to the need for the evaporation it would prevent. Planting of trees around a dam may assist in wind flow provided that these trees do not take up more water from the dam by their natural transpiration.

- Various covers of thin plastics have also been suggested but these would only have application for small to medium reservoirs. Free-floating covers and fixed-support covers, including shade cloth, have been suggested but there is a problem with the water quality and ecology below the covers if sunlight and air is totally blocked.

- Use of a layer of aquatic plants such as water hyacinth has also been suggested. Whilst it is doubtful that the transpiration rates from the leaves of these plants would be less than evaporation rates, there is also the massive problem of oxygen-depletion within the water itself.

- Other forms of individual, free-floating covers include the use of large numbers of floating mats, plastic circles and even wood pieces and sealed glass bottles. One commercial brand uses recycled rubber tires filled with polystyrene to make them float. These appear to have some success on smaller reservoirs.

See also:

https://www.e3s-conferences.org/articles/e3sconf/pdf/2019/23/e3sconf_form2018_05044.pdf
(An academic review from Russia on different ways of reducing evaporation)

https://core.ac.uk/download/pdf/11036429.pdf
(A good summary of similar methods from Australia)

https://www.sheepconnectsa.com.au/water-security/water-sources/controlling-dam-evaporation
(A good article which lists the main products and methods for reducing evaporation from farm dams)

https://ramblingsdc.net/Australia/EvapReduc.html
(An excellent website by a local farmer who has trialed a number of methods on his dam and has settled on a rubber tire system)

Final Remarks

1. Available freshwater for drinking, agriculture and industry only makes up about 2.5% of the world's water supplies and most of this is locked up in the world's ice caps and in the permafrost.

2. With increased population, more intensive agriculture and industry as well as climate change, many of the world's rivers are decreasing in volume.

3. Of the available freshwater, 30% lies underground in sub-artesian basins or pressurised artesian basins. The ability of rock to hold water depends upon the nature of the rock's porosity, its amount of pore spaces, and its permeability, its ability to allow water to flow through it.

4. Many countries have low reserves of surface and sub-surface water and must resort to desalination – the physical removal of salt and other impurities – from seawater or other forms of impure water. Methods of desalination include distillation, ion exchange, reverse osmosis and electrodialysis.

5. Many large urban areas now recycle waste water and sewerage using water treatment plants which are able to return freshwater to the environment.

6. Some water for agriculture can be saved by using intensive hydroponic systems with recycling of drained water.

More Applications

1. Consult local government maps or websites about the potential for flooding in your area.

2. If a homeowner, devise a flood plan if necessary, ensuring that doors can be sandbagged or sealed. Both homeowners and apartment dwellers should make provision for lack of drinking water, food and lighting if flooded.

3. Conserve water at all times by fixing dripping taps, having minimal showers and hand-watering gardens.

4. Take a bucket into the shower to collect waste water. Use grey water if suitable on gardens.

5. Homeowners could also install rainwater tanks connected to the roof. Apartment owners could discuss the collection of rainwater with the Body Corporate or Owners Association. It may be possible to install collecting systems on the roof and channel rainwater down internal pipes to small tanks on each floor.

6. Property owners with drought problems might consider installing water tanks near shallow dams for storing water during rainy seasons rather than allow for evaporation of dams during the dry. Explore the provision of underground tanks in drainage hollows.

7. Property owners using underground water should also campaign against excessive use of artesian reservoirs, especially the use by large companies and the abuse of water rights.

8. Coastal communities and individuals may consider the use of desalination for additional water either as a community installing a commercial plant or individually by constructing a home system. This could range from a simple kettle/stove arrangement or larger a glass or sheet plastic structure.

More Applications Continued

9. If water or space for a home garden is limited, then consider growing vegetables using hydroponics, either in a closed greenhouse if space is appropriate or as a vertical garden on an apartment balcony or front room.

10. Campaign for local water conservation and storage and support organisations which foster clean water projects in developing countries.

Further References

Ashton, K. with Heckler, A & Jones, C., 2012. Water for Life – Investigating Water as a Global Issue. Melbourne: Geography Teachers' Association of Victoria. ISBN 978-1-876703-23-3.
https://www.globaleducation.edu.au/verve/_resources/Water_for_Life_web.pdf

Australian Bureau of Meteorology, 2018. Water in Australia 2016-17. Melbourne: Bureau of Meteorology. ISSN: (Print) 2206-7809 (Online) 2206-7817.
http://www.bom.gov.au/water/waterinaustralia/files/Water-in-Australia-2016-17.pdf

Chellaney, B., 2013. Water, Peace and War: Confronting the Global Water Crisis. Lanham, Maryland: Rowman & Littlefield Publishers, Inc. ISBN: 9781442221406 (electronic).

Cooley, H., Gleick, P.H., Wolff, G., 2006. Desalination, With a Grain of Salta California Perspective. Oakland California: Pacific Institute. ISBN: 1-893790-13-4.

Gude, V.G. (Edit), 2018. Renewable Energy Powered Desalination Handbook: Application and Thermodynamics.
Oxford (UK): Butterworth-Heinemann. ISBN: 9780128152447.

Colonel Daniel Rivière, 2015. The Thirsty Elephant – India's Water Security Challenges: A Test for Regional Relations over the Next Decade. Canberra, ACT: The Centre for Defence and Strategic Studies (CDSS).
https://www.defence.gov.au/ADC/Publications/IndoPac/RiviereIPSPaper.pdf

Roberto, K. 2003. How-To Hydroponics, 4th edition. New York: The Futuregarden Press. ISBN - 0-9672026-1-2.
http://www.agriculture.uz/filesarchive/HowToHydroponicsRobert2003.pdf

United Nations, 2009. Facts on Water Resources - A summary of the United Nations World Water Development Report 2.
Brussels: Greenfacts.
https://www.greenfacts.org/en/water-resources/water-resources-foldout.pdf

Van Loosdrecht et al., 2016. Experimental Methods in Wastewater Treatment. London: IWA Publishing. ISBN: 9781780404752 (eBook). https://experimentalmethods.org/wp-content/uploads/2018/01/Experimental-Methods-in-Wastewater-Treatment.pdf

Vickers, A., 2001. Handbook of Water Use and Conservation: Homes, Landscapes, Industries, Businesses, Farms 1st Edition. Amherst, Massachusetts: WaterPlow Press. ISBN-10: 1931579091.

Williams, J. & Swyngedouw, E. (Edits.), 2018. Tapping the Oceans: Seawater Desalination and the Political Ecology of Water. Cheltenham, United Kingdom: Edward Elgar Publishing Ltd. ISBN10 1788113802.

Chapter 12: The Way Ahead

12.1 Introduction

This chapter is about the future. Some of the ideas presented here and possible scenarios for the next few decades are based on aspects of technology and science which has been tried and tested for many years. Some techniques and processes are now available or are in the early stages of their development and have great potential for the future. Others are more in the realm of possible developments; things which are possible but are yet to come as common aspects of society. There is the hope that readers may use their talents to develop some of these potential ideas for the benefit of the rest of humanity in the future.

12.2 Back to the Future

This is not an appeal to go back to the good old days when life seemed to be simpler and global warming was only in the minds of a few well-informed people and environment issues referred only to endangered wildlife. Rather, it is an appeal to look at some of the simpler approaches to life which could now be further extended into making the future world more secure and livable. For example, in the post-Second World War years, everything was in short supply so more items were recycled and used with less waste. Milk, soft drinks, beer, wine and spirits were all sold in glass bottles which could be returned to a convenient place to be re-used or recycled. Food was sold in bulk and not over packaged in plastic. There also seemed to be more cooperation within and between communities rather than the isolation due to the individualism advocated by the media.

In the near future, the problems which need to be confronted and overcome concern those associated with:

- global warming due to the increased use of fossil fuels such as coal and oil products;

- increase in population and the ever-increasing demand for basic resources such as water, food and living space;

- increasing demand for energy and other resources needed for maintenance of living as well as future development;

- pollution on land, air and in the world's waterways and oceans due to an increasing amount of industrial and domestic wastes; and

- poor social interaction and cooperation at the local and international level with lack of initiative from government and international bodies in dealing cooperatively with the world's problems.

Future survival will depend upon cooperation amongst the world's developed nations and the use of real, pragmatic initiative. Innovation and positive, cooperative social change will also be the key to the development of future social order. New processes and technologies which will help to reduce the environmental problems now facing the planet will need to be developed from existing systems or invented. The beginnings of such change are already here with an evolution in thinking in terms of living <u>with</u> the environment rather than simply using it.

Indigenous peoples knew about this basic principle and some tribal groups are still allowed to practice these environmentally-friendly ways of life. This does not mean that modern urbanized societies should become tribal, nor live in isolated communes and generally try to avoid the rest of the world. What is needed is a change in social thinking so that people can live cooperatively with each other and with the environment. The world's great environmental protection and research agencies are on the right track. So too are those world leaders who see global warming as a serious and current treat and are prepared to put party politics aside and get on with leading change.

In looking forward to the near and distant future, societies and nations should consider such possibilities as:

- Conservation of natural resources, especially water, clean air and natural environments with the planned provision for reducing deforestation and planting more trees.

- More effective food production using available space both by traditional agricultural and grazing methods as well as using new systems more suitable to the changing world. Such changes could include using available urban spaces for agriculture as well as farmland further afield. Urban agriculture could be carried out within, between and on top of city buildings so that agriculture returns to a full local community base. Agricultural methods such as the extended use of intensive hydroponics, large-scale commercial farming of fish and other aquatic life and other practices which are not as sensitive to climate change as traditional methods would be an advantage.

- Continued development of alternatives to energy supply and transportation systems using other power sources such as hydroelectricity, solar power, and other non-pollution systems. The development of such systems, however should not include continued hidden polluting factors as part of the manufacturing processes used in their development.

- More effective use of wastes as a source of raw material and energy rather than simply disposable rubbish. Here too, one must be warned about hidden disadvantages such as the production of methane gas in organic waste burial. This gas may be a valuable raw material but gas emissions of methane would only add to the problem.

- Use of processes which can completely absorb gas emissions should continue to be developed. There is still a role for coal and other fossil fuels including waste-bed methane provided that the manufacturing and energy processes in which they are used have a zero-emissions outcome. The technology for zero emissions from fossil fuels is already available but the cost is high and the motivation for its use so far is low. The use of algae ponds, burial in stable saline aquifers and sealed systems which use the waste gases as resources in manufacturing are still to be developed as the standard commercial practice.

12.3 The Urban Context

Greenhouse gases which contribute to global warming mainly come from the major industrialized countries and from developing countries where burning of forests for land clearance has been tolerated. However, all countries in the world are affected by the problems associated with global warming. These include sea level rise, ocean acidification, melting of ice caps and glaciers, habitat destruction and the increased spreading of disease.

In poor rural communities, life is still relatively simple and the people are probably more adaptable to climate change within limits which still permits their use of the land. Where climate change has reduced their living standards, mass migration will have to be an alternative putting more pressure on other nearby communities, especially those in already crowded cities.

The problems which will be encountered more in an increasingly warmer environment with an increasing population which requires more living space would be associated with:

- living space and energy;
- water;
- temperature control; and
- transportation.

12.4 Living Space

Almost all of the great commercial cities of the world are overcrowded - with people, traffic and commerce. All compete for the basic requirements of life such as clean air and water, food and a space to live or conduct business.

The world's population has grown considerably in recent decades. In 1950 there were 2.5 billion people on the planet, 7.7 billion by 2019 and the United Nations predicts that by the end of this century there will be 11.2 billion. Increasing life expectancy and falling child mortality in most countries are the major factors in these increasing population numbers, however this is counterbalanced by the falling trend in fertility rates. The trend of couples having fewer children has brought population growth to an end in many countries and that will possibly bring an end to population growth globally. For some projections on the future birth rate by countries, see also:

https://ourworldindata.org/future-population-growth#projections-of-the-drivers-of-population-growth

An interesting website is that of the World Population Clock which gives a current world population based upon computer estimates of growth. See this at:

https://www.worldometers.info/world-population/

In rural economies, traditionally families would be large as there would be a need for maintenance of the economy once one generation has become elderly. In the cities however, over population and lack of space and job opportunities limits and sometimes prevents an increase in local population. According to the United Nations, over 55% of the world's populations live in cities. Some of the most populated cites are given in the next table:

CITY	COUNTRY	POPULATION	POP. DENSITY (per km^2)
Tokyo	Japan	37,435,191	21,556
Delhi	India	29,399,141	23,893
Shanghai	China	26,317,104	3,814
Sao Paulo	Brazil	21,846,507	8,055
Mexico City	Mexico	21,671,908	14,616
Cairo	Egypt	20,484,965	31,399
Dhaka	Bangladesh	20,283,552	42,659
Mumbai	India	20,185,064	20,634
Beijing	China	20,035,455	1,267
Osaka	Japan	19,222,665	5,200

Table 12.1 Some of the world's most populated cities. Figures will vary depending upon the region taken as the city's boundaries.

Such statistics often do not include the immediate surroundings of each city which may also have high population densities. It is no surprise to note that some of the biggest cities are in countries which have very high national populations such as China with 1,428 million in 2019, and India with 1,369 million people. An important factor is the population density which gives an idea of how many people are crammed into the area bounded by the city limits – sometimes geographical features such as mountains, rivers or parts of the sea.

Figure 12.1: On the Bund – the riverside walk in Shanghai, China.

Figure 12.2: Traffic in New Delhi, India.

The need for energy is going to be a large-scale social issue, especially in urban areas. Already many city homes have solar power to add to or even replace utility-supplied electricity. Energy storage is the main problem here, with alternative energy-generation systems still being unreliable or unsuitable for a 24-7 production. New lithium batteries are useful but one must look at the cost, use of non-renewable resources and the pollution involved in their manufacture and ultimate disposal.

Energy use in domestic, commercial and industrial workplaces is becoming a problem of supply and demand. In countries with a national electricity grid, cities are dependent on local suppliers. They in turn, will rely on a smooth change-over from the use of fossil fuels to more renewable energies in the future. Nuclear power may be seen as an interim measure until the renewable energies are at a level which can meet future demand.

Innovative changes in the living and working environment would also be an appropriate area of the future use of energy. This would mean a dramatic reduction in the waste of energy, especially in the home and in industries where there may be a reduction in the use of electricity through the use of more productive processes.

Many buildings today rely upon the massive use of air conditioning to bring temperatures to a satisfactory level of comfort. This will probably continue into the future but more efficient use of the air conditioning systems may have to become the norm. Continual operation of air conditioning over a 24-hour period and at a low level such as the 21^{0C} often set in many buildings, may no longer be appropriate. Moreover, other practices, long used in hot countries in past times, may become more common. The use of green plants within the home, office, factory and institution is not new but perhaps innovative ideas can be developed to replace the cost and detrimental heath problems with air conditioning. Early single houses had wide verandahs, windows which could be shut or opened depending upon the outside temperatures and high ceilings to allow good air flow. In the future, architects and builders may look back to the past and build homes more suited to the local environment.

The need for high-rise units is also apparent in a world with a rapidly-increasing population, but even here there could be changes to reduce the need for extreme energy use in air-conditioning. These changes could include: through-ventilation using ducted systems; indoor gardens in alcoves facing the sun; and water circulation for indoor plants which would improve humidity in air-conditioned buildings. However, with increased air conditioning and ducted humidifiers, one must consider water-borne dangers such as mold, fungus and Legionnaires' disease. Gardens on balconies and rooves and indoor greenhouses are now becoming more common and one day may be extended to most buildings.

Green spaces within cities are also becoming a major feature of new town planning and urban development. They are becoming a very important feature for recreation and air circulation as well allowing for some absorption of the higher concentrations of carbon dioxide in the cities. Indoor plant systems and green spaces would supplement more robust systems such as internal building ventilation with filter systems. The use of solar panels on city buildings to power air conditioning, lighting and machinery within the city would also assist in reducing carbon dioxide levels generated at the national electricity grid.

See also:

http://filelibrary.associationsites.com/aia/collection/Walk_the_Walk/Supporting_Docs/architectsandclimatechange.pdf
(Brief article on Architecture and climate change)

https://www.researchgate.net/publication/41892454_Climate_Change_and_Architecture_Mitigation_and_Adaptation_Strategies_for_a_Sustainable_Development
(Professional paper on Architecture and climate change)

http://www.wilsonarchitects.com.au/powerlink/
(A good example)

12.5 Decentralization and Satellite Cities

Decentralization of large urban populations is not a new concept, but cities continue to grow. This growth, especially at the city's margins, is due to the migration of people from outlying areas due to lack of employment and increasing climate problems such as drought, floods and sea level rise.

Decentralization often has not been successful because their new urban sites:

- Do not offer sufficient job opportunities because the area does not have the potential to be exploited as a place of commerce, extraction of raw materials, growth and harvesting of crops or the raising of livestock.

- Are not attractive to people for their resettlement and often only offer minimal or no infra-structure such as schools, hospitals, roads and recreational facilities.

- Are developed in places which are unsuited to a larger population because of their physical or climatic conditions.

- Are developed in places where there are cultural conflicts between local peoples or because the site is more important as a place of environmental concern.

- Are poorly planned and financed such that living conditions and infrastructure needed for a viable city are inadequate.

- There is inadequate high-speed transportation between the satellite cities and major ports, administration centres, sources of raw materials and other people.

With an ever-increasing number of people moving to greatly overcrowded cities, local authorities are having difficulty in finding both space and jobs for the new arrivals. With the failure to keep up with the required infrastructure and utilities in current cities, decentralization will probably become a major necessity in future years.

The following is an hypothetical idea for a satellite city located in a suitable region:

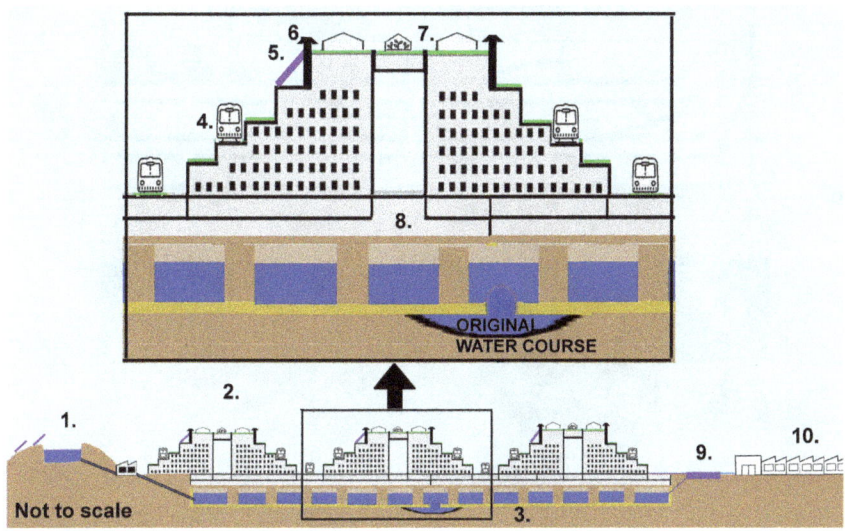

Figure 12.3: A possible concept for a satellite city (partly shown). Note the following features:

1. Pumped hydroelectricity generation
2. Large-sized high-rise buildings
   5 to 10 stories only with apartments & shops
3. Underground cistern for water storage
4. Light rail loop system on buildings
5. Solar panels on all available roof areas
6. Through ventilation chimney
7. Green spaces for gardens, orchards and recreation parks
8. Underground spaces for storage, utilities and carparks
9. Sewerage & water treatment plant
10. Industrial precinct

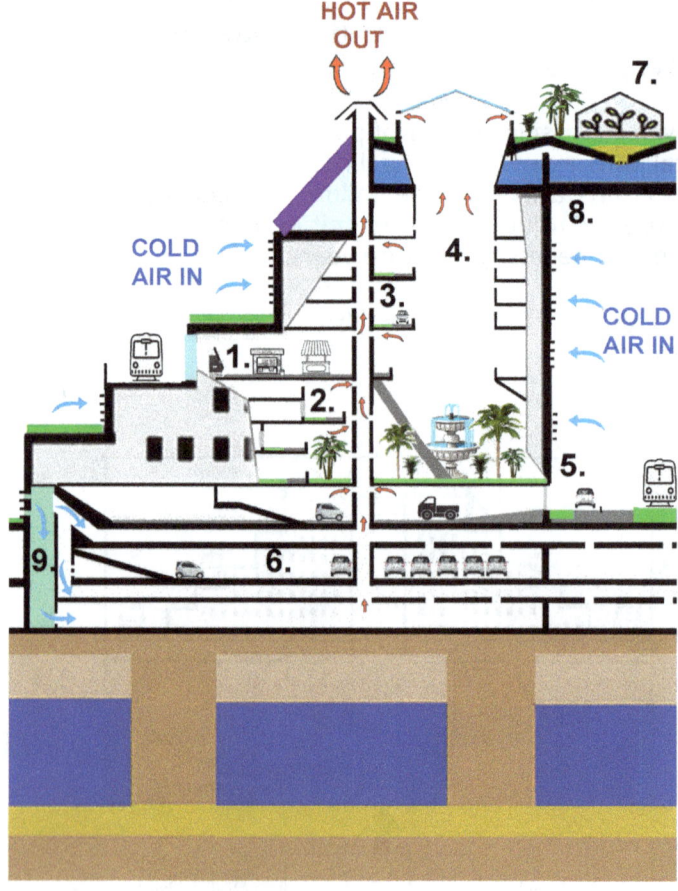

Figure 12.4: An idealised cross-section of the satellite city showing:

1. Light Rail (LR) station with shops
2. Two-story apartment with garage
3. Internal road for small electric or hybrid cars and pathway
4. Large open atrium for light and ventilation - with gardens
5. Exit to primary road of two lanes each way and LR in centre.
6. Underground carpark for bigger cars, trucks etc.
7. Green space on roof with greenhouses, parks, gardens.
8. Underroof cisterns which drain green spaces and which drain to basement cisterns (connections not shown)
9. Air filters on all cold air inlets with solar-powered pumps.

The main features of such a hypothetical model includes:

- Clusters of large buildings, each being one block of the city and being of a squat nature of only five to ten stories high. They would contain internal open spaces such as glassed-in atria, escalators and small roads such as those found in elevated carparks. Each building would contain apartments on the upper floors with shops, additional car parking and some light industry on lower floors. Apartments would have a small garage space for a small electric or hybrid car which would open internally to a small spiral road system and walkway around the atria. Some current cities have large amounts of wasted space between buildings. This is because the cost of land in cities is high and often the site was once that of a previous small area development. Consequently, modern city buildings are usually very tall and thin structures.

- Light rail or monorail systems would operate at several levels as well as at ground level. They could be constructed on the terraces of the buildings; each being connected to the next building by an overpass allowing the rail system to form as several looped routes. These would provide the main transportation around the city with private cars being used for travel outside the city precinct. Car parking facilities would be strictly limited and would be situated below each building, not on the roadside. Commercial vehicles would operate out of appropriate loading zones in the basement areas.

- Satellite cities would be connected to the outside through high-speed electric trains operating on traditional rail networks, as monorail systems or even high-speed maglev systems. Rail stations would occupy the first floor of an outer building with access to the internal light rail system.

- Each building would be constructed on top of a vaulted water cistern similar to those often found in some ancient Greek cities. Supporting structures could be made from local hard rock quarried to make space for the city. Ideally, this would be sited on an existing minor waterway which would flow into the cistern and then overflow out into the old water course downstream from the city. Provision would be made to ensure that this efflux is large enough to cope with any seasonal or rare flood.

Figure 12.5 Ancient water cisterns or sterna in the ruins of Delos, Greece. Many ancient ruined houses here have stern which are still filled with water.

- The basements of each building would contain the various facilities for the building's operations such as utilities and control panels. They would also offer storage facilities on a group or rented basis for residents as well as additional car spaces for larger vehicles and loading zones for commercial vehicles.

- Wherever possible, the roof of each building complex will be used as green space, for providing ventilation through atria openings and for through-ventilation chimneys. Green spaces may consist of commercial or private glasshouses for vegetables, small orchards or as open areas of tree-filled parks for sport and recreation.

- Whilst not shown on the diagram, glass windows should be able to be opened to allow natural ventilation to assist the through-ventilation system. Through-ventilation is achieved by pumping cool, filtered air from below the building up through ventilation shafts into apartments, the atria and other open spaces with hot air being passed out of the roof through a series of chimneys. With appropriate engineering, this upwards ventilation should be natural and require no power. All apartments and other spaces would have ducted fresh air coming in through ventilators at floor level and hot air removed through ceiling ventilators. Windows facing the sun can also be coated with a thin-film photo-voltaic material which would produce electricity as well as providing shading.

Figure 12.6 Portcullis House, London has a through ventilation system as shown in the previous diagram. There is also a large glass atrium, shops and gardens inside.

- Below each green space would be a draining cistern which would hold sufficient water for the plants but allow good drainage via pipes to the living spaces below. They would then finally drain into a water treatment plant for purification and recycling into the main storage cisterns below the buildings.

- Sewerage and grey water from each building would also be run into the water treatment plant and then recycled into the main cisterns or used externally in local agriculture in fields surrounding the city.

- Additional energy for the city could be provided by a solar farm/pumped hydroelectricity system on nearby hills or in the quarry which used to obtain stone for building parts of the city.

- Heavy processing, agricultural and industrial operations would occur just outside of the city with appropriate environmental systems in place to reduce emissions and waste. These areas would also have road and rail access to the city and to other external centres of trade and manufacture by the high-speed rail network and highways. Waste products from the city and industrial precinct would also be recycled or destroyed using efficient high-temperature furnaces which would also meet some of the energy needs of the city.

There could be many similar designs for satellite cities but the basic need remains that of depopulating currently over-crowded cities whilst revitalizing regional centres with new citizens, infrastructure, industry and advanced agriculture.

12.6 Retrofitting Cities

Satellite cities could become a necessity of the future, but it will take considerable government incentive, innovation and drive from both governments and private enterprise to have them constructed. In the meantime, and for the future, large urban populations will remain in the over-crowded cities of today. The continuation of current cities is also necessary as most of them were built on old trade routes which are still a major part of the networked world economy. However, it will be difficult for these cities to continue to function as they do now with increasing population and in a world of global warming and its consequences. The cities and their citizens will have to adapt to future change both physically and socially.

Of course, most cities are evolving as new technology comes into use and increased populations put more pressure on local governments to improve the situation. For example, China has embarked on a modernization program to replace low-rise slums with new high-rise blocks of units. Many cities have also developed light rail systems, some driverless, to take commuters to major centres such as airports and sea ports and other trading centres. However, with increased global warming and its consequences, town planners will need to adopt new and innovative measures to meet these challenges. These new measures will naturally take some time due to costs, manpower available, technology required and social reaction. They will require thorough pre-planning followed by an achievable and systematic sequence of long-term development. Many cities have already embarked on such planning and change. Such a sequence could include:

1. **First Stage: preparation** for any immediate consequences of global warming. These may include: the threat of inundation due to sea level rise; increased cyclone (hurricane) activity with strong winds and flooding; drying up of water supplies and failure in agriculture and grazing; and the encroachment of wildfires.

For those cities which have already experienced such disasters, such preparation has probably occurred to some extent but with forecasts of increased hazards due to increased temperature rise, such preparations need to be reevaluated.

In cities on a coastal strip, the consequences of sea level rise would mean further coastal erosion and inundation. This is especially true of those cities which are at the mouths of rivers where flooding can also occur from rain in the hinterland as well as flood surges during combinations of king tides and high off-shore winds. The use of sea dykes, flood gates and an increase in height of river levees may be options for the future. In addition, current high-rise buildings which may have the potential for flooding should look to changing the uses of their lower and street levels for temporary storage rather than to house vital electrical utilities and permanent, locked car parks.

Cities on the margins of current storm, tornado and cyclone (hurricane) belts may find that global warming of the oceans nearby may bring storms to their region as well. The lessons learned from previous cyclones (hurricanes) in well-known regions should now be applied to cities on the margins of these belts. See:

https://www.cyclocane.com/
(Interactive world cyclone/hurricane map. Zoom in for cities)

Hot, dry seasons may also pose a wildfire (bushfire) hazard, especially to cities which are spread over a large area with hilly terrain covered in leafy suburbs and near forested areas e.g. cities in California and northwestern United States, southwestern Canada, eastern Australia. southern Europe and South Africa. See:

https://firms.modaps.eosdis.nasa.gov/map/#z:7;c:148.0,-36.3;d:2020-01-18..2020-01-19
(An interactive world map of fire zones. Move to specific cities and zoom in)

Cities with potential wildfire hazard could look to the creation of buffer zones of open space, water features or extensive sprinkler systems at these margins. Additional provision would also need to be made so that city fire authorities have adequate equipment and hydrant points to fight internal city outbreaks.

2. **Second Stage: reduction in the amount of traffic**, especially private cars, into the central business district of the city. This can be achieved with the introduction of well-considered systems such as:

    a. A city **congestion tax** based on a tag or GPS system whereby individual motorists who enter a designated city zone would pay a fee for entry. This has been tried in several large cities including London and Singapore. This could also be supplemented the use of private ride-share arrangements whereby employees working nearby in the city and who live in the same dormitory suburb could roster days when one driver picks up the others to transport them into the common area of employment.

    b. Use of commercial ride-share companies such as Uber or taxi services on a regular basis with at least 3 or 4 commuters meeting each morning and afternoon at a common pick-up point i.e. similar to (a) above except using a commercial vehicle instead of a private one.

    In the ideal scenario, cars with only one occupant, private car parking on city roads and randomly in commercial parking stations would be a thing of the past. Where places of employment could not provide off-street parking for their employees within a salary package, then individual car parking would have to be done at a shared cost within high-rise parking stations in which each space is permanently rented for specific times for specific individuals. Street parking would be limited during working hours to the current loading zones and short-term drop-off points for vehicles with congestion tax tags. Naturally the current model of late-night and entertainment parking out of business hours could still operate. This may include a limitation of the hours of operation of the congestion tax outside of business hours and with a return to random parking in parking stations and the usual restriction in street parking.

    Other than changes to signage and installation of detectors systems similar to those now operating on most tollways could be applied to a city precinct congestion tax. Prior to this there would have to be a massive public relations and advertising campaign to show the city car commuters that such a tax is necessary.

3. **Third Stage: improve public transport** so that commuters do not have to use their own cars. In many cities, public transport is still inadequate during peak hours. Buses, trams and subway trains always seem to be

overcrowded and inadequate in numbers at such times. Some cities currently have efficient subway, or metro, systems capable of moving large amounts of people in a short period of time. Tokyo, Singapore, New York and London, just to name a few. These systems, like any proposed future public transport system, are electric and in the future, there would have to be an appropriate expansion of the power grid to meet such demands.

Many cities once had tramway networks which were efficient at moving large amounts of people but were deemed then as hindering the flow of motor vehicle traffic in the city and inner suburbs. These were often removed and replaced by busways, push bike lanes or simply tarred over as roads. There has been in recent years an upsurge in the introduction of light rail systems and the discovery that they can operate within city centres and move people around in larger numbers than cars or even buses. The frequency of services as well as the number of carriages or seats is also a major factor. To reduce competition with motor vehicles, light rail tracks could be built on old tramway thoroughfares, in secondary streets where motor traffic is kept to service vehicles, along raised tracks on the edges of waterways or even overhead on raised structures or even on connected building terraces.

Public transport could take several, integrated forms such as bus-rail-ferry networks as is the case in many cities now, but with an increased global warming and congestion, the main aim is to reduce motor vehicle access within cities. Where cities have spread over large areas with dormitory suburbs and their regional shopping and trade centres forming miniature satellite zones, there should also be provision for commuters to drive and park at public transport hubs or park-and-ride facilities. These should also be interconnected by secondary ring systems of other public transport.

4. **Fourth Stage: retro-fit existing buildings with solar panels** for the production of electricity. This could make use of the tops of smaller roof area buildings which often have wasted space apart from roof-top apartments and lift and air conditioning units. These panels could be installed at the shared cost of the building's tenants and used against their normal power use. Alternatively, the panels could be installed by a utilities company on a rental basis with payment going to the Body Corporate or Tenants' Association to offset the usual rates or fees imposed on the building.

In addition, some buildings which have a large glass exterior facing the sun may also consider coating the windows of the building with photovoltaic film for both electricity generation and reduction of glare. This could be an expensive option however, and tenants would need to consider the cost – benefit ratio in coating their windows with such films.

See also:

https://www.solarquotes.com.au/panels/photovoltaic/thin-film/
(Types of thin film systems available)

https://www.kashanu.ac.ir/Files/Content/thin%20film%20solar%20cells.pdf
(Comprehensive article on the application of thin film voltaics)

https://biblio.ugent.be/publication/4238935/file/4238983.pdf
(Basic overview of thin film voltaics)

https://ocw.tudelft.nl/wp-content/uploads/Solar-Cells-R5-CH7_Thin_film_Si_solar_cells.pdf
(Technical article on thin film solar cells)

https://arena.gov.au/assets/2019/09/cutting-edge-reserch-in-pv-arena-funded-projects.pdf
(Some current research in solar cell development)

5. **Fifth Stage: increase green spaces** within the city with the planting of more trees. This may include the development of gardens, community farms, water features and recreational fields being developed over old, disused spaces such as demolished building sites, old quarries, disused industrial land and along river sides. This idea is not new and many cities have ongoing beautification plans already in place.

As an additional innovation, some cities have begun to look at the tops of buildings as potential green spaces for small gardens and recreational parks. These would only be suitable for those buildings which have considerable flat roof space and have the structural integrity to support the sub-structures needed for the planting and maintenance of large trees and gardens. Some smaller garden units could also be grown on terraces of buildings and even down their sides. Even relatively small balconies could be used for small potted plants and vertical hydroponic gardens.

Figure 12.7: The new city centre of Hangzhou, China showing an emphasis on the use of green space (Photo: Public Domain).

See also:

https://www.projex.com.au/incredible-rooftop-gardens-from-around-the-world/
(Examples of rooftop gardens around the world)

https://www.wikihow.com/Create-a-Rooftop-Garden
(Construction a rooftop garden at home)

https://www.growinggreenguide.org/wp-content/uploads/2014/02/growing_green_guide_ebook_130214.pdf
(A guide to creating rooftop gardens in Victoria, Australia. Local but many good ideas)

https://www.niu.edu/communiversitygardens/_pdf/Vertical-Rooftop-Gardens.pdf
(Good article on growing vertical gardens)

https://www.academia.edu/25927056/ROOF_GARDENING
(A book which can be downloaded with detail on rooftop gardens)

Figure 12.8: Building gardens, Singapore.

6. **Sixth Stage: adequate water supplies** in dryer regions where there may also be a case for the provision of water for rooftop gardens, community gardens on the ground and to supplement general water storage. Water could be collected from the roofs of small structures such as garden sheds, shelters, lift or stairwell buildings or other structures which are often found on top of buildings. This water can be collected in the usual way and stored in water tanks on the roof near the garden or structure. However, with water weighing one kilogram for every litre, a modest 3000L tank would weigh 3000 kilograms or 3 metric tonnes, so the supporting structures below the tank would need to be strong enough to hold the weight.

For gardens below the roof, such as those on terraces, large balconies and on the ground, buildings may be retrofitted with several smaller tanks on different levels which would collect water from any existing roof drainage system. Alternatively, for buildings with a complex roof structure full of air conditioning ducts, solar panels and elevator buildings, rain-catching structures may have to be added. These would probably be simple awnings or roof panels which would cover the existing structures but be anchored strongly to the main building to be able to resist high winds. Drainage off these roof structures would go via pipes, internally or externally, down to smaller water tanks on each of the garden levels. This would be a major plumbing project which would have to be shared by the tenants of the building, but the benefit would be more useable water which could even be used as potable water by incorporating appropriate filtration and sterilization systems.

Figure 12.9: Diagram showing a possible retrofit for water collection and supply to roof gardens and internal spaces:
1. Rooftop collecting surface
2. External drainage pipes to terraces
3. Internal water tanks on each level
4. External drainage pipes
5. Mains water drainage.

These stages are not absolute but one possible logical sequence in retrofitting existing buildings within an established city. Many of these stages have already been implemented in local plans of some cities and it is possible to complete several stages concurrently as the implementation of one stage will also allow for the development of another.

12.7 Transportation - on Land

Transportation is also another major issue in greenhouse emissions. Current transportation still generally relies on road and sea systems which use fossil fuels. Re-development of light rail and high-speed inter-city and country rail using electricity generated by methods other than fossil fuels would be a better alternative for land transport. With such development there would be less need to work in the central business district of cities and massive decentralization of industry would also help to decrease city populations.

On the road, electric vehicles are already making an appearance and the use of compressed hydrogen gas as a fuel has great potential. Already, some petroleum companies and automobile associations are already establishing charging and hydrogen refueling stations, but it will take time before petroleum-based vehicles are replaced completely. To date, the options for road transport include:

- Hybrid electric-petroleum vehicles with several mainstream brands already on the road;

Figure 12.10: Current well-known petrol-electric hybrid car (Photo: Public Domain).

- Electric vehicles using lithium rechargeable batteries as the only power source;

Figure 12.11: Fully electric town car (Photo: Public Domain).

- Internal combustion engine vehicles which use compressed hydrogen gas as a fuel and requiring only small modifications to some current vehicles. When hydrogen gas is burnt with air, the main waste is steam, however with incomplete burning there may also be some oxides of nitrogen as well. See also:

  https://www.energy.gov/sites/prod/files/2014/03/f11/fcm03r0.pdf
  (Hydrogen gas used in internal combustion engines - details)

Figure 12.12: Prestige road car using compressed hydrogen gas (Photo: Public Domain).

- Electric vehicles using compressed hydrogen gas as a fuel in fuel cells which create electricity for the motor.

- Electric vehicles using overhead power lines which are engaged continuously or as required using a pantograph to connect the vehicle to the wires. This is not new technology as electric trolley buses have been used in many cities for a long time. Innovations may involve the use of rechargeable batteries which are used in older streets not wired for electricity.

- Other hybrid combinations using combinations of fuel cells, batteries and hydrogen power.

In urban transportation, hydrogen-fueled vehicles would greatly assist the reduction in vehicular pollution, especially the carbon monoxide produced by current internal combustion engines. Over small distances such as within cities, overhead electrification for light rail, buses and some commercial vehicles would be a relatively simple development. Moreover, in time, with the new trend for many automobile companies to manufacture electric vehicles, their price should drop considerably. Associated with policies for reduced city traffic and the provision for on-site charging stations, electric vehicles as city transportation becomes more applicable. This would be further enhanced with the introduction of fast-charging outlets at employment centres, parking stations and at home.

Charging centres are probably best set up by existing fuel companies as they have had long experience in roadside services. Moreover, there would be only marginal conversion required for the use of hydrogen gas as a fuel. Electric recharging could be done using common power from a national grid network or from solar voltaic farms on site, especially in more remote service centres. Currently regular battery charging could take some time, especially at home or office charging points, but with at such places time would not be a major factor. Fast charging centres are now opening with full charges being done in about 30 minutes with new systems being developed that would be comparable to the current petrol refueling times. Another innovation for quick service with electric vehicles would be the use of interchangeable battery packs. This would mean that a vehicle fitted for quick replacement as well as recharging could stop at designated service centres and quickly replace a depleted-charge battery with one that is fully charged. This would only take a few minutes and the new battery could even be dispensed from a vending machine at the service centre at all hours of the day.

One new innovation which has been trialed in several cities, including Ipswich, Australia is the **Autonomous Vehicle (AV)** or self-driving car. These are fitted with a range of local sensors and regional GPS data to enable the vehicle to be driven within certain zones without a human driver. Within a city with reduced transport zones, such vehicles could be programmed to pick up several passengers along a designated route to and from work places. Autonomous vehicles also have great potential for service vehicles such as taxis, ambulances, delivery vehicles and transportation for the elderly and handicapped. Of special benefit would be the use of autonomous vehicles in long distance haulage along set routes.

See also:

https://infrastructure.org.au/wp-content/uploads/2017/09/AV-paper-FINAL.pdf
(Details about the future of autonomous vehicles)

https://assets.kpmg/content/dam/kpmg/au/pdf/2019/australia-future-transport-and-mobility-report.pdf
(General article about the future of transportation in Australia)

https://assets.kpmg/content/dam/kpmg/xx/pdf/2019/02/2019-autonomous-vehicles-readiness-index.pdf
(2019 autonomous vehicle readiness index)

Long distance travel and freight haulage poses more problems but the technology is not new, only the longer distances and the need for regular recharging stations for electric road transport and hydrogen fuel service stations have to be overcome.

See also:

https://na.chargepoint.com/charge_point
(Interactive map of world electric charging stations)

https://openchargemap.org/site
(Another interactive map of world charging stations)

Figure 12.13: Hybrid prime mover for a semi-trailer road haulage.

In countries where extensive electrified rail systems have been available for many years, the use of high-speed passenger services and freight haulage has not been a problem. In larger counties such as Australia, Canada, the United States and Russia which have relied upon diesel locomotives to operate these services over incredibly long distances, overhead electrification may be expensive. The use of hybrid electric-hydrogen systems to power the locomotives may be the answer.

For general information See also:

https://www.greenvehicleguide.gov.au/pages/Information/HydrogenVehicleInformation
(General green vehicle guide)

https://www.energynetworks.com.au/news/energy-insider/refuelling-your-car-with-hydrogen/
(Guide to refueling vehicles with hydrogen gas)

https://www.shell.com.au/make-the-future/cleaner-mobility/the-great-travel-hack/season-one/fuels-vehicles.html#vanity-aHR0cHM6Ly93d3cuc2hlbGwuY29tLmF1L21ha2UtdGhlLWZ1dHVyZS9jbGVhbmVyLW1vYmlsaXR5L3RoZS1ncmVhdC10cmF2ZWwtaGFjay91c2EvZnVlbHMtdmVoaWNsZXMuaHRtbWw
(Good site from the Shell Company about new power sources)

12.8 Transportation – at Sea

Carbon emissions from international shipping are around 0.9 gigatonnes (Gt) of CO_2 which account for about 3% of total global emissions. Growing demand for shipping could see these emissions increase to 1.7 Gt by 2050 if current systems continue to operate. Shipping is regulated by the International Maritime Organisation (IMO), a UN agency with 174 countries as members. In 2018 the IMO agreed to reduce carbon emissions from ships by 50% by 2050 compared to those of 2008. They also recommend pursuing efforts to phase out carbon emissions consistent with the Paris Agreement on climate change, despite shipping not being covered by it.

Currently, most commercial ships at sea use diesel oil or liquefied natural gas (LNG) as the primary fuel source. These fuels can be used in a variety of propulsion units such as diesel engines or hybrid diesel-electric generators to power electric motors linked to propeller or water jet systems. There are also a few large naval ships, such as aircraft carriers and submarines and a limited number of merchant vessels and ice-breakers which are nuclear powered. Unfortunately, ships using fossil fuels as their primary fuel source still continue to put considerable amounts of carbon dioxide and oxides of nitrogen (NO_X) and sulfur (SO_X) into the atmosphere.

Trials have been carried out on a variety of environmentally-friendly propulsion systems including the use of fuel cells using liquefied natural gas (LNG) to power electric motors. The *Viking Lady* is a 92-metre-long vessel of 6100 gross tonnage which is used to supply gas and oil platforms and runs on liquefied natural gas which significantly reduces NO_X and CO_2 emissions. Using such a dual system, it is claimed that the Viking Lady would reduce emissions by 30% to 50%, especially with the installation of batteries to store excess electrical power. One step beyond this would be to use compressed hydrogen or ammonia gas instead of the LNG as the fuel. Another power source would be the use of hydrogen gas to power marine gas turbine engines, either directly for propulsion through the ship's propellers or coupled with generators to produce electricity which would then be used to drive the propellers.

Improvements in efficiency of operations such as reducing and optimising ship speed, improvements in ship design and wind assistance technology could deliver reductions of 30-55%. However, the cost here would be the need for ship owners to move away from traditional oil fuels and change to more environmentally-efficient propulsion. This may mean however, a change from using large, fast but expensive container ships to many smaller, slower and less-expensive ships more frequently on direct but well-established shipping lanes. Within certain parameters, and well-defined navigation routes, some of these ships may be autonomous and under control from one or several ports.

See also:

http://www.energy-transitions.org/sites/default/files/ETC%20sectoral%20focus%20-%20Shipping_final.pdf
(Future concepts of shipping)

https://www.ijcaonline.org/archives/volume178/number25/koikas-2019-ijca-919043.pdf
(Autonomous shipping)

https://commons.wmu.se/cgi/viewcontent.cgi?article=1601&context=all_dissertations
(Technical research paper on the future of shipping and autonomous ships)

https://www.rolls-royce.com/~/media/Files/R/Rolls-Royce/documents/customers/marine/ship-intel/aawa-whitepaper-210616.pdf
(Good articles on autonomous ships from Rolls Royce)

Another concept which may be used in the future, is the use of auxiliary wind ships. This is another turn towards some of the ideas of the past. By the beginning of the 20th century, coal-fired, then oil-fired steam ships had established their superiority as the main mode of shipping. However, at the same time, cost-effective sailing ships had also developed as the logical evolution to the clipper ships of the mid-19th century. Square rigged ships which had previously been the main-stream form of marine transport began to give way to the more manoeuvrable fore-and-aft or schooner-rigged ships which could sail closer to the wind. These used well-established sea lanes which followed the consistent winds across the ocean. Many of these winds were subsequently called **Trade Winds** and many of the world's great seaports were developed at coastlines which could best make use of these winds. Global warming may change the intensity of some of the world's wind patterns but essentially the location and direction of major winds should stay the same.

Most shipping routes also make use of the world's great ocean currents which are driven by the surface winds. Future shipping would need to also make use of such currents to save fuel.

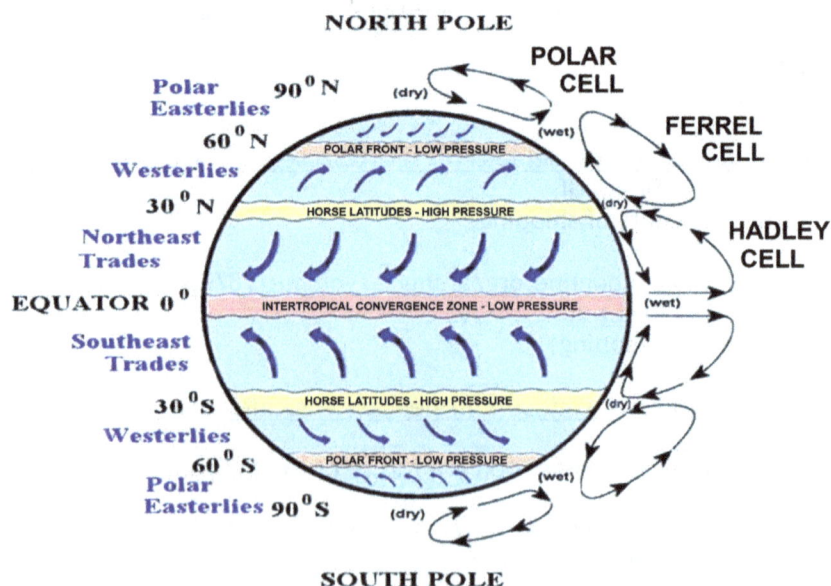

Figure 12.14: Diagram showing the main winds of the world. Note the Coriolis Effect deflecting the winds due to the rotation of the Earth.

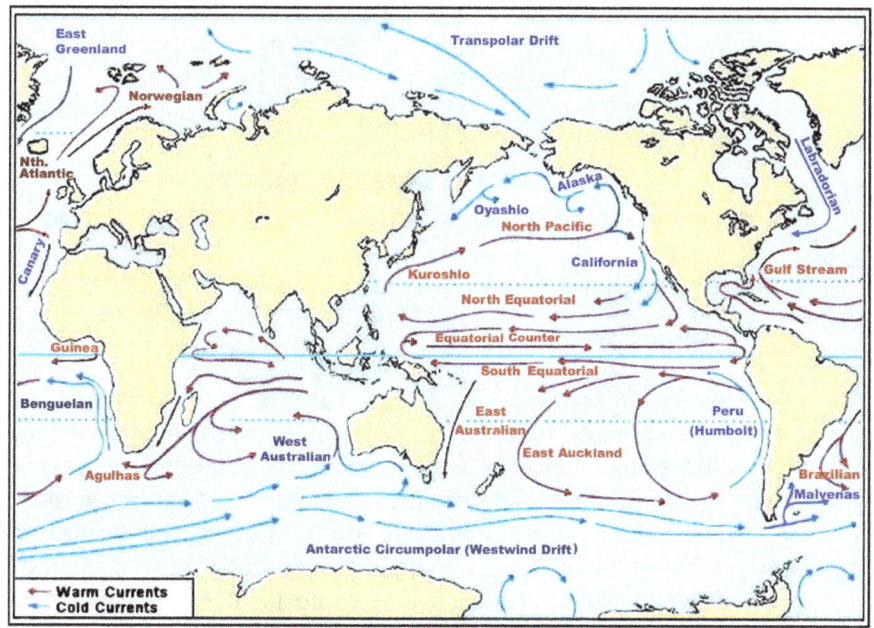

Figure 12.15: Diagram showing the main ocean currents of the world

See also:

https://www.nnvl.noaa.gov/weatherview/index.html
(World's winds from NOAA - use pointer to move Earth)

There are a number of wind-assisted designs for use with major shipping being trialed or on the drawing board. These include:

- **Traditional sail concept**, including the use of hard and soft sails to catch the wind in a way similar to the traditional use of sails in the past. This could include both square-rigged and schooner-rigged ships but unlike the windjammers of the past, sail and rigging handling would be fully automated and computer controlled to get the best performance from the wind along the intended route of the ship. Ships could be fitted with dynamically-shaped sails which could be rotated from the fore-and-aft rig of a schooner to a square sail format depending upon the wind direction. Masts could be designed so that they could be hydraulically tilted and lowered in case of high winds. In this case and when entering port or tight channels, the ship would use its auxiliary power sources. Moreover, if the sails were also made from or coated with flexible photovoltaic film, electrical power could be generated for supplementary battery charge and propulsion.

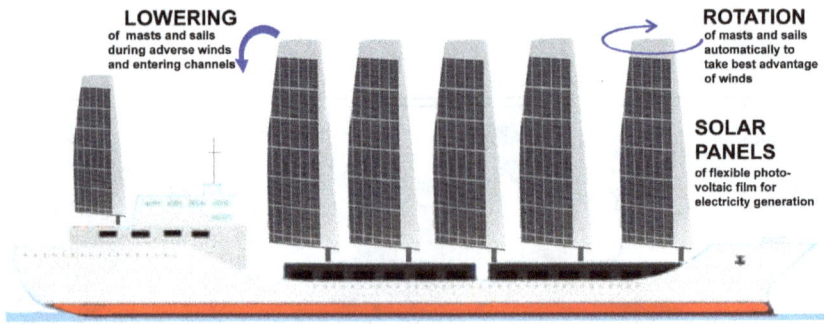

Figure 12.16: Diagram of a cargo ship under sail. Note the flexible solar panel sheeting on the outside of the sails to provide addition power to batteries when the sails are lowered.

- **Flettner rotor concept** which makes use of a spinning cylinder standing on the deck of the ship. This would use the **Magnus effect** to create propulsion. The Magnus effect is named after Gustav Magnus (German: 1802 - 1870), and it explains that there is a force generated when a fluid flow such as wind passes over a rotating body. This force will be at right angles to both the direction of flow and the axis of rotation and could be used to power the vessel. German engineer Anton Flettner (1885-1961), was the first to build a ship that attempted to tap this force for propulsion, and ships using his type of rotor are sometimes known as Flettner ships. The first ship with Flettner rotors was used in 1920. When the wind hits the spinning cylinder the airspeed is higher on one side of the cylinder creating a force directed forward. This system could also be retro-fitted to existing ships, especially container ships and bulk carriers. Several ships have already been fitted with rotors to supplement and reduce their use of fossil fuels. These include the oil tanker *Maersk Pelican* and Viking Line's passenger ship M/S *Viking Grace*.

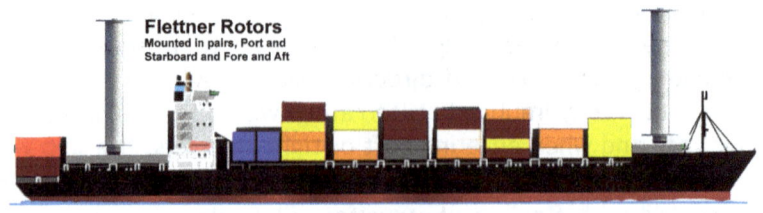

Figure 12.17: A container ship powered by Flettner rotors.

- **Kite sails**, similar to those used for kite surfing but on a larger scale could be connected to the bow of the boat and used when there as an almost constant wind blowing from astern. Ships need hardly any modification for this arrangement and it does not take up space on deck when not in operation.

See also:
https://www.vos.noaa.gov/MWL/apr_09/skysails.shtml
(Research and limitation of kite sails)

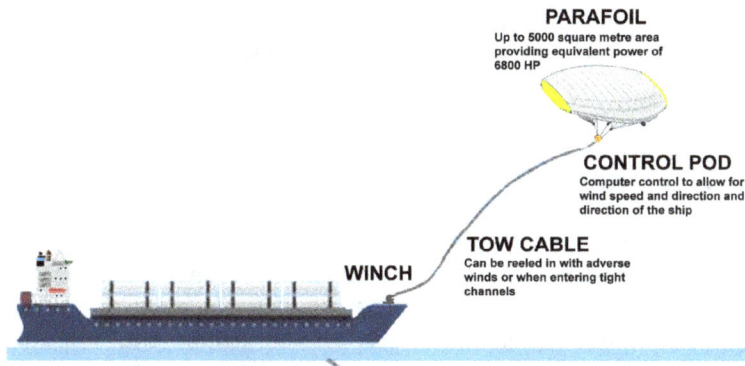

Figure 12.18: A container ship using an employable parafoil kite as a supplementary power source.

In addition to the use of sails and rotors, some ships may supplement or even replace other propulsion systems using solar-powered electric motors in conjunction with battery storage. Whatever the future holds for maritime transportation, propulsion systems will need to be free of fossil fuels and rely upon a mixture of other alternative energies such as hydrogen fueled turbines, wind assistance, nuclear power and solar power.

See also:

https://unctad.org/en/PublicationsLibrary/rmt2019_en.pdf
(Detailed review of current marine transport by the UN)

https://www.ecomarinepower.com/en/aquarius-eco-ship
(Good site about on-going research including a good video)

https://www.irena.org/-/media/Files/IRENA/Agency/Publication/2015/IRENA_Tech_Brief_RE_for-Shipping_2015.pdf
(Comprehensive review of future wind potential)

https://www.stg-online.org/onTEAM/shipefficiency/programm/06-STG_Ship_Efficiency_2013_100913_Paper.pdf
(Good technical article about Enercon-Eship 1 and of wind power in general)

https://www.itf-oecd.org/sites/default/files/docs/decarbonising-maritime-transport-2035.pdf
(very detailed report on marine zero emissions by 2035)

12.9 Transportation – in the Air

There are fewer alternative choices of power when it comes to air transportation and most still rely on the use of aviation gasoline for propulsion. The two main parameters of flight have always been lift and propulsion: lift to get the aircraft off the ground and propulsion to move it through the air from place to place. Both generally require a power source to provide the energy required. In all cases, weight and air resistance are also been important factors. In the early days, lighter-than-air balloons obtained lift using hot air, then explosive hydrogen gas and more recently helium gas. Propulsion was provided by mechanical means, simple steam or internal combustion motors using fossil fuels. These evolved into aerodynamically-shaped blimps which were totally filled with hydrogen gas but had no rigid structure. By the beginning of the 20th century, rigid airships or **dirigibles** appeared. Often called 'Zeppelins' after the pioneer engineer Count Ferdinand von Zeppelin (German: 1838 - 1917), these airships seemed to hold great potential for future air long-distance travel at a time when fixed-wing aircraft where in their infancy.

Following their unfortunate use as aerial bombers in the First World War, several successful variations were made for luxury air travel. The *Graf Zeppelin* LZ 127 (*Luftschiff Zeppelin 127*) was a German passenger-carrying dirigible which flew from 1928 to 1937 to many parts of the world including regular flights between Germany and Brazil and an historic long-distance flight around the world in 1929. It was given lift by hydrogen gas in 16 large internal gas bags made from cotton and lined with goldbeater's skin, a thin membrane made from the intestines of oxen which prevented massive gas loss. The airship was propelled by five Maybach internal combustion engines using **Blau gas** as fuel. This gas, named after its inventor, Dr. Herman Blau (German: 1871 - 1944) was similar to coal gas and had the same density as air. On board the airship, it was contained in pressurized cylinders and was vaporized upon release to the engines. The *Graf Zeppelin* made 590 flights totalling almost 1.7 million kilometres and was operated by a crew of 36. It carried 24 passengers in luxury with cabins similar to sleeper cabins found on luxury trains. It also and had a well-appointed dining room, smoking room and an observation deck.

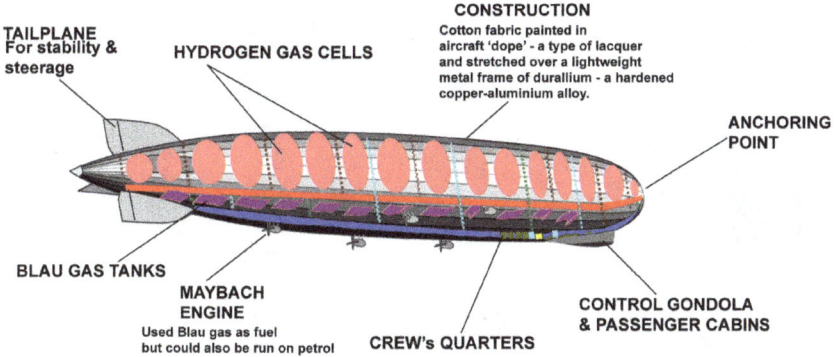

Figure 12.19: A diagram showing a cutaway view of the Graf Zeppelin dirigible which operated efficiently from 1928 to 1937 when it was withdrawn from service (Photo: Public domain and modified).

Other countries, notably the United States and Great Britain also embarked on large dirigible construction. The British built the bigger *R101* and *R102* to be luxury airships but scrapped their plans when the *R101* was wrecked during a storm in France during its maiden voyage on 5 October 1930, killing 48 of the 54 people on board.

The United States Navy built the *Shenandoah*, *Akron* and the *Macron* to be forward scouting platforms for their ships and lift was provided by the inert gas helium. The Macron and the Akron were designed as flying aircraft carriers and each carried five small fighter biplanes.

The end of the romance of lighter-than-air airships came with the crash and explosion of the German zeppelin, the *Hindenburg* as it was landing at Lakehurst, New Jersey on May 6, 1937. Whilst many of the passengers and crew survived the horrendous, fiery crash, this and the wrecking of other airships, including the three American Navy zeppelins, spelled the end of the giant airships. During the period of their development and service, heavier-than-air fixed-wing aircraft became larger, safer and more popular. Using firstly fossil fuel aviation gasoline (avgas) and later kerosene-based fuels in jet aircraft, modern-day airliners contribute greatly to the world's greenhouse gases due to emissions of carbon dioxide, nitrogen oxides, particulates and other emissions.

Figure 12.20: The Hindenburg and a US Coastguard seaplane over the Airship hanger at Lakehurst, New Jersey, 1936 (Photo: US Coast Guard).

See also:

https://www.biologicaldiversity.org/programs/climate_law_institute/transportation_and_global_warming/airplane_emissions/
(Good general article on aircraft emissions)

Modern aviation for both passenger and freight has increased greatly over the last twenty years. Pollution in the upper atmosphere due to high altitude aircraft accounts for about 2% of total global emissions, mainly from high volume international travel. There is no doubt that the need for aviation will continue and this is expected to rise in the future. Accordingly, there will be a greater need to reduce or remove the use of fossil fuels used in jet engines. Whilst there may be some improvement in emission reduction as aircraft and engine design become more efficient, the problem with combustion of hydrocarbon fuels still remains.

For short duration local flights, the development of light weight aircraft with propellers driven by electric motors powered by lithium ion batteries is well underway. The main problem here, however is that the energy-to-weight ratio of these batteries is still very much greater than that of using hydrocarbon fuels. That is, in a small aircraft, much of its weight is taken up by the power source rather than passengers and freight. This also limits the range of the aircraft between battery charges, but some companies envisage that with further development in light weight batteries, international travel in electrically-powered aircraft may be possible in the near future.

Figure 12.21: NASA's concept STARC-ABL concept hybrid-electric propulsion aircraft (Photo: NASA).

See also:

https://mashable.com/feature/electric-airplanes-future-flight/?utm_cid=a-seealso
(Some information about electric aircraft)

https://www.nasa.gov/centers/armstrong/features/Major-Milestones-for-NASAs-Electric-X-Plane.html
(Details and video of NASA's electric aircraft)

https://www1.grc.nasa.gov/aeronautics/electrified-aircraft-propulsion-eap/eap-for-larger-aircraft/
(NASA's development of larger electric aircraft)

There are also developments in the use of biofuel and hydrogen gas as fuels for aviation. Some argue that aviation biofuels are carbon neutral because they are derived mainly from plants e.g. alcohol produced from fermentation of sugar cane. Others suggest that whilst biofuels come from renewable sources unlike fossil fuels, they still produce carbon dioxide gas emissions when used in engines. Never-the-less, aviation biofuels have been blended with normal aviation gasoline for several years on some international flights. See:

https://en.wikipedia.org/wiki/Aviation_biofuel
(Scroll down to the list of flights using biofuels)

Hydrogen-powered aircraft can be fuelled either by the combustion of the compressed, liquid hydrogen in a jet engine, or other kind of internal combustion engine, or can be used to power a fuel cell to generate electricity to power a propeller. However, if used in a jet engine, liquid hydrogen would need about four times the volume to provide the same amount of energy to an engine compared to kerosene-based jet-fuel. In addition, hydrogen gas is highly volatile and would need to be contained in strong, pressurized containers precluding it from being in the wings, as with conventional fuels. Because of this, any liquid hydrogen aircraft designs would need to store the fuel in the fuselage, leading to a larger fuselage length and diameter than a conventional fuelled aircraft. This larger fuselage size would cause more surface friction drag and wave drag over the fuselage reducing the speed of the aircraft or requiring more energy for the same speed. On the other hand, hydrogen is about one-third of the weight of kerosene jet-fuel for the same amount of energy so a hydrogen-fuelled aircraft would have about one-third of the fuel weight and give greater range.

Another concept for the future of aviation is the return of the airship. Whilst there has not been as much research into airships as with fixed-wing aircraft, the airship may be useful where air transportation is needed in remote areas where a large landing strip cannot be built. Airships have several advantages over conventional fixed-wing aircraft:

1. Lift is provided by the gas within the envelope and so there is no need to expend fuel to maintain flight.

2. Low-speed manoeuvrability which will allow them to land on only a small area, especially in remote areas. Hybrid airships, which have a fuselage shaped like a wing which also provides some extra lift would only need a small runway.

3. High potential lifting power with some of the largest designs theoretically being able to carry about the same cargo weight as some fixed-wing cargo aircraft.

4. High cargo volume if used as a heavy lift airship, as it would have to be very large, and so would be able to accommodate a correspondingly huge cargo bay.

However, there are still some major technical problems, notably with the nature of the lifting gas. Initially hydrogen gas was used in some of the early zeppelins but this proved to be dangerous as in the case of the *Hindenburg*. Helium gas, whilst heavier than hydrogen, is inert and non-explosive and would be the preferred lifting gas. Unfortunately, it is rare and much more expensive and it is also in great demand in industry and in a variety of medical processes. Helium gas forms within the Earth as a result of radioactive decay, with some of it escaping to the surface and some being contained within oil traps. A few natural gas wells in the United States are the main source of helium and this is extracted from the oil by fractional distillation. Recent discoveries have also indicated additional probable reserves under the Rocky Mountains in North America and in the East African Rift. The main supplies of commercial helium gases come from:

United States	55%
Qatar	32%
Algeria	6%
Australia	3%
Russia	2%
Poland	1%
Others	1%

Apart from the difficulty in obtaining helium for their lifting gas, airships also have problems with the weather. Their size and low speeds make them unmanoeuvrable in strong winds which was the main cause of destruction of the earlier airships. Never-the-less, there is still strong interest in the commercial applications of airships, especially hybrid models which combine a flattened wing shape which provides some lift at speed with the lifting power of the gas inside the envelope.

The successor of the original Zeppelin Company in Friedrichshafen, Germany, has also reengaged in airship construction, building semi-rigid airships consisting of a cylindrical gas bag which supports an extended gondola powered by small engines. Apart from having a greater payload than simple gas-filled blimps, their main advantages are higher speed and excellent manoeuvrability in light air. So far, they have been limited to short distance flights for advertising and joy flights but interest in bigger airships continue with both NASA and Lockheed-Martin developing hybrid airships for commercial purposes.

In the future, hybrid airships may become very useful in their role as heavy lift or even luxury passenger airships in the same vein as the *Graf Zeppelin*. They may be powered by rotatable, ducted electric or hybrid fuel cell/combustion motors and have supplementary power drawn from an upper surface skin of flexible photovoltaic thin film panels. Buoyancy and gas wastage could be controlled by having a two-way gas compression system which releases gas into the gas bags when extra buoyancy or change in trim is required but then compressed back into cylinders in the lower part of the hull when less buoyancy is required. Buoyancy may also be controlled by an internal **balloonette** inside one or several of the main gas bags. This smaller internal balloon can be inflated or deflated to adjust buoyancy as required. The passenger compartment held below or with the lower hull of the airship may also be detachable and able to be replaced with cargo containers in a freight configuration. For passenger transfer, one or two small electrically-powered aircraft could also be carried as air taxis. These could be carried in an internal space near the rear of the airship in the same way that aircraft were carried in the US Naval airships. Composite materials including light but strong bamboo might also be used in their construction. Whilst slow and subject to weather conditions they may offer a new era in aviation transport.

See also:

https://ntrs.nasa.gov/archive/nasa/casi.ntrs.nasa.gov/20170001665.pdf
(A NASA report on airship potential and design)

https://ubibliorum.ubi.pt/bitstream/10400.6/5473/1/3744_7425.pdf
(A good, detailed thesis on the future of airships - scroll past the Spanish introductions)

https://apps.dtic.mil/dtic/tr/fulltext/u2/a568298.pdf
(More on hybrid airships)

https://www.airshipsonline.com/airships/interior/R100Interior.htm
(Inside the airship R100, built in 1925 like a 'floating hotel')

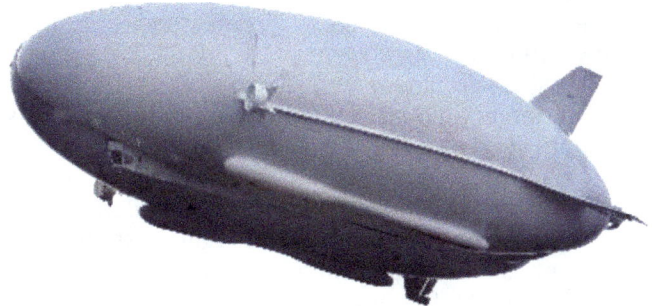

Figure 12.22: An artist's impression of a hybrid airship (Photo: NASA).

Whatever the future holds for commercial aviation, it seems likely that the dominance of fossil fuels will be phased out in favour of electricity, hydrogen or biofuel or probably a hybrid of a more environmentally-friendly fuel and electrical power.

See also:

http://www.energy-transitions.org/sites/default/files/ETC%20sectoral%20focus%20-%20Aviation_final.pdf
(A future for aviation)

https://www.oecd.org/sd-roundtable/papersandpublications/49482790.pdf
(A good summary of the future of aviation within a greener world)

https://www.aerosociety.com/media/5962/3-hybrid-power-in-light-aircraft.pdf
(Scientific details about hybrid electric/fuel aircraft from Cambridge University)

12.10 Construction and Materials

Currently, the construction industry is dominated by materials such as concrete, steel, aluminium and glass. Copper for electrical wiring and some construction is also a major material. Stone and brick have been used for centuries for some modest high-rise buildings but materials such as timber are more common in low-rise construction and interiors of larger buildings.

Concrete is responsible for about 4-8% of the world's carbon dioxide emissions and it is one of the most widely used substance on Earth. Concrete is made from cement, gravel, sand and may also use steel as a reinforcing material. The raw materials for cement, limestone rock, gravel and sand are non-renewable materials and must be mined, removing or destroying part of the local environment in the process. The mixing of concrete also requires a large amount of fresh water which is also becoming short in its supply in some places.

Cement is really a binding material which is needed to be added to other material such as sand, gravel or crushed rock or aggregate to make concrete. **Portland cement** is the most commonly used cement in the construction industry. It is made by heating limestone rock or dead coral, both forms of calcium carbonate ($CaCO_3$), with other materials such as clay to 1,450 °C in a kiln. This process is called **calcination** and it removes carbon dioxide from the calcium carbonate to form calcium oxide, or quicklime.

limestone		quicklime		carbon dioxide
$CaCO_3$	$\xrightarrow{1,450\ °C}$	CaO	+	CO_2

This is then chemically combined with the other materials in the mix to form calcium silicates and other compounds forming a hard substance, called clinker which is then ground with a small amount of gypsum to make ordinary Portland cement (OPC). When mixed with water and aggregate, an irreversible chemical reaction hardens the cement. Whilst there can be some reduction of from 40% to 95% of carbon dioxide emitted in this process by the addition of fly ash or slag waste, both come from the burning of coal in power stations and manufacturing of steel respectively. However, a mixture of fly ash, slag and water reacted with alkaline fluids such as the hydroxides or silicates of sodium and potassium, can be used to form a slower setting concrete, called **geopolymer concrete**, when mixed with sand and/or aggregate. Sodium hydroxide can be manufactured by the electrolysis of seawater which also produces chlorine gas for bleaching. Other materials could be used instead of fly ash, including volcanic ash and its hardened rock variety tuff which could be powdered. This has good properties as a concrete without the need for heating and the release of large amounts of carbon dioxide. Mixed with volcanic ash and using seawater, a concrete similar to that used by the ancient Romans called **opus caementicium** for their long-lasting concrete constructions could be used in modern constructions.

See also:

https://www.nature.com/news/seawater-is-the-secret-to-long-lasting-roman-concrete-1.22231
(Recent tests on Roman concrete)

https://www.fhwa.dot.gov/pavement/concrete/pubs/hif10014/hif10014.pdf
(A short summary about geopolymer concrete)

Figure 12.23: Street in Pompei destroyed in 79AD by the volcano Vesuvius (seen beyond the end of the street). Note the walls of bricks and Roman cement and the paved road made from slices of natural hexagonal columns of basalt rock also from the volcano.

Such **hydraulic cement**, which not only hardens by reacting with water but also forms a water-resistant product, was not available in ancient China and so could not be used as a mortar in the many brick buildings which they built, including the Great Wall. Around 500 AD however, they found that sticky rice soup could be mixed with slaked lime, to make an inorganic–organic composite mortar that had more strength and water resistance than basic lime mortar. The **slaked lime**, or calcium hydroxide was made by heating limestone to form calcium oxide which was then slaked with water. This sticky rice mortar had high adhesive strength, sturdiness, waterproofing capability, and would rival modern cement in strength. Modern chemists have identified amylopectin, a type of complex carbohydrate, found in rice and other starchy foods, to be responsible for the sticky rice mortar's strength and durability. The bricks of the Great Wall and many other Chinese structures were essentially large, mud bricks made by baking shaped mud strengthened with straw in ovens. Put together with the sticky rice mortar, these brick structures were strong enough to resist earthquakes and keep the buildings intact for centuries.

Figure 12.24: Part of the Great Wall showing the large mud bricks commented together with sticky rice mortar.

Steel, aluminium and other metals, must be extracted and refined from their ores; often oxides of the metal or other metal compounds which must be first turned into oxides. Usually they are reacted with other non-renewable materials and heated or processed using large amounts of energy and produce considerable amounts of emission products.

Steel is highly processed iron which has had much of its carbon impurity removed. Iron is produced in a blast furnace using iron ore, mainly the oxide called haematite (Fe_2O_3), along with calcium carbonate as limestone ($CaCO_3$) and coke, a form of carbon (C) derived from the heating of high-grade coal in the absence of air. The mixture is heated to about 1200 ^{0}C and the complex series of reactions reduces the iron oxides to molten iron called pig iron. There is also a waste mixture of silicates called slag and hot furnace gas emissions, mainly carbon dioxide and some sulfur dioxide, produced in the processes. A simplified reaction is:

$$\underset{\text{haematite}}{2Fe_2O_3} + \underset{\text{coke}}{3C} \xrightarrow{1200\ ^0C} \underset{\text{iron}}{4Fe} + \underset{\text{carbon dioxide}}{3CO_2} + \text{wastes}$$

Producing the coke from coking coal also produces ammonia, coal tar and coal gas which are useful but also give off some harmful emissions such as carbon monoxide and hydrogen sulfide depending upon the efficiency of the process and the nature of the original coal.

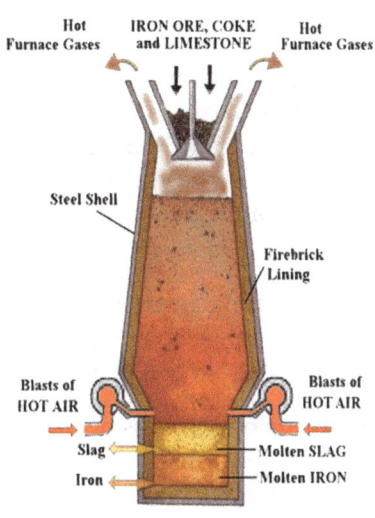

Figure 12.25 & 12.26: Blast furnace at Port Kembla, New South Wales, Australia (left), and a simplified diagram of a blast furnace.

Once the iron has been made, it is remelted and treated with a blast of oxygen to remove some of the carbon as carbon dioxide gas to form steel.

The steel industry has improved its environmental footprint considerably over the last 50 years such that from 1960 to 2015, global steel industry decreased its energy consumption per tonne of crude steel produced, by around 60%. However, it is still an energy intensive industry and accounts for about 7.0% of global CO_2 emissions. In the future, steel will continue to be a major material required in building and manufacture and the industry is mindful of the role that steel-making has in producing greenhouse gas emissions. Some of the ways that these emissions may be reduced might include:

- Recovery of carbon dioxide and other gases emitted from the blast furnace. It would be technically feasible using available capture technology to decarbonise the blast furnace gases before it is used for heat recovery. Applying CO_2 capture could add some 15-20% to the cost of steel production but this would be probably acceptable considering the benefits of CO_2 removal. Unfortunately, this captured CO_2 could not be used within the steel process on site and would need to be transported for subsequent use elsewhere or geological storage.

- Using more recycled scrap iron in the steel-making process. In 2016, the amount of scrap iron and steel available globally was about 700 million tonnes (Mt). This availability is expected to reach about 1 billion tonnes by 2030 according to the World Steel Association. This not only assists in the reduction of gas emissions but also preserves some of the natural resources required for iron and steel making.

- Using electricity as an initial source of heat in both the making of iron and of its processing into steel. This would come from renewable sources such as hydroelectricity, wind generators or solar voltaic cells or most likely a combination of these sources.

- Using electrolysis by passing an electric current directly through molten iron oxides to precipitate the iron out onto positively- charged electrodes or **Metal Oxide Electrolysis (MOE)**. This was a research project reported in Scientific American of May 2013, but very expensive platinum or iridium electrodes were used as these metals can withstand the high temperatures of 1600 ^{0}C. The breakthrough would be to use much cheaper chromium alloys that can also do the job but a commercial-scale demonstration is yet to appear. See also:

https://steel.org/~/media/Files/AISI/Public%20Policy/9956factsheet.ashx

- Using charcoal made from wood, or wood waste. As the source of reducing carbon. This would be a return to the very first methods of producing iron but it would only be considered as an environmentally-friendly process if the wood needed for the process could be grown faster than it is being burned. It may be questionable whether this use of wood would be environmentally sustainable, especially if there was a huge shift in land use which could displace food production. The use of charcoal in steel-making has been trialled to some extent in New Zealand with some success.

- Replacing coke with hydrogen gas which should be technically possible with the reconfiguration of existing technology and some new installations. If hydrogen gas is used, then the main by-product of steelmaking is water, H_2O, rather than CO_2. It is reported that the world's largest steelmaker, Arcelor Mittal in Germany is building a demonstration plant that will use hydrogen gas produced using electricity from offshore wind farms. The hydrogen will reduce ore into iron pellets, which will then be melted into liquid metal but the company says commercialisation is still a long way off and there will need to be a way of reducing the current high cost of such production.

- A similar approach would use gaseous reducing agents—such as natural gas, syngas or hydrogen to flash heat a falling stream of iron ore particles in just a few seconds within a reaction chamber. The reactor design can be incorporated into existing production chains through the feed process, or else replace the blast furnace outright without the need for an extensive system overhaul. Not only would this be effective in reducing energy consumption and costs, but also the system produces no CO_2 emissions.

See also:

https://www.worldsteel.org/en/dam/jcr:96d7a585-e6b2-4d63-b943-4cd9ab621a91/World%2520Steel%2520in%2520Figures%25202019.pdf
(Steel production globally in 2019 from World Steel)

Aluminium is processed from its ore, bauxite which is a complex oxide-hydroxide of aluminium formed from the mass weathering of feldspar minerals in rocks. It exists on or near the surface in huge quantities in many tropical countries including Australia:

Country	Mine production Mt	Reserves x 1000 Mt
Guinea	45,000	7,400,000
Australia	83,000	6,000,000
Vietnam	2,000	3,700,000
Brazil	36,000	2,600,000
Jamaica	8,100	2,000,000
China	68,000	1,000,000
Indonesia	3,600	1,000,000

Table 12.1: Bauxite production and reserves for the top countries in millions of tonnes (Mt)

The complex ore is treated by several steps until it becomes the fine white powder of aluminium oxide or alumina, Al_2O_3. This is then dissolved in molten cryolite or sodium aluminium fluoride - Na_3AlF_6, and the molten solution is then refined by electrolysis in a purpose-built cell in a process known as **Hall-Héroult** electro-refining. This process requires enormous amounts of electricity and gives off the toxic gas hydrogen fluoride (HF) as well as carbon dioxide, perfluorocarbon (PFC) gases and particulates. The HF is reacted with sodium hydroxide to form sodium fluoride which is a useful salt in the production of pharmaceuticals and the particles are filtered out.

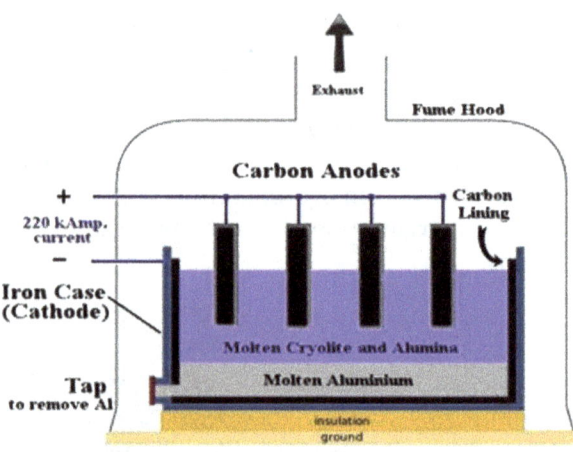

Figure 12.27: A simplified diagram of the Hall–Héroult electro-refining process

The aluminium industry has long sought to develop a process which would have reduced or zero greenhouse gas emissions. Steps towards such a goal might include:

- Revising the current operating systems with greater efficiency with reduced emissions.

- Use of renewable sources of electricity such as hydroelectricity, wind turbines or solar voltaic farms.

- Increased use of recycled scrap aluminium within the smelting process.

- Using new processes which would reduce or eliminate emissions. Such a process is being trialled in Canada, called *Elysis*, which uses an inert anode, rather than the carbon anodes used in conventional aluminium smelting. Instead of releasing carbon dioxide, the process emits only oxygen and the pilot plant will run on hydropower, further cutting the process's emissions.

See also:

http://www.world-aluminium.org/media/filer_public/2013/01/15/fl0000329.pdf
(A brief but detailed look at the future of aluminium)

https://www.aluminum.org/sites/default/files/Aluminum_The_Element_of_Sustainability.pdf
(Aluminium as a sustainable material)

https://www.nrel.gov/docs/fy01osti/29340.pdf
(An overview of aluminium in the future)

Copper, one of the first metals exploited by humankind as bronze, an alloy of copper and tin. It is vital to construction and the electrical industries. It has proven to be the most practical metal for electrical conduction and circuits, however in some applications electrical conduction may be replaced by fibre optics using LASER transmission.

Like aluminium, copper is produced from a complex ore which must be reduced to its oxide before being treated and then purified using electrolysis. One of the most common ores of copper is chalcopyrite, copper iron sulfide $CuFeS_2$, which is roasted in air to copper sulfide:

chalcopyrite	oxygen	copper I sulfide	iron sulfide	sulfur dioxide gas
$2CuFeS_2$ +	O_2 ⟶	Cu_2S +	$2FeS$ +	SO_2

The iron and copper sulfides are further reacted in oxygen to form iron oxide, copper and sulfur dioxide. The impure copper is dissolved in sulfuric acid to form a copper sulfate solution which is then electrolysed to get pure copper. The sulfur dioxide gas liberated in the first stages of the process is converted to sulfuric acid and then used in the electrolysis process. Cooper at 99% pure is formed on the cathode, or negative, plate in the electrolysis process and impurities such as arsenic, selenium, tellurium, gold and silver collect at the anode or positive plate. These then fall to the bottom of the electrolytic cell as a slime which can be further processed to extract any of the other valuable metals. Arsenic, for example is removed by roasting the mixture in air and extracting it as arsenic trioxide which can be further processed and used in agriculture, medicine and electronics to name a few.

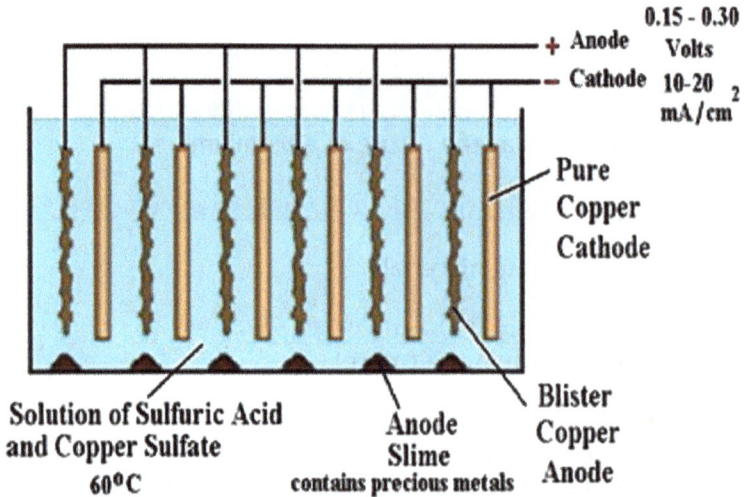

Figure 12.28: A simplified diagram of copper electro-refining process

See also:

https://sustainablecopper.org/wp-content/uploads/2018/07/ICA-EnvironmentalProfileHESD-201803-FINAL-LOWRES.pdf
(A report on the environmental profile of copper)

http://cfsites1.uts.edu.au/find/isf/publications/muddetal2012copperemissionscleanenergy.pdf
(Greenhouse gas emissions from copper mining)

Glass is a non-crystalline material with widespread use in windows, tableware, containers, optics and construction and optoelectronics as glass fibre. The most familiar type of glass, silicate glasses, are made from pure white quartz sand or silica and other compounds. The most common glass is soda-lime glass, composed of about 75% silica (SiO_2) as sand, sodium oxide (Na_2O) derived from heating sodium carbonate (Na_2CO_3), calcium oxide (CaO) or lime, from heating calcium carbonate. Other minor additives such as lead oxide can be added to produce lead crystal glass.

Along with concrete, steel and aluminium, glass as sheeting forms a major part of building construction.

Figure 12.29: An example of Egyptian glass of the 5th Century A.D. (Photo: Public Domain)

Following combining of the appropriate raw materials and the silica, the mixture is melted in gas fired furnaces. After melting and removal of bubbles, the glass is formed as flat sheets or blown and moulded into shapes.

Most of the greenhouse gas emissions come from the removal of carbon dioxide from the raw materials and from any fuel used in firing the furnaces. This can be reduced by increasing the amount of recycled glass used in the process and by using electrical furnaces with the power coming from renewable sources.
See also:

https://recycleglass.com.au/glass-lifecycle/carbon-footprint/
(Report on the carbon footprint of glass-making)

https://www.nrel.gov/docs/fy01osti/29342.pdf
(Glass as a material of the future)

It must be remembered that all of these useful materials come from finite i.e. non-renewable sources. Whilst there has been considerable effort in the resources industry to reduce or eliminate greenhouse gases and conserve energy, there is still the problem of obtaining the raw materials.

Some processes have attempted to reduce the amount of raw resources needed by reducing the size of the finished product yet retaining the structural strength as in steel fabrication. Concrete can also be made more economical by using reinforcement material which has been sourced from man-made materials such as fly ash and even bamboo fibres. All major construction materials can also be made using recycled materials such as:

- Crushed and powdered recycled concrete which can be used as aggregate. Whilst this is a major reduction in using natural gravel and sand, it cannot be rendered back into hydraulic setting cement.

- Recycled iron from a great variety of sources both industrial and domestic. A major component of the steel-making process is scrap steel and, in some situations, the furnace may be charged with up to 75% of scrap steel. Moreover, some of the waste products of the steel-making industry such as the slag waste can be used in other industries such as making concrete.

- The recycling of aluminium cans is well known in domestic recycling systems. These and other used aluminium frames and structures are recycled in the original aluminium smelting process. Aluminium is very suitable for recycling as it forms a thin coating of its oxide on its surface when used in construction and this protects the metal below. Some countries such as Brazil recycle over 85% of their produced aluminium cans.

- Glass, apart from some specialized products, is a material which can easily be recycled in the manufacturing process. Moreover, it has long been a traditional material for containers such as jars and bottles, especially for milk, beer and wine. These items can be recycled, cleaned and reused.

See also:

https://www.fhwa.dot.gov/pavement/sustainability/hif16013.pdf
(About recycling concrete)

https://www.worldsteel.org/en/dam/jcr:16ad9bcd-dbf5-449f-b42c-b220952767bf/fact_raw%2520materials_2019.pdf
(Recycling of scrap iron and steel)

http://www.world-aluminium.org/media/filer_public/2013/01/15/fl0000181.pdf
(Recycling aluminium report)

Timber is another construction material which has resurfaced as one with great potential for the future. Used from the beginnings of civilization, timber plays a major role in the construction of low-rise buildings, the interiors of high-rise units as well as a great variety of furniture and other products. In China, tall pagodas and temples were constructed from timber with flexible joints to reduce the vibrations due to earthquakes. In modern times however, excess logging of old-growth natural forests has led to a world-wide shortage of hardwood and softwood plantations do not seem to be keeping pace with demand.

With the desire to use renewable products, especially in construction, timber used in new and innovative ways may be a way of replacing many of the non-renewable materials in construction. This is particularly true of new laminated timber sheets which can be reformed into lightweight but strong columns and beams for use in structures instead of hardwood or concrete slabs.

The Mjøstårnet building in Norway consists of 18-stories made out of **Cross-Laminated Timber (CLT)** with some concrete slab flooring to stabilize the structure. Cross-laminated timber consists of several layers of timber slices glued with a resin so that the grain of one timber slice is at ninety degrees to the next and so on. The slices are then hydraulically pressed to make solid structural wood panels for floors, walls and roofs. These can be manufactured in a variety of sizes and shapes and so can be customised to suit specific needs. The Mjøstårnet building was completed in March 2019 and is the world's tallest timber building at 85.4 meters tall. It contains apartments, hotel, swimming pool, office space and a restaurant. Other timber laminate buildings of even bigger dimensions are planned for the United States, Japan and elsewhere. It is claimed that the normal fire potential of timber is reduced by the compact nature of the laminated wood which usually chars when subjected to flame, protecting the internal structure from burning.

The potential of CLT in the future of construction however, depends upon the continual re-supply of renewable timber resources. Many timbers can be used for CLT including softwoods such as spruce and hardwoods such as birch. Low-grade hardwoods can also be used. In the future, plantations of timber suitable for CLT will need to be developed to supplement those already in existence.

In addition to the traditional laminates used in current and possibly future building, bamboo also shows great potential for the future. It has also been used traditionally for buildings in Asia and has also been used for scaffolding in high-rise construction. Some research has also been done in using it as a reinforcement in concrete, but whilst bamboo fibres and rods are very strong, the timber itself is subject to attack by water and mould.

See also:
https://crosslamtimber.com.au/SBR%20Cross%20Laminated%20Timber.pdf
(A good article on CLT)

https://e360.yale.edu/features/as-mass-timber-takes-off-how-green-is-this-new-building-material
(New uses of timber in building)

https://www.awc.org/pdf/education/mat/ReThinkMag-MAT240A-CLT-131022.pdf
(More on the future of wooden buildings)

https://www.apn-gcr.org/resources/files/original/1654f846a58279adea4aeb44a881321b.pdf
(More about bamboo)

https://www.globalforestwatch.org/map/country/NZL?mainMap=eyJzaG93QW5hbHlzaXMiOnRydWV9&map=eyJjZW50ZXIiOnsibGF0IjotNDEuOTg3OTUwODg2Njc4MTUsImxuZyI6MC4xMTk0MzgxNzUwMDAwMjc1MX0sImJlYXJpbmciOjAsInBpdGNoIjowLCJ6b29tIjoxLjExNjk5MzkzNzg0NDMyMzksImNhbkJvdW5kIjpmYWxzZSwiYmJveCI6W119
(An excellent interactive world map of forest resources. Click on a country to obtain local data about forests and plantations)

https://www.dpti.sa.gov.au/__data/assets/pdf_file/0009/293688/Environmentally_Sustainable_Building_Materials.pdf
(Building materials which can be environmentally friendly)

With a major demand for dwindling resources in the future, especially resources which now must be made using environmentally-friendly mining, processing and fabrication, architects and builders are faced with several options:

- Using existing high-rise buildings in more efficient ways. This might mean the conversion of old office blocks, storage buildings and the like into accommodation or other purposes. There has already been a move to do this amongst some of the older stone and brick buildings of the 19th and 20th centuries.

- Building structures out of recycled materials such as brick and stone to modest heights. Concrete can be recycled to make more aggregate which can be used with the new forms of geopolymer cement and reinforcing material. Glass would be used less as a feature and more of designated window space as in older buildings with less need to use aluminium frames. More wall-to-glass ratio would also provide better insulation.

- Using modern materials such as cross-laminated timber in conjunction with geopolymer concrete for high-rise buildings of modest heights.

- Continuing to construct low-rise housing where appropriate out of recycled brick, stone and timber.

- Considering new options such as underground structures built into the sides of hills, in old quarries and the like so that they are above water channels and they are well-drained with ventilation and illumination provided by side windows or ducted skylights.

Away from the urban zones, hopefully now limited in area by decentralization, rural communities would make best use of local materials as well as the new modern timbers and concrete.

Whatever the architectural and construction scenario will take in the future, recycling and the use of renewable resources will be dominant processes in construction of living space and general infrastructure.

12.11 Permaculture

Permaculture is also not a new concept, deriving its origin from 1978, and refers to an holistic approach in agriculture where one works with natural processes rather than replacing them. David Holmgren in his *Permaculture: Principles and Pathways Beyond Sustainability* outlined twelve basic principles in applying permaculture in crop and livestock production:

1. *Observe and interact* the local natural features and then adapt methods of production accordingly.
2. *Catch and store energy* by developing systems that collect resources at peak abundance for use in times of need.
3. *Obtain a yield* by ensuring that best practice leads to productivity.
4. *Apply self-regulation and accept feedback* by looking at the positive and negative aspects of the work and discarding what does not work.
5. *Use and value renewable resources and services* by making the best use of natural resources in a sustainable and conservative way limiting the reliance on non-renewable resources.
6. *Produce no waste* by making use of all the resources that are available.
7. *Design from patterns to details* by evaluating and observing patterns in nature and society which can then be used to form the basis of system designs.
8. *Integrate rather than segregate* by working with nature so that suitable relationships will develop which work together and provide support.
9. *Use small and slow solutions* rather than attempting to maintain large, often unwieldy projects which make better use of local resources to produce more sustainable outcomes.
10. *Use and value diversity* so that there are fewer threats as there are in mono-cultures and take advantage of the local nature of the environment.
11. *Use edges and value the marginal* of the local environment as it is often the interface between various ecologies where there is greater potential for development and change.
12. *Creatively use and respond to change* by carefully observing natural systems in time and space and then intervening at the right time.

These principles generally suggest that the wastage, which occurs in current food production with crops and livestock can be reduced with greater productivity. This means that if the primary producer makes better use of natural systems, especially concerning the type of crops, livestock and methods used, then more food can be produced with less environmental change. However, not all permaculture principles may work in large-scale food production and in every case. For example, in Australia, the soil types and ecology are more suitable for its unique native animals such as the various species of kangaroo. It was totally unsuited for hooved animals such as sheep and cattle which were introduced by the first European settlers. Consequently, much of the native ecology was destroyed and the land left exposed for soil erosion by wind and water. Now, two hundred years later, it would not be feasible to replace the huge sheep and cattle industries with one based on native animals such as the kangaroo.

Many ancient civilizations practiced natural techniques which could be classified as permaculture. For example, many cultures such as those in the Pacific Islands cultivate crops more suited to their environment and often the local village garden is located within the nearby local forests as forest gardens. Many nomadic tribal peoples also planted specific crops in areas which were more suitable for each crop. These could be harvested later at times of maturation. The Incas, masters of crop engineering, constructed terraced gardens, sometimes around a large natural depression so that each type of crop was able to be planted at appropriate climatic conditions of altitude and water requirement. Even on flat surfaces, plants can be grown as height layers with a descending mixture of taller fruit trees, shrubs and vegetables occupying the same space. Moreover, a variety of domestic animals such as bees, chickens and even dairy cattle can be raised in a similar environment provided that they support that environment.

Figure 12.30: Incan terraces still in use near Arequipa, Peru. The volcano, El Misti in the background.

On a larger scale, such practices have been used in many agricultural communities these include:

- Minimal or no-till agriculture in which the new crops are planted within the remains of the old without the need for disturbing the soil by tillage. This not only retains moisture in the soil, but the past vegetation remains also add to the fertility of the soil.

Figure 12.31: Newly-planted soybean plants are emerging from the residue left behind from a prior wheat harvest (Photo: USDA Natural Resources Conservation Service).

- Drip or limited irrigation which applies smaller amounts of water directly to the base of crops rather than mass spraying which allows considerable water waste due to runoff.

- Use of natural organic fertilizers rather than chemical fertilizers which are often over-rich in ammonia and nitrogen products which are carried by runoff into waterways and contribute to mass algae blooms which deplete oxygen content.

- Use of more drought-resistant crops and livestock to counter increasingly more arid conditions but also allowing for less water consumption.

- Use of water-gathering technology and additional reservoirs such as tanks or sub-surface cisterns and collection of water from fog and coastal mists as in the western Andes of Peru and Chile and also in southwestern Africa.

- Crop rotation, especially including legumes to replace nitrogen in the soil.

See also:

https://holmgren.com.au/wp-content/uploads/2013/02/HolmgrenAdelaidePublic.pdf
(A presentation on Permaculture by David Holmgren who, along with Bill Mollison, coined the term 'Permaculture' in 1978)

https://www.crmpi.org/What_is_Permaculture_files/permarticle.pdf
(A simple guide to the nature of permaculture)

https://files.holmgren.com.au/downloads/Essence_of_Pc_EN.pdf
(A more detailed look at the concept of permaculture)

http://hautrive.free.fr/permaculture/Facilitators-Handbook-For-Permaculture.pdf
(Detailed notes on permaculture for developing countries)

https://permateachers.eu/wp-content/uploads/2014/04/Permaculture%20Design%20Course%20Handbook.pdf
(Notes on permaculture design)

https://www.usda.gov/sites/default/files/documents/urban-agriculture-toolkit.pdf
(An Urban Agriculture Toolkit from the US Dept. of Agriculture)

12.12 Waste and Recycling

Most of the industrialized countries produce large amounts of wastes which are often burnt, added to land-fill, dumped at sea or even exported to less-developed countries. The amount of urban waste being produced is growing faster than the rate of urbanisation itself. According to the World Bank, by 2025 there will be 1.4 billion more people living in cities worldwide, with each person producing an average of 1.42kg of waste per day – more than double the current average of 0.64kg per day.

Despite pushes for recycling and responsible resource use globally, waste is still a major problem virtually everywhere. Annual worldwide urban waste is estimated to more than triple, from 0.68 to 2.2 billion tonnes per year. More commonly known rubbish or trash, **Municipal Solid Waste** (MSW) consists of everyday items such as product packaging, grass clippings, food scraps, and newspapers. Most countries struggle with the problem of waste disposal. Many poor island nations in the Caribbean, the Pacific region and Asia have difficulties because of their lack of disposal facilities and their limited space in which to bury or otherwise dispose of their waste.

There is considerable variation in statistics as to which countries produce the most waste. Some of the poorer countries rate very highly because of their lifestyle, small populations, dependence on packaged foods and lack of facilities in disposing of the waste. Larger, industrialized countries will rate very highly depending upon their population, reliance on package foods, wasteful life-styles and limited recycling. The following table gives a set of statistics (from Statista - https://www.statista.com) of the most wasteful countries in kilograms per head of population in the year 2018. There are many such rank-order lists which vary greatly depending upon criteria used.

COUNTRY	POPULATION (million)	WASTE Kg/capita/year
United States	327.2	2.58
Canada	37.59	2.33
Australia	25.72	2.23
Germany	82.9	2.11
South Africa	58.78	2.00
France	66.99	1.92
United Kingdom	66.87	1.79
Japan	126.5	1.71
Saudi Arabia	34.14	1.30
Mexico	About 126	1.24
South Korea	51.41	1.24
Brazil	About 210	1.03
China	1,427.65	1.02
Russia	146.70	0.93
India	About 1,369	0.34

Table 12.2: Showing some statistics of the main countries with high amounts of solid wastes.
(from: https://www.statista.com/statistics/689809/per-capital-msw-generation-by-country-worldwide/)

Figure 12.32: Typical waste landfill. This practice is now banned in many countries (Photo: Public Domain).

One must actively look for and choose materials or products which come from renewable raw materials requiring little greenhouse gas emissions in their manufacture and which can also be recycled. A return to a more comprehensive and productive human-based urban recycling can be put into practice immediately. Recycling of current products made from non-renewables should be a priority. Car tyres, all forms of plastics, timber, organic wastes and the like could all find a new use in a renewable society. One needs to think away from one's limited perspectives e.g. bamboo could be used as scaffolding instead of valuable steel pipe in building construction in Western cities in the same way that it has been used for a long time in Asia.

A close examination of life before the introduction of synthetics made from petroleum products in the 1930s would be an advantage to see how natural materials could once again become the main supply of raw materials. Much of the clothing, other cloth products, containers and the like which today are often made from synthetics could be replaced with renewable plant and animal products such as linen, cotton and wool as they were in the past.

In the 1950's, before plastic containers and the widespread use of packaging, recycling and reuse was more common and considered the usual practice in society. The most common items which were reused or recycled were:

- Glass bottles for milk, soft drink (soda pop), beer, wine and spirits. Empty milk bottles were left on the doorstep to be returned to the milk company for washing and reusing and glass soft drink bottles were returned to the shop to claim a small cash deposit. Beer, wine and spirit bottles were collected by charities and then sold to glass companies.

- Clean newspapers, cardboard and other paper items were also often collected by charities for resale.

- Most fruit and vegetables were sold unwrapped with any perishables placed in paper bags which could then be reused. Purchasers would also have their own reusable carry-bags in which the bulk items would be placed.

- Staple items such as flour, rice and powdered milk could be purchased in bulk with the buyer supplying their own container, often a metal carry-can, or the powder packed in paper bags. In most cases these items could also be purchased in pre-packaged cloth or paper bags.

- Clothing and household goods were usually purchased unwrapped but were often wrapped in paper by the shop assistant.

- Packaged fast food was generally unknown. Any take away was wrapped in paper, often recycled or simply carried by hand.

Waste paper and plastic from over-wrapped products often finds its way in the streets as thoughtless people of throw away generations simply discarded their wastes. This philosophy or way of behaving has become a major social issue in modern society. Whilst local authorities seek to remove the street litter by providing waste bins and collection systems, there is still the attitude that unwanted materials should be simply discarded and then they are no longer a personal problem. This attitude also is applied to larger items such as furniture, white goods and many other household products.

Plastic containers, bags, household items, toys and wrapping continue to be a major concern. In general, most plastics are not biodegradable and last a long time. Ultimately, much of it ends up in the ocean where it accumulates as masses of floating debris in the central circular ocean currents or **gyres**, or it is washed up upon beaches. Such material then becomes a major environmental hazard to all marine life which may ingest the material as pieces or as broken-down micro-plastic pellets, or become tangled within it.

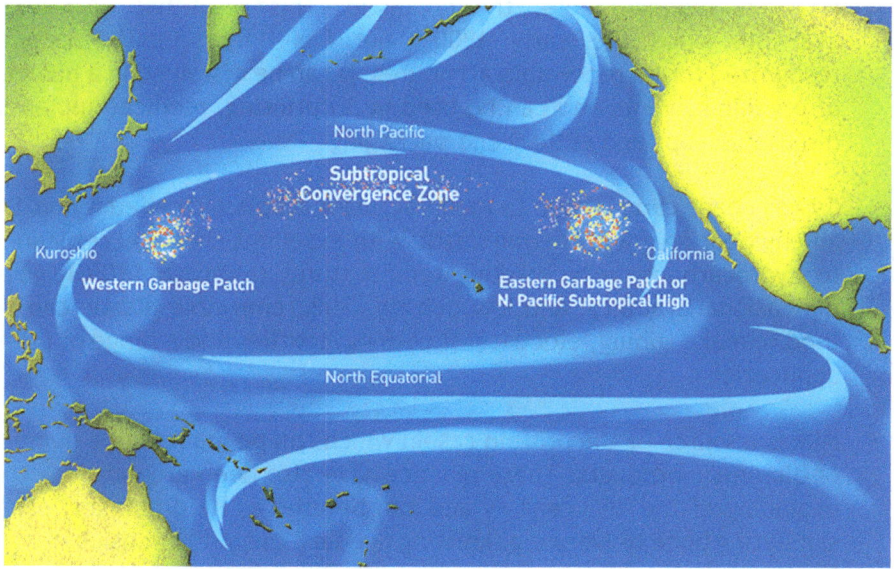

Figure 12.33: Map showing the Pacific Garbage Patches of accumulated plastics and other material (Photo: NOAA)

Plastics are usually made from petroleum products, so their source material is a non-renewable fossil fuel which has very limited time as a resource; probably less than 50 years at current usage. There are however, alternative plastics or **bioplastics** and other similar materials which may be friendlier to the environment because they can be made from natural, renewable sources such as:

- Wood can be used to make cellulose acetate bioplastics which are derived from the acetylation of the plant cellulose with acetic acid made from grape vinegar. Cellulose acetate and some of its derivatives can be spun into textile fibres such as acetate rayon, or moulded into solid plastic parts such as tool handles or cast into film for photography or food wrapping. Developed in the early 20th century, its use diminished with the rise of the petrochemical industry. It is biodegradable and it is claimed that it can be reused by converting it into a thermal plastic which then can be remoulded. Similarly, bamboo fibre is now becoming a major source material for a range of goods including clothing.

- Corn starch or sugar cane pulp can be used to make Polylactic Acid (PLA), a biodegradable, thermoplastic. This resin has also been available since the early 1900's but has only recently become recognized for its renewable attributes. It can be dissolved in a number of organic solvents and can be used in 3D printing, medical implants, disposable cups and cutlery and packaging.

- Castor beans can be used to manufacture a bioplastic called Nylon 11, an engineering grade thermoplastic with excellent chemical resistance to hydrocarbons, wide working temperatures (-40^{0}C to 130^{0}C), good dimensional stability, and low density. This renewable plastic can be used to make tubing, electrical sheathing, textiles and a range of other applications.

- Milk is a source of the protein casein which can be made into a number of bioplastic products. These plastics have also been around since the beginning of the 20th century and can be used as hard plastics, often simulating bone or ivory e.g. knitting needles, buttons, utensil handles, food wrapping and other applications. It is very biodegradable.

When considering the nature of modern bioplastics, one must be able to distinguish between the terms 'biodegradable' and 'renewable'. Whilst these products are sourced originally from natural, renewable raw materials, the finished product may yet still be unsuitable for the environment in the long term. The used item still must be collected to be used or disposed of properly. Some bioplastics may not be truly biodegradable, simply degradable and break down to smaller, harmful pellets of organic sludge. True biodegradable products decompose by the action of microbes or even sunlight to produce water and carbon dioxide. Only some bioplastics can be reused by further industrial processing into other materials.

See also:

https://marinedebris.noaa.gov/info/plastic.html
(Plastics in the ocean - from NOAA)

http://www-g.eng.cam.ac.uk/impee/topics/RecyclePlastics/files/Recycling%20Plastic%20v3%20PDF.pdf
(Renewable plastics)

https://ec.europa.eu/environment/integration/research/newsalert/pdf/FB1_en.pdf
(More on the future of plastics)

http://www.trendingpackaging.com/the-most-common-eco-friendly-alternatives-for-plastic-packaging/
(Substitutes for petroleum-based plastics)

The development of biodegradable and renewable products from renewable raw materials will be of the greatest importance for the future. Unfortunately, the fact still remains that societies, especially urban societies, continue to produce large amounts of wastes which must be removed from the environment. These municipal solid wastes (MSW) include a variety of organic and inorganic material:

WASTE MATERIALS	AVERAGE PERCENTAGE %
Paper	27
Food scraps	15
Yard trimmings	14
Plastics	13
Rubber & textiles	9
Metals	9
Wood	6
Glass	4
Other	3

Table 12.3: Breakdown of common urban waste as MSW in the United States in 2013 (Data: U.S. Environmental Protection Agency)
https://www.statista.com/statistics/689809/per-capital-msw-generation-by-country-worldwide/)

There is a strong view that much of this can be recycled. Waste should be seen as a valuable, potential resource to be collected, separated and converted into useful material.

Currently there is a move towards recycling of waste materials with sorting beginning at the household level with rubbish bins for separate items. Hopefully this will continue with even more selectivity such as paper, plastics, vegetable matter and glass being different categories. At the refuse station were waste is taken there should also be more thorough separation with removal of recyclable materials to specific recycling companies. Whilst many countries do this by labor-intensive hand sorting, there are now robotic-artificial intelligence machines appearing in the industry.

Separated material such as metals, paper and cardboard and glass are relatively easy to recycle but plastics have become the dominant waste problem. Some plastics cannot be recycled by current means and others claim to be able to be recycled. The following chart gives a quick review of some of the main plastic product types and their ability to be recycled:

Symbol	Polymer Type	Examples	Recyclable	Recycled Into
PETE 1	Polyethylene Terepthalate PETE or PET	Water and soda bottles; Many food oil containers.	Yes. Easiest to Recycle	Egg cartons, carpet, T-shirts.
HDPE 2	High Density Polyethylene HDPE	Milk & juice cartons, Detergent containers.	Yes	Toys, traffic barriers trashcans.
V 3	Polyvinyl Chloride PVC	Plastic pipes, shrink wrap, frames, roofing & decking.	Very difficult and expensive to recycle	Notebooks, shoe soles, construction material, boat bumpers, automotive floor mats.

![4 LDPE]	Low Density Polyethylene LDPE	Disposable diaper **liners**, cable sheathing, cellophane wrap	Recycled except cellophane wrap	Plastic 'timbers', trash can liners, shipping envelopes, floor tiles.
![5 PP]	Polypropylene PP	Drinking straws, yogurt containers, butter and margarine tubs	Not commonly recycled	Signal lights, bicycle racks, trays, battery cables, ice scrappers when recycled.
![6 PS]	Polystyrene (Styrofoam) PS	Egg cartons, cups, plastic forks, spoons, knives	Not commonly recycled	Egg cartons, foam packing, light switch plates

Table 12.4: Some common plastics and how they can be recycled.
(https://www.recyclereminders.com/recycling-plastics?gclid=EAIaIQobChMIv9-EypKy5wIVgo2PCh0dHgMDEAAYAiAAEgIAOvD_BwE)

Continued research is on-going into 100% recycling and at least one company claims to be able to convert all plastics into basic liquid fractions which could then be recycled into new products:

Online Video 12.1: A new process designed to recycle all plastics.
Go to: https://www.youtube.com/watch?v=MTgentcfzgg

Burning waste in incinerators has been a major method of removing household, industrial and medical waste for a long time, especially in countries where there is little space for landfill, the other traditional means of removing waste. Incineration is a disposal method in which solid organic wastes are subjected to combustion so as to convert them into residue and gaseous products. It has been claimed that high-temperature incineration can reduce the volumes of solid waste by 80 to 95 percent. Many new plants use the waste-to-energy systems which create high temperatures which can destroy many harmful biological hazardous wastes as well as providing heat to make steam for electricity generation. However, whilst it claims to be less polluting than landfill, incineration is controversial as it still produces gaseous emissions including some organic compounds such as dioxins, as well as the oxides of carbon, nitrogen and sulfur. Never-the-less it is also claimed that ash from these incinerators is also useable and that modern incinerator plants use filters which can trap hazardous gases and particulate **dioxin**.

Figure 12.34: Waste-to-energy incinerator plant at Thun, Switzerland (Photo: Public Domain)

In a modern, take-away and throw-away society, food has become a waste product. Many modern urban societies have a major problem with obesity which was uncommon before the late 20th century. Food servings at home and in restaurants have become traditionally over-large whilst in the streets and in many countries, people continue to be malnourished and starvation is a real problem. Whilst this problem is recognized and some public charities and government agencies are attempting to redress it, the main issue about excessive food processing and disposal continues to be a major social and health issue. In general. Food waste should follow a hierarchy of processes:

https://ecosystems.psu.edu/youth/sftrc/environ-series/rnr-mat
(General look at the use of renewable materials)

https://www.ellenmacarthurfoundation.org/assets/galleries/ce100/CE100-Renewables_Co.Project_Report.pdf
(Good general article)

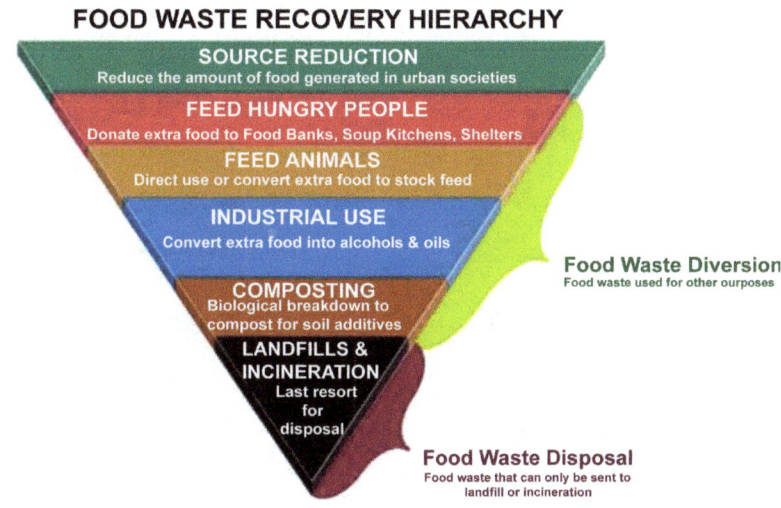

Figure 12.35: A hierarchy of food use (Photo: modified from Public Domain)

12.13 Socio-political Consequences

So far in this book the scientific causes and effects of global warming and subsequent climate change has been emphasized. However, the effects of global warming have and will continue to have profound social, emotional and political ramifications. Some of the main effects of global warming which will influence these factors are:

- rise in sea level;

- increased intensity and frequency of storms and flooding;

- increased desertification with reduced food production;

- increased frequency and intensity of wildfires; and

- increased competition for dwindling resources, especially freshwater.

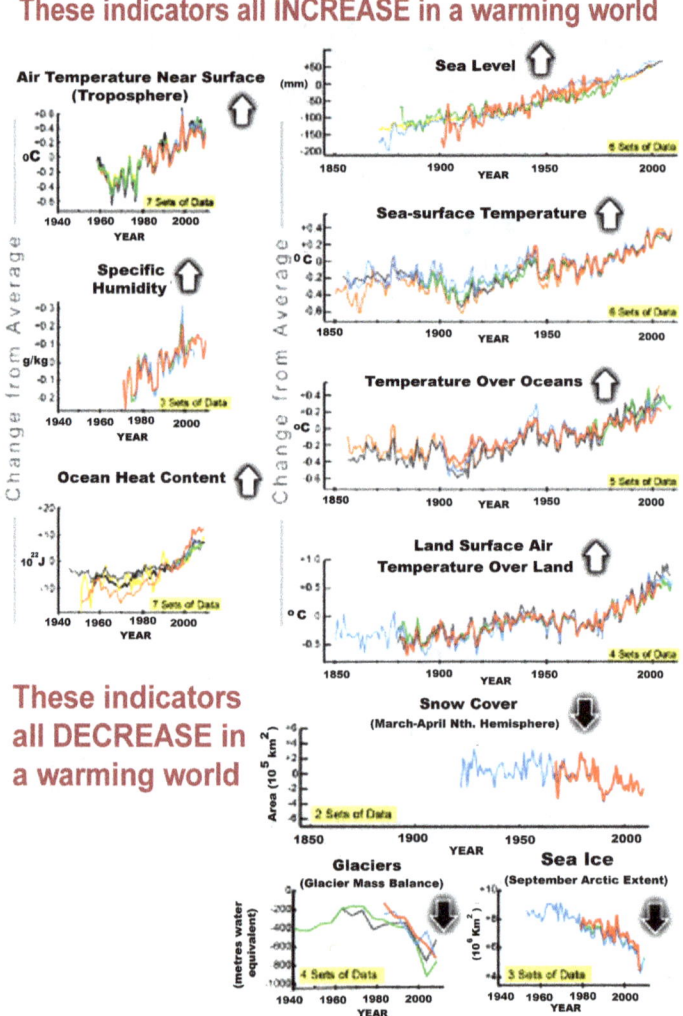

Figure 12.36: Changes in indicators with global warming (Photo: modified Data: NOAA)

These consequences will affect people differently in different parts of the world as they begin to occur to the extent of disrupting local societies. For example, sea level rise will affect those communities living in low-lying coastal regions, coastal river plains and on oceanic islands. In many parts of the world, such as Bangladesh and the central Pacific Islands, the effects of sea level rise are already impacting on societies with problems of storm and wave inundation and rising salty water in the underlying water table which destroys crops.

Superimposed on many of these countries are the seasonal storms which come as a result of ocean warming and sudden rise and movement of moisture. The regular monsoons and tropical cyclones, also called typhoons or hurricanes, which come each year to countries in the tropical zone of the Earth will probably extend further south and north from the tropics into the sub-tropical zones which are yet to experience the deluge of rain, high winds flooding and damage formally only experienced by their tropical neighbours. Moreover, it is predicted that these events will last for a longer period of time because of the new extent of warmer oceans.

Meanwhile, in countries with large arid zones, desertification of the margins of these zones is leading to loss of arable land and livestock production. Lack of rainfall is the main contributor here with large areas of countries such as Australia, many of those in North Africa and the Middle East, the western coastal strip of South America and parts of central Asia all being prone to extensive drought. The reduction in the local production of food from crops and livestock, as well as to the lives of the local people will come as a direct response to the sudden reduction in the availability of freshwater from rains or streams from snowmelt. Water security and the competition for water will become a source of major conflict between nations which share major water supplies from the world's major rivers. Disputes between Egypt, Sudan and Ethiopia have already occurred over the building of dams and introduction of irrigation projects which will greatly reduce the water supply of countries downstream. The Intergovernmental Panel on Climate Change of the United Nations (IPCC) predicts that by 2050, more than a billion people in central, south, east and south-east Asia and between 75 million and 250 million in the African region are expected to be affected from freshwater shortages due to climate change.

With current trends in global warming, it is predicted that there will also be major problems with health worldwide. Whilst some northern countries will receive some warmer climates, the general effect worldwide will be mainly detrimental. The risks posed by climate change will most likely come from:

- Direct effects due to increased temperature such as heat waves, increased air pollution and more frequent and violent weather phenomena.

- Changes in ecological systems such as malnutrition due to decreased crop and livestock yields, increased diseases carried by bacteria and insects such as mosquitos.

- Indirect consequences relating to impoverishment, displacement, resource conflicts (e.g. water), and post-disaster mental health problems.

The World Health Organization (WHO) reports suggests that rising temperatures and unpredictable rainfall due to climate change since the 1970s has claimed over 140,000 deaths with about another 60,000 deaths each year due to weather-related natural disasters mainly in developing countries.

In the years to come there will be greater need for action plans within nations to combat the specific threats posed by global warming. Stable and wealthy countries such as Australia, those in Europe and North America will need to continue to meet threats caused by resource shortages, natural disasters such as drought and wildfires, severe storms and flooding as well as to assist in supporting poorer nations with their specific needs such as sea level rise in the Pacific, drought in Africa and many others. There will also be the need for diplomatic assistance from the United Nations and richer nations in the many political, social and military disturbances which are likely to occur as the Earth becomes warmer.

See also:

https://postconflict.unep.ch/publications/GN_Renewable_Consultation.pdf
(Notes from the United Nations on renewable resources and conflict)

https://postconflict.unep.ch/publications/UNEP_Sahel_EN.pdf
(A case study of problems in North Africa)

https://www.consilium.europa.eu/media/30862/en_clim_change_low.pdf
(A detailed paper on climate change and international security)

https://unfccc.int/resource/docs/publications/impacts.pdf
(Impacts of climate change on developing countries)

https://www.ipcc.ch/site/assets/uploads/2019/08/Fullreport-1.pdf
(A paper from the IPCC on global warming and land use)

http://www3.weforum.org/docs/WEF_Global_Risks_Report_2019.pdf
(A global risks report from the World Economic Forum)

Final Remarks

1. In the future there will be many and diverse problems associated with increased global warming and climate change with more extremes of those already locally experienced.

2. Some of these problems on a global scale may include: increase in population with an ever-increasing demand for basic resources such as water, food and living space; increasing demand for energy; continued pollution of land, air and in the world's waterways and oceans due to an increasing amount of industrial and domestic wastes; and problems with social interaction and cooperation at due to the lack of initiative of governments in dealing cooperatively with the world's problems.

3. An increasing population, especially in the less-developed nations will but pressure on the need for living space, food, water and transportation.

4. Many of the world's future problems can be overcome with a change in thinking about living space and transportation with the possible development of decentralization to specifically-built satellite cities and changes to current cities limiting population, transportation whilst enhancing water conservation and increased green spaces.

5. The push to replace petroleum-based fuels in cars, ships and aircraft with more environmentally-friendly and renewable fuels should continue. New power sources such as electricity, hydrogen fuels and wind power will reduce greenhouse emissions and reduce global warming.

Final Remarks Continued

6. A new perspective is needed in the search for construction material in the future. Recycling and modern methods of processing of many non-renewable materials such as metals, glass and some solid building materials can be enhanced by new applications of renewable materials such as wood as well as composite materials.

7. New applications of some older agriculture and livestock practices but on a larger scale using permaculture can be supplemented by hydroponics and urban gardens.

8. Waste should be reduced through more efficient social practices, especially in the use of resources such as food and water. Solid wastes, especially plastics should be seen as potential raw material for recycling and conversion to useful material which in turn could also later be recycle.

9. Major social disruption may occur due to global weather extremes, shortages of natural resources and energy and increased local and international competition for living space and these resources. More cooperation and assistance will be required to prevent conflict and provide help to those in need.

More Applications

1. Look at the local environment whether it be rural, urban or suburban and make a complete and written audit of the potential hazards, problems and shortages which may occur assuming increased temperatures with local possible consequences such as sea level rise, more storms, more extended droughts, more competition for energy and increased population and waste. Using the guides in previous chapters, develop strategies to cope with change.

2. Such strategies may include: installation of water tanks; development of home gardens; specific waste recycling including the use of grey water; and more cooperation with cooperation and the use of vehicles using friendly fuels. This may require community efforts, especially with urban dwellers living in high-rise units.

3. Reduce the amounts of waste in food intake, water use, energy use, packaging and waste materials in general. Recycle organic wastes as compost and encourage local collection of metals, glass bottles, renewable plastics and building materials such as bricks.

4. Encourage local and national politicians to legislate for actions which will reduce greenhouse emissions and support the environment. This may include writing to local members, peaceful protest and voting only for candidates who support good future practice. One must be cautious, however in the hidden agendas of some politicians and their parties.

5. Support reliable and well-known charities and support organisations in the drive to help less-advantaged communities at home and abroad, especially during and after extreme disasters such as fire, flood, famine and forced migration.

More Applications Continued

6. Readers who have some power as politicians or as entrepreneurs should use their positions to encourage or develop devices, processes or systems which will enhance future living space, food and water use and transportation.

7. Encourage more cooperation in society by a change in attitude from being an isolated individual to belonging to community groups which have positive aims and real action in supporting the community and the environmental.

8. Use social media in a positive way to contact those individuals or groups which actively assist the community and the environment (CAUTION: there are many false and negative social media groups who do little other than social disruption).

9. Heat waves will cause stress and even death if extreme so look at ways of cooling the house or flat. House owners can insulate the ceiling space with fibre glass wool or similar, install roof ventilators and skylights in the surface of the roof and put ventilation holes under the eaves of the roof:

10. High-rise dwellers have little choice except fans and air conditioning but use those appliances with good power rating and a timer switch to use them only as required. In dry climates, evaporative AC units are satisfactory or even hang wet cotton sheets across the open windows.

Further References

Bows-Larkin, A. 2015 All Adrift: Aviation, Shipping and Climate Change Policy. Abingdon, UK: Taylor & Francis. DOI: 10.1080/14693062.2014.965125.
https://www.tandfonline.com/doi/pdf/10.1080/14693062.2014.965125

California Environmental Protection Agency. 2016. Recent Research on Climate Change: A bibliography with an emphasis on California.
https://oehha.ca.gov/media/downloads/climate-change/document-climate-change/climatechangecaliforniabibliography2016.pdf

Chopra, K.L, Paulson, P. D. & Dutta, V. 2004. Thin-Film Solar Cells: An Overview. IN Progress in Photovoltaics: Research and Applications. Prog. Photovolt: Res. Appl. 2004; 12:69-92. DOI: 10.1002/pip.541.
https://www.researchgate.net/publication/244973238_Thin-Film_Solar_Cells_An_Overview

Dharmadasa, I.M. 2012. Advances in Thin Film Solar Cells. Boca Raton, FL: Taylor & Francis Group. ISBN 13: 978-9-81436-412-6
https://www.kashanu.ac.ir/Files/Content/thin%20film%20solar%20cells.pdf

Gardiner, S.M. et al. (edits.). 2010. Climate Ethics: Essential Readings. New York: Oxford University Press. ISBN: 978-0-19-539962-5.

Holmgren, D. 2002. Permaculture: Principles & Pathways Beyond Sustainability. Holmgren Design Services. ISBN 978-0-646-41844-5.

Scott, P.T. 2018. Riches from the Earth. Brisbane: Felix Publications. ISBN: 978-0-9946432-6-1.

Scott, P.T. 2018. Through Sea and Sky. Brisbane: Felix Publications. ISBN: 978-0-9946432-2-3.

Smerdon, J. 2018. Climate Change: The Science of Global Warming and Our Energy Future. New York: Columbia University Press. E-ISBN: 978-0-231-54787-1.

Tull, Kerina. 2019. Development, Climate and Environment: An Annotated Bibliography.
https://assets.publishing.service.gov.uk/media/5da5e89ee5274a392e9c9468/644_Climate_and_Environment.pdf
(An excellent list of links to global warming matters)

Glossary and Index

ablation – the melting of glaciers at their end or snout · 33
aerosol - a suspension in air of fine solid particles, liquid droplets or another gas. · 185
Air Quality Index (AQI) - general indicator of air pollution from a number of instruments monitoring a variety of air pollution parameters. Usually an AQI under 50 is low pollution. · 190
Aleutian Low - the variation in the strength of a semi-permanent area of low atmospheric pressure off the Aleutian Islands in the north eastern Pacific. · 131
algorithms – computer programs used to analyse data. · 38
alpha particles (α) - consist of a helium atomic nucleus of two neutrons and two protons. · 216
alpha track detectors - passive radioactivity detector in which a plastic film becomes etched by the alpha particles that strike it. · 189
anaerobic – pertaining to conditions without air. · 79
aluminium – light-weight metal processed from bauxite ore which formed from the mass weathering of feldspar minerals in rocks. · 391
Andean-Saharan Ice Age - occurred between 450 and 420 million years ago due to a change in patterns of glaciation in what is now present-day Sahara Desert as well as in present-day South America in the Andes Mountains. · 19
anode - negative electrode of an electrical cell or battery. · 247
anomaly – a variation from normal. · 131
anthropogenic - pertaining to climate change, fueled by the activities of humankind · 16, 30, 68
aphelion – the position of the Earth furthest from the Sun. · 40
aquicludes – rocks which will not hold water. · 314
aquifer – rocks which will hold water underground and allow it to pass through. · 95, 314
aquitards – rocks underground that will only allow a slight passage of water through them. · 314
artesian – underground water confined between two aquitards in water and dissolved gases are held under pressure. · 314
atomic number – the number of protons within the nucleus or centre of an atom. · 213
ATP - the biochemical adenosine triphosphate which is stored in living cells as a source of energy. · 70
atmosphere – the envelope of air surrounding the Earth. · 19
Autonomous Vehicle (AV) - or self-driving which are fitted with a range of local sensors and regional GPS data to enable the vehicle to be driven within certain zones without a human driver. · 370
axial precession - the wobbling effect of the Earth's axis. · 42
balloonette - small balloon, often inserted inside another, larger gas bag to provide some control over buoyancy. · 384
base flow – lowest flow rate of water in a stream. · 199
Beringia - the land where the continents of Asia and North America were joined during at least part of the great ice ages. · 21, 23

beta particles (ß⁻) - high-speed, negatively-charged electrons. Positrons (ß⁺) have the same mass as an electron but with a positive charge.   216

Biochemical Oxygen Demand (BOD) - amount of dissolved oxygen needed by organisms to break down organic material present in a given water sample at certain temperature over a specific time period.   195

bioplastics – plastic materials made from natural, renewable sources such as wood fibre, corn starch and others.   407

biosphere - that part of the Earth's environment containing living things.   38

Blau gas – an artificial gas made from distillation of oil and used in lighting and as a fuel in early airships.   379

Bradfield project – rejected scheme which proposed the collection and transfer of excessive water from the north coast of Queensland into the dry interior of the Murray-Darling River system in Australia.   332

BTEX - is a mixture of benzene, toluene, ethylbenzene and xylene with sodium hypochlorite, hydrochloric acid, cellulose, acetic acid and disinfectants used in fracking.   94

buoyancy - is an upward force exerted by a fluid such as water or air that opposes the weight of an object immersed in it.   312

Carbon Cycle - the series of processes by which carbon compounds are recycled in the environment.   275

catalysts – substances used to promote chemical reactions.   96

catchment area - the land surface which collects the rain for a river system.   156

cathode - positive electrode of an electrical cell or battery.   247

chain reaction – rapid, successive nuclear reaction as the neutrons from released by the fission of one atomic nucleus collides with other nuclei in other atoms causing more fission, more neutrons, radioactivity and large amounts of heat.   213

charcoal liquid scintillation devices - measure radiation levels such as those from radon by absorbing it or its products onto charcoal which can then be analysed in the laboratory.   189

Christmas tree – multi-outlet pipe and valve attached to the end of an oil drill pipe when oil starts to flow.   89

climate-change modelling – use of computers to examine and predict what will happen to the climate in the future.   38

climate forcings – are influences which can cause climate change and can be both natural and man-made.   38

climatology – the study of climates   16

coalification - the process which then turns compressed vegetation into coal.   79

Coal-seam Gas (CSG) - found within cracks and cavities of sub-surface coal seams and containing about 95-97% methane gas (CH4) with a little carbon dioxide, ethane and nitrogen.   93

Colorado River - with headwaters are in the Rocky Mountains, it flows south-west into Utah, Arizona, Nevada and California before entering Mexico and into the Gulf of Mexico near Baja California.   305, 337

Concentrated Solar Power (CSP) - uses mirrors or lenses to concentrate a large area of sunlight onto a small area to generate steam to run electrical turbines.   242

concrete - composite material composed of fine and coarse aggregate, such as rock or pebbles, bonded together with a fluid cement that hardens over time.   386

congestion tax - a **tax** paid by vehicle users to enter a restricted area, usually within a city centre, as part of a management strategy to relieve traffic **congestion** within that area.   362

Continuous Radon Monitors (CRMs) – powered devices which measure radon emissions over a time period.   189

control rods - used to prevent the chain reaction happening too quickly in a nuclear reactor and are often made from an alloy of silver and cadmium or boron mixed with iron or carbon as they are good at absorbing neutrons.   218

COP-24 is the 24th Conference of the United Nations Framework Convention on Climate Change which met in Katowice, Poland Between 2nd and 15th December 2018. There 197 countries signed of on guidelines to implement the Paris Agreement of 2016.   10

copper - salmon-coloured metal used since ancient times and made from its ores and refined electrochemically.   393

Coriolis effect - the apparent force on air masses across the surface of the Earth causing deflection to the right in the Northern Hemisphere and to the left in the Southern Hemisphere due to the Earth's rotation.   140, 374

cracking – when larger oil molecules are broken down into more useful, smaller molecular fractions.   96

critical mass - amount of nuclear fuel to start a chain reaction and is of about 52 kilograms of U-235.   215

Cross-Laminated Timber (CLT) - is an engineered wood product, similar to plywood made by gluing thin sheets of timber at right angles to each other.   397

Cryogenian Ice Age - began about 720 million years ago following the breakup of the ancient supercontinent Rodinia.   19

current bedding – angle lines seen in sedimentary rocks such as sandstone formed by deposition of sediment by currents.   27

darcy (D) - permits a flow of 1 cm^3/sec of a fluid with viscosity of 1 millipascal second (mPa·s) under a pressure gradient of one atmosphere per centimetre (atm/cm) acting across an area of one square centimetre (cm^2). It is <u>not</u> an S.I. unit.   314

Darcy's law - states that the rate at which a fluid flows through a permeable medium, such as an aquifer, is directly proportional to the drop in the elevation between two places in the medium (the head) and inversely proportional to the distance between them. 313

daughter products - the new isotopes formed when an atom is split or when an isotope undergoes decay. 213

decay chains (nuclear) - series of changes in unstable radioactive elements to less radioactive elements and finally the stable element, lead. 213

deforestation - is the removal of a forest or stand of trees from land which is then converted to a non-forest use. 279

dendroclimatology - the study of tree growth rings used to give a measure of time as growth respond to climatic changes with different amounts of growth each year. 28

deuterium - natural isotope of the element hydrogen having an extra neutron in its nucleus. 217

dioxin - highly toxic and persistent chemical compound, 2,3,7,8-tetrachlorodibenzo para dioxin (TCDD) which causes problems with reproduction, development, and the immune system and lead to cancer. 412

Dipole Mode Index (DMI) - an indicator of the east-west temperature gradient across the tropical Indian Ocean, linked to the Indian Ocean Dipole. 133

Direct Ethanol Fuel Cell (DEFC) - fuel cell in which ethanol is fed directly into the cell to produce electricity. 248

dirigibles - airship with a rigid framework containing gas balloons. 378

distillation - method of obtaining freshwater from impure or saltwater by boiling or evaporation and collecting the resulting vapour which is cooled and condensed into freshwater. 319

Doha Amendment - second agreement after the Kyoto Protocol with 37 countries agreeing to have binding targets. 9

Doppler effect - an increase (or decrease) in the frequency of sound, light, or other waves as the source and observer move towards (or away from) each other. 147

drawdown (of a well) - the drop in the level of water when it is being pumped out of the well or bore 317

eccentricity - usually referring to the Earth's orbit, is its elongation and variation from a circle. 40

ecosystem - a community of living organisms in conjunction with the nonliving components of their environment, interacting as a system or cooperative unit. 113

Electret ion detectors - device for measuring radiation using an electrostatically-charged Teflon disc which produces an electrical current when struck by an ion generated from radon decay. 189

El Niño-Southern Oscillation (ENSO) - an irregularly periodic variation in winds and sea surface temperatures over the tropical eastern Pacific Ocean, affecting the climate of much of the tropics and subtropics. 50, 131

energy – the ability to do work by applying a force over a distance and is measured in **joules** (J).   218

Enhanced Geothermal Systems (EGS) - use high-pressure water injected deep within hot crystalline rocks to produce steam to drive turbines to make electricity.   249, 284

erratics – large, isolated boulders dropped from a glacier as it retreated.   17, 18

eutrophication - an excessive nutrient load in water, particularly dissolved compounds of phosphorus and nitrogen which results in an overgrowth of algae and micro bacteria.   195

evapotranspiration – loss of water from the leaves of plants.   75

extant – organisms which are still surviving today.   51

extinct – when an organism ceases to exist as a species.   51

extremophiles – organisms which can live in extreme conditions of temperatures, pH etc.   198

feedbacks – are data from natural processes of climate change giving predictable results used in verifying computer models.   38

Fenno-Scandinavian Ice Shield - the Northern European ice sheet of the Quaternary glaciation.   21

fermentation - the chemical breakdown of substances such as sugars by bacteria, yeasts, or other microorganisms, typically involving effervescence and the giving off of heat.   248

ferromagnesian minerals - minerals dark in colour due to high percentages of iron and magnesium ions e.g. hornblende, augite and biotite found in many igneous rocks.   202

fire ecology - focuses on the origins of wildfires fire and their relationship to the environment, both living and non-living.   165

fire storm or localised **fire tornado** - when the sudden updraft of hot air brings in strong, swirling winds at ground level, further increasing the fire's intensity.   160

Flettner rotor concept – marine propulsion system which makes use of a spinning cylinder standing on the deck of the ship.   376

flood stage - when the height of water flowing over the river's banks are of sufficient magnitude to cause widespread inundation.   156

foraminifera are tiny marine, single-celled animals which secrete calcium carbonate shells or tests. These live in the ocean today and are very sensitive to temperature change.   25

Forest Fire Danger Index (FFDI) - fire-warning colour coded system to alert communities about the current threat of wildfires.   169

fossils - are the remains of living things which existed a long time ago and have been preserved in rocks and as impressions.   24

fracking or hydraulic fracturing - used to crack up the coal seam and surrounding rock to release c o a l s e a m gas.   94

fractional distillation - a process whereby c r u d e o i l is heated at the base of tall fractionating towers and the vapours rise, condense and removed from the tower at different cooling levels.   96

fuel cells – produce electricity using various electrodes and hydrogen gas as a fuel.   102

fuel rods - cylinders containing nuclear fuel such as uranium inserted into a reactor core to begin fission. 215

Gaia hypothesis - questionable idea that the Earth as a total environmental system may have the capacity to heal itself of all of the negative changes which have occurred due to humankind's activities. 54

gamma radiation (γ) - high energy electromagnetic radiation similar to X-rays but more penetrating and dangerous. 216

geopolymer concrete - type of **concrete** made by combining source materials rich in silica and alumina such as fly ash, ground granulated blast furnace slags with strong alkali solutions such as potassium hydroxide (KOH) or sodium hydroxide (NaOH). 386

geo-sequestration - the burial of waste gas emissions underground, usually in old saline gas wells. 104

geosphere - the rocky part of the Earth's surface and interior. 38

glacial periods - times when the ice caps have extended towards the Equator giving continental ice shields and extensive glaciers. 17

glass - non-crystalline material made from pure white quartz sand or silica and other compounds. 394

Global Warming Hiatus - episode from 1998 to 2012 when there seemed to be a pause in the increase in global air temperatures. 31

Graf Zeppelin (*Luftschiff Zeppelin* 127) - was a German passenger-carrying dirigible which flew from 1928 to 1937 to many parts of the world including regular flights between Germany and Brazil. 378

greenhouse effect - warming due to the capture of heat radiation by the atmosphere. 67

Great Oxygenation Event (GOE) - a period from about 2.0 to 2.4 billion years ago when there was increased atmospheric oxygen with a decrease in methane which was oxidised into carbon dioxide and water. 18

Gulf Stream - the great surface ocean current sweeping up the North Atlantic. 48

Günz - a glacial event of the latter Pleistocene in Europe. 21

gyres - is any large system of circulating ocean currents, particularly those involved with large wind movements and are caused by the Coriolis effect 407

half-life - time taken for the radioactivity of the element to decay to half of its original level of radioactivity or the original mass of the isotope. 213

harvesting - the process of gathering a ripened crop from prepared fields on a seasonal basis or the harvesting of fish, prawns and other animal life. 286

Hall-Héroult process - an electrochemical method of making aluminium from a molten mixture of its oxide. 392

heavy water - consist of water molecules (H_2O) in which the hydrogen part is a natural isotope called deuterium, which has an extra neutron in its nucleus. 217

High Efficiency, Low Emissions (HELE) – new generation of coal-fired power plants with lower gas emissions — 103

hindcasting – the process of testing and validating computer models by using known data from the past. — 39

Holocene Interglacial – a period of warming which began about 11,500 years ago — 17

homeothermic – animals which maintain a constant body temperature e.g. birds and mammals. — 193

Horizontal Axis Wind Turbines (HAWT) – wind electricity turbine with its rotating shaft parallel to the level of the land. — 239

Huronian Ice Age – or the Makganyene glaciation, a period of cooling between 2.5 and 2.2 billion years ago — 18

hydraulic cement – technical name for common cement which hardens by reacting with water and forms a water-resistant product. — 387

hydraulic conductivity (k) - is a measure of a material's capacity to transmit water and is defined as a constant of proportionality relating the specific discharge of a porous medium under a unit hydraulic gradient in Darcy's law. (see **Darcy's Law**). — 313

hydrogeology or groundwater hydrology - the study of movement of water in the ground through permeable rock units. — 313

hydrology - the study of water, especially **hydrodynamics** or the specific study of moving water. — 156

hydrosphere – pertaining to the waters of the Earth. — 38

hydrostatic head - height difference between where water may enter permeable rock and where it is able to flow out. — 312

ice age - an extended period of time when the Earth's poles and nearby areas are covered with continental ice sheets and are often called glacial periods. — 17

Indian Ocean Dipole (IOD) - an irregular oscillation of sea surface temperatures in which the western Indian Ocean becomes alternately warmer (positive phase) and then colder (negative phase) than the eastern part of the ocean. — 133

Industrial Revolution - the transition to new **manufacturing** processes in Europe and the United States, from about 1760 to sometime between 1820 and 1840 — 79

isobars - lines on a weather map (synoptic chart) joining places of equal air pressure. — 145

isotopes - variations of a particular chemical element which differ in neutron number, and consequently in mass. — 24, 26, 213

integrated gasification combined cycle (IGCC) - uses a high-pressure gasifier to turn coal and other carbon-based fuels into pressurized syngas. — 107

interglacial periods or simply interglacials are periods of warming between ice ages. — 16, 17

intergranular porosity - the space between grains with a rock which will hold fluids. — 308

iridium - is a natural chemical element found only in the Earth's mantle and extra-terrestrial objects such as meteors and asteroids. 52

isotopes - different types of the one element having different atomic weights due to extra neutrons in their nucleus or centre e.g. Oxygen-16 and Oxygen-18 27

Karoo Ice Age - or the Late Paleozoic Ice Age, occurred from 360-260 million years ago associated with major mountain building during the formation of the supercontinent of **Pangea.** 19

kelly – hexagonal rotating stage into which drill pipes are secured. 89

Kyoto Protocol an agreement to legally bind developed countries as signatories, to agree on targets to reduce greenhouse gas emissions 9

lacustrine – pertaining to lake environments. 79

lag time - the time between rainfall and river flooding at a certain point. 156

Last Glacial Maximum (LGM) - near the end of the Pleistocene, between 24,000-18,000 years ago was the last of most recent glacial events when ice sheets and glaciers were at their thickest and the sea levels at their lowest. 21

Laurentide Ice Shield, covered most of Canada and extended into the United States 21

lightning – the sudden discharge of the static electricity which builds up by the friction of the ice particles and other droplets in the clouds of the central part of the thunderheads. 136

lignocellulosic biomass - non-edible, low-value plant waste consisting of lignin and cellulose fibres. 248

liquid petroleum gas (LPG) – common gas made from distilling petroleum and used extensively. Largely replaced town gas. 95

lithium-ion battery (LIB) - a rechargeable battery in which a lithium compound is used as the material at the positive electrode and typically graphite at the negative electrode. 246

Magnus effect - explains that there is a force generated when a fluid flow such as wind passes over a rotating body and that this force will be at right angles to both the direction of flow and the axis of rotation. 376

mass extinctions - sudden disappearances of species of organisms due to catastrophic events and involving large parts of the entire Earth. 51

Maunder Minimum - a period between 1645 and 1715 during which sunspot activity was generally exceedingly rare. 43

Metal Oxide Electrolysis (MOE) – the production of pure metals by passing electric current through a molten mixture of the metal's oxides. 390

microbursts - sudden concentrated pockets of downdraft of air in the atmosphere. 136

microsiemens (μS) – measure of electrical conductivity being one thousandth of a sieman which is equivalent to one ampere of electrical current per volt of electrical pressure. 196

Milankovitch Cycle – natural variations of the Earth's orbit which have been used to correlate long-term climate changes. — 40

Mindel – a glacial event of the latter Pleistocene in Europe. — 21

Moderate Resolution Imaging Spectroradiometer (MODIS) – device on several NASA satellites which scans the Earth using thirty-six spectral bands, or groups of wavelengths. — 146

moderator – substance used to slow down the neutrons of the chain reaction in a nuclear reactor. — 217

mole – is defined as exactly $6.02214076 \times 10^{23}$ particles of a substance or its atomic or molecular weight in grams. — 185, 197

Murray-Darling system covers a large basin in the eastern part of Australia draining around one-seventh of Australia and covers about 1,061,469 km² of flat, low-lying inland country which receives little rainfall. — 302

NASA – National Aeronautics and Space Administration is the United States' agency for satellite and space activities. — 29

Neolithic Age – or new stone age which began around 12,000 years ago and ended as civilizations started to rise around 3500 BC. — 60

nephelometric turbidity units (NTU) – measurements of the turbidity or cloudiness of water by the amount of light which can pass through the turbid liquid. — 199

neutron moderators – materials such as graphite used to slow down the neutrons with a nuclear reactor and thus control the chain reaction. — 215

neutrons – atomic particles in the centre of atoms which have no electrostatic charge. — 213

NOAA – National Oceanic and Atmospheric Administration is the United States' agency which monitors the atmosphere and oceans. — 29

Non-Proliferation of Nuclear Weapons (Treaty) of 1970, is an international treaty whose objective is to prevent the spread of nuclear weapons and to promote cooperation in the peaceful uses of nuclear energy, and to further the goal of achieving nuclear disarmament. — 221

North Atlantic Deep Water (NADW) – deep ocean current formed when the Gulf Stream sinks at the Arctic. — 48

NOx – collective term for gaseous oxides of nitrogen. — 101, 179

nuclear fission – the splitting of an atom's nucleus using sub-atomic particles such as neutrons. — 213

nuclear winter – concept that global cooling could occur much like the effects of cloud from nuclear bombs. — 46

obliquity – the angle of the Earth's axial tilts in relationship to its orbital plane or the flat surface prescribed by its orbit. — 41

Ogallala aquifer – an aquifer in the southwestern United States underlying parts of Nebraska, Kansas, Oklahoma and Texas. — 335

opus caementicium – ancient Roman cement made for their long-lasting concrete constructions using volcanic ash as a major ingredient. — 386

Organic Rankine Cycle (ORC) - a closed-loop thermodynamic cycle for the generation of electric and thermal power using multiple sources, such as renewables (biomass, geothermal energy, solar energy), traditional fuels and waste heat from industrial processes, waste incinerators, engines or gas turbines.   250

osmosis – a process whereby water and some mineral salts pass through a thin membrane due to differences in salt concentration called osmotic pressure.   322

overbank flow - a localised flow of water over a river's banks into another waterway.   156

Pacific Decadal Oscillation (PDO) - recurring pattern of ocean-atmosphere climate variability centred over the mid-latitude Pacific basin. The PDO is detected as warm or cool surface waters in the Pacific Ocean, north of 20°N. Measured as an Index.   131, 132

Palaeocene-Eocene Thermal Maximum (PETM) – period of global warming as Greenland pulled away from Europe and the associated volcanic events put a considerable amount of carbon dioxide into the atmosphere.   44

parallel circuit - electrical circuit with each cell connected to the next using the same terminal so that anodes are connected to each other as are the cathodes.   246

palaeoclimatologists – scientists who study ancient climates.   21

palaeontology - the study of past life   24

paludal – pertaining to swamp conditions.   79

palynology - is a precise science which studies small particles such as dust and pollen.   25

particulate matter – any small pieces of material in air or water.   183

peak (of a flood) - the extent of the potential maximum height of the flood at any given point.   156

perihelion – the position of the Earth nearest to the Sun.   40

permaculture - an holistic approach in agriculture where one works with natural processes rather than replacing them.   399

permafrost is ground including rock and soil, with a temperature that remains at or below the freezing point of water 0 °C for two or more years   119, 179

permeability – a general term for the ability of a rock to allow the passage of fluids through it.   311

permeability (intrinsic) - is the specific permeability of a rock and represents the void capacity through which a fluid can flow and the fluid flowing through it. It's unit is the square metre.   313

pH – measure of acidity being minus the logarithm (power of 10) of the hydrogen ion concentration.   197

photo bioreactors - containing algae which removes carbon dioxide from waste gas emissions and produce oxygen by photosynthesis.   105

photoelectric effect is the emission of electrons or other free carriers when electromagnetic radiation, like light, hits a material.   242

photons – particles of light and other electromagnetic radiation.   243

photovoltaic effect - produces a voltage or electrical pressure when light strikes a sensitive plate. The main distinction between this and the **photoelectric effect** is in this process, the electron is ejected out of the material (usually into a vacuum) and in the photovoltaic effect the excited charge carrier is still contained within the material.   242

picocuries (pCi) – is 10^{-12} of a curie (Ci), which is a unit of radioactivity equal to 3.7×10^{10} decays per second.   189

plate tectonics - the theory that the Earth's surface is covered by many large, moving plates with the continents upon them.   44

poikilothermic – animals whose body temperatures changes with that of the environment.   193

porosity (Φ) - the ability of a rock to hold fluids such as water, mineral solutions, oil and natural gases, because of the many pores or interstitial spaces between its grains   308

Portland cement – most common cement made by heating limestone rock or dead coral, both forms of calcium carbonate ($CaCO_3$), with other materials such as clay to 1,450 ^0C in a kiln.   386

potato blight - a virulent disease of that crop which spread across Europe and into southern Ireland where it caused over one million deaths.   289

power - the use of energy over time and is measured in **watts** (W).   218

primary air pollution – that which come directly from its source,   177

pulse ionization chambers – powered radioactivity measuring device which works by small electrical charges forming a current when the radioactive particles strike gas molecules in an evacuated chamber.   189

Pumped Hydro Energy Storage (PHES) – system of distributing electrical power using pumps to fill a dam during low electricity demand and then releasing water to make hydroelectricity during peak demand.   234

pyrolysis - slow heating of wood and other substances in the absence of oxygen.   64

pyrophytes - plants which have adapted to tolerate fire because it has become part of their natural environment.   165

Quaternary Ice Ages – from about 2.58 million years ago at the beginning of the Pleistocene Epoch leading to the spread of ice sheets from the Artic into Europe as far as France and Germany and into North America past the Great Lakes.   20

quanta (of light) – light acts like particles, called **photons**, rather than continuous waves, which can knock electrons out of metals if the photons had sufficient energy.   243

quantum mechanics - determines the properties of matter such as atoms, molecule and electromagnetic radiation in terms of small particles and their dual properties as waves.   243

RADAR - is an acronym for RAdio Detecting And Ranging, is used locally and nationally for detecting objects including those which are weather-related.   147

radiometric dating - is used to date materials such as rocks or carbon by comparing the abundance of a naturally occurring radioactive isotope within the material to the abundance of its decay products, which form at a known constant rate of decay. 28

radon gas (Rn) is a radioactive inert gas which comes from the radioactive decay of natural deposits of radioactive substances such as the compounds of uranium and thorium. 179

ranks (of coal) - different sequential forms of coal with varying amounts of components. 80

recharge zone - where water enters an aquifer. 315

Saffir-Simpson scale - used to classify hurricanes in the Atlantic and Northern Pacific regions using a five-point scale based on the maximum velocity of wind recorded in any one-minute time span. 141

Sahul - landmass in the Southern Hemisphere where the Australian mainland, New Guinea, Tasmania and many smaller islands were joined together. 22

scientific method - the systematic process and philosophy used in studying any task or proposed question or hypothesis. 4

scintillation detector - measures radiation as points of light on a sensitive screen when it is struck by radiation. 189

seams - layers of coal within a sedimentary sequence. 82

sea surface temperature gradient (SSTG) - the surface water temperature change over distance across the ocean. 130

Secchi disk - is a black and white quartered disk about 20 cm diameter which is dropped in the water attached to a cable to measure the clarity of the water. 200

semiconductor - a material such as the silicon which conducts electricity only under certain conditions. 243

shelf cloud - an ominous dark bank of low cloud coming in front of a squall. 136

silviculture - the science of forest management. 279

series circuit - electrical connection with the negative terminal of one cell is connected to the positive terminal of the next cell and so giving an addition of the voltage of each cell with the amount of current flow equal to one cell. 245

slaked lime - or calcium hydroxide made by heating limestone to form calcium oxide which was then 'slaked' or quenched with water. 387

smog - intense air pollution. The word and is a contraction of the words smoke and fog. 178

sodium battery - is a wet-cell battery that uses a reaction with salt water electrolyte, air, a magnesium oxide cathode and a carbon anode to produce electricity. 247

Solar electricity is the production of domestic and industrial electricity using light energy from the Sun to produce direct current (DC). 242

Solar Thermal Energy (STE) - uses solar collectors for heating water. 240

Southern Annular Mode (SAM) or the **Antarctic Oscillation (AAO)** 134
involves the oscillation north or south of the regular low-pressure westerly winds which flow around Antarctica.

Southern Oscillation Index (SOI) - is computed from fluctuations 131
in the surface air pressure difference between Tahiti in the Pacific and Darwin on the Indian Ocean.

SO_X - general formula for any oxide gas of sulfur e.g. SO_2, SO_3 etc. 372

steel - highly processed iron which has had much of its carbon 388
impurity removed by subjecting it to a blast of oxygen gas.

sticks – lengths of hollow drill pipe used in oil drilling. 89

Sunda – ancient peninsula of Asia extending south to Wallacia. 22

supercapacitors (SC) - ultra-capacitors, are high-capacity electric 247
charge storage devices which can be more quickly recharged than batteries but then can only be used for a shorter time.

supercell is a thunderstorm characterized by the presence of a 137
deep, persistently rotating updraft.

sustainability - the dynamic equilibrium between natural inputs 285
and outputs, modified by external events such as climatic change and natural disasters of any natural system.

syngas - or synthetic gas is a mixture of carbon monoxide and 95
hydrogen made by combining powdered coal with oxygen and steam while being heated and pressurized.

tektites - small broken fragments typical meteorite impacts. 53

Thermohaline Circulation (THC) - the great conveyor belt of 48
ocean currents above and below the surface.

Tidal Stream Generators (TSG) – electricity generators using 236
turbines placed in streams or narrow tidal inlets.

timber – products from the use of trees and wood. 397

tipping point - an irreversible change after which a return to 16
previous conditions would be uncertain.

tornado is a rapidly rotating column of air that is in contact with 139
both the surface of the Earth and a cumulonimbus storm cloud.

town gas – derived from the distillation of coal and sold to 95
consumers for heating. Once used extensively.

Trade Winds - winds blowing steadily towards the equator from 374
the north-east in the northern hemisphere or the south-east in the southern hemisphere and are named because they were used by early sailing vessels for trading between ports.

transboundary pollution - which enters the environment in one 193
place but travels a great distance to have an effect in a place far from the source of pollution.

Tropical Atlantic SST Dipole refers to a cross-equatorial sea 134
surface temperature (SST) pattern that has a period of about twelve years and occurs around 10-15 degrees of latitude off of the Equator.

tropical cyclones - intense low air pressure systems which develop 140
in the hotter months and which are known by this name in Australia and New Zealand; **typhoons** in South-east Asia; and **hurricanes** in the Americas.

troposphere - the lowest layer of Earth's atmosphere where nearly all of the weather conditions take place. The total average height of the troposphere is 13 km. 136

tsunami – an ocean wave caused by an earthquake with a vertical displacement below the sea. 223

Uniformitarianism, Principle of - states that the processes observed today in the formation of rocks and their structures were the same that caused them in the past. 26

United Nations Environment Program (UNEP) - is a programme of the United Nations that coordinates the UN's environmental activities and assists developing countries in these activities. 8

United Nations Framework Convention on Climate Change (UNFCCC) – is a body of the UN seeking to stabilize greenhouse gas concentrations in the atmosphere. 8

vector borne diseases - those transmitted by insects and other animals which carry parasites which cause the disease e.g. some mosquitoes carry the malaria parasite. 120, 153

Vertical Axis Wind Turbines (VAWT) – electrical turbine with its rotating shaft inclined at right angles to ground level. 239

viscosity – ability of a fluid like water or oil to flow, and concerns the frictional effect due to the rock and the inherent nature of the fluid as a sticky substance. 312

vog or volcanic **smog** - widespread haze around volcanic areas such as Hawaii. 178

Volatile Organic Compounds (VOCs) - those compounds which come from living things, especially from trees which can emit volatile chemical vapours on warm days. 179

vortex - the bottom of the rotating current such as in a tornado within a cumulonimbus cloud. 139

Wallacea - an archipelago of islands to the north-west of the ancient continent of Sahul in the Southern Hemisphere. 22

Wallace Line – a boundary line which separates the animal ecologies of Asia and Australia between what are now the islands of Bali and Lombok in Indonesia is the. This was 22

water budget - a means for evaluating availability and sustainability of a **water** supply by noting the balance of the rate at which **water** flows into and out of a watershed. 306

Water Cycle - also known as the **hydrologic cycle**, describes the continuous movement of water on, above and below the surface of the Earth. 275

waterspout – sudden uplift of water due to the passage of a tornado. 139

weather front when a wall of air of one temperature meets a mass of air of another temperature e.g. a warm front when warm, moist air moves into cooler air. 135

World Meteorological Organization (WMO) - an agency of the United Nations dedicated to international cooperation and 8

coordination on the state and behaviour of the Earth's atmosphere.
Würm - a glacial event of the latter Pleistocene in Europe. 21

About the Author

Dr. Peter Scott is an award-winning teacher of Earth Science of over forty years' experience in both Secondary and Tertiary Education. He holds a Bachelors' Degree, two Masters' Degrees and a Doctorate including many years on his own research in exploration geology, research methods and sociology. He has travelled the world extensively including Australia, Antarctica, North and South America, Europe, Asia and Africa to view global warming effects at first hand. This includes: travelling down the Antarctic Peninsula; climbing to glaciers in New Zealand, European Alps, the Andes and Canada; canoeing down the Amazon and travelling across many of the world's great oceans. He is the author of many books on Earth Science and the environment as well as two works of fiction set in South America and Antarctica.

Dr Scott at Paradise Bay, Antarctica 2011.

Other Books by the Author

ADVENTURES IN EARTH SCIENCE (2017) covering Geology. Oceanography, Meteorology and also Astronomy. Over 800 pages, 1200 illustrations (mostly taken by the author in many exotic places on seven continents), and over 30 video links on skills and virtual excursions to many parts of the world. Suitable for Senior Secondary Schools, Junior College and Universities and there are accompanying TEACHERS' GUIDES and LABORATORY MANUALS.

Other books in the ADVENTURES series are also available in electronic format which can be purchased from Kindle and can be used on any other electronic devices such as PCs and iPad using the free Kindle App. Books in this series (2017) are also available in print form (A5 – novel size) from Felix Publishing, Australia (info.felixpublishing@gmail.com) and include:

EXPLORATION SCIENCE Field Geology & Mapping

FOSSILS – LIFE in the ROCKS. Fossils and past

RICHES from the EARTH Minerals & Energy

ROCKS – BUILDING the EARTH

| A DANGEROUS PLANET Volcanoes & Earthquakes | CHANGING the SURFACE Erosion & Landscapes | THROUGH SEA and SKY Oceanography & Meteorology | BEYOND PLANET EARTH An Introduction to Astronomy |

In 2019, the **ADVENTURES in EARTH and ENVIRONMENTAL SCIENCE** series was released as an A4 paperback and an eBook. It consists of a two-volume textbook with accompanying TEACHERS' GUDES and LABORARY MANUALS. The textbook is designed to follow the Australian Environmental Earth Science national syllabus for Senior Secondary Schools and has been written for easy reading and includes over 2000 photographs and illustrations and many video links.

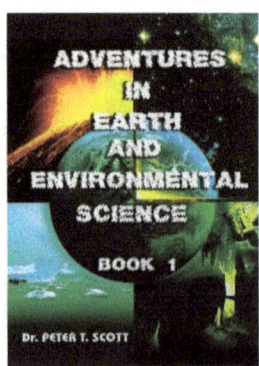 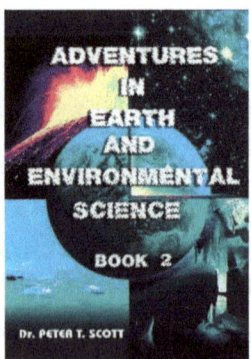

Enquiries are welcome at: info.felixpublishing@gmail.com

www.ingramcontent.com/pod-product-compliance
Lightning Source LLC
Chambersburg PA
CBHW070028040426
42333CB00040B/896